Recent Advances in Differential Geometry and Related Areas

CONTEMPORARY MATHEMATICS

821

Recent Advances in Differential Geometry and Related Areas

AMS Special Session
Recent Advances in Differential Geometry
April 1–2, 2023
Virtual

Tenth Congress of Romanian Mathematicians Special Session
Recent Advances in the Geometry of Submanifolds
June 30–July 5, 2023
Pitești, Romania

Bogdan D. Suceavă
Adara M. Blaga
Marian Ioan Munteanu
Cezar Oniciuc
Shoo Seto
Gabriel-Eduard Vîlcu
Lihan Wang
Editors

AMERICAN MATHEMATICAL SOCIETY

2020 *Mathematics Subject Classification.* Primary 14D07, 53A07, 53A55, 53B20, 53B35, 53C15, 53C18, 53C25, 53C26, 53C43.

Library of Congress Cataloging-in-Publication Data

Names: Suceavă, Bogdan, 1969- editor | Blaga, Adara M., 1978- editor | Munteanu, Marian Ioan, 1973- editor | Oniciuc, Cezar, 1972- editor | Seto, Shoo, 1986- editor | Vîlcu, Gabriel-Eduard, 1975- editor | Wang, Lihan, 1986- editor | AMS Special Session on Recent Advances in Differential Geometry (2023 : Online) contributor | Special Session on Recent Advances in the Geometry of Submanifolds (2023 : Pitești, Romania) contributor

Title: Recent advances in differential geometry and related areas : AMS Special Session on Recent Advances in Differential Geometry, 2023 Spring Eastern Sectional Meeting, meeting virtually, hosted by the American Mathematical Society, Meeting # 1185, April 1–2, 2023 : the Special Session on Recent Advances in the Geometry of Submanifolds at the Tenth Congress of Romanian Mathematicians, Pitești, Romania, June 30–July 5, 2023 / [edited by] Bogdan D. Suceavă, Adara M. Blaga, Marian Ioan Munteanu, Cezar Oniciuc, Shoo Seto, Gabriel-Eduard Vîlcu, Lihan Wang.

Description: Providence, Rhode Island : American Mathematical Society, [2025] | Series: Contemporary mathematics, 0271-4132 ; volume 821 | Includes bibliographical references.

Identifiers: LCCN 2024062350 | ISBN 9781470475864 paperback | ISBN 9781470480202 ebook

Subjects: LCSH: Geometry, Differential–Congresses | AMS: Algebraic geometry – Families, fibrations – Variation of Hodge structures | Differential geometry – Classical differential geometry – Higher-dimensional and -codimensional surfaces in Euclidean n-space | Differential geometry – Classical differential geometry – Differential invariants (local theory), geometric objects | Differential geometry – Local differential geometry – Local Riemannian geometry | Differential geometry – Local differential geometry – Hermitian and Kählerian structures | Differential geometry – Global differential geometry – General geometric structures on manifolds (almost complex, almost product structures, etc.) | LCGFT: Conference papers and proceedings

Classification: LCC QA641 .R428 2025 | DDC 516.3/6–dc23/eng/20250522

LC record available at https://lccn.loc.gov/2024062350

Contemporary Mathematics ISSN: 0271-4132 (print); ISSN: 1098-3627 (online)

DOI: https://doi.org/10.1090/conm/821

---

10 9 8 7 6 5 4 3 2 1 30 29 28 27 26 25

Contents

Preface vii

Bibliography ix

Harmonicity and biharmonicity of quadratic maps between spheres
RAREŞ AMBROSIE 1

A survey on submanifolds of nearly Kähler spaces
MATEO ANARELLA 23

Expressions of circles in a complex hyperbolic space by trajectories on tubes around complex hyperplanes
YUSEI AOKI and TOSHIAKI ADACHI 35

On conservation laws for polyharmonic maps
VOLKER BRANDING 49

Chen inequality for statistical submanifolds in Kenmotsu statistical manifolds of constant ϕ-sectional curvature
SIMONA DECU and GABRIEL-EDUARD VÎLCU 67

Thurston geometries in dimension four from a Riemannian perspective
MARIE D'HAENE 77

Recent results in Ricci and Yamabe solitons in contact geometry
AMALENDU GHOSH, RAMESH SHARMA, and RAHUL PODDAR 87

Understanding of constant p-mean curvature hypersurfaces in the Heisenberg groups
HSIAO-FAN LIU 99

Open problems on compact constant mean curvature surfaces with boundary
RAFAEL LÓPEZ 105

On holomorphic conformal structures associated with lattice polarized K3 surfaces
ANDREAS MALMENDIER and MICHAEL T. SCHULTZ 119

Triharmonic curves in the 3-dimensional Sol space
S. MONTALDO and A. RATTO 135

A note on Weingarten surfaces foliated by helices
ANA IRINA NISTOR 145

Characterizations of rotational biconservative hypersurfaces
Álvaro Pámpano 159

Einstein warped-product manifolds and the screened Poisson equation
Alexander Pigazzini, Luca Lussardi, Magdalena Toda, and Andrew DeBenedictis 173

Preface

In preparing this volume, our first thought was to present the reader with a useful tool, a reference that could inspire and inform. We are all aware that in the last two decades various techniques produced new results in the realm of differential geometry, and many of them focused on the geometry of submanifolds, where one of the most important questions, the Willmore Conjecture, was solved by F. Codá Marques and A. Neves (Ann. of Math. (2), **179** (2014), no. 2, pp. 683–782). This invites a reflection on the new trends and research directions that contemporary differential geometry might and will take in its natural organic development. Some of the most important advances have been obtained by using techniques in geometric analysis, while some others have been using more classical methods, which are still of interest. Motivated by these new evolutions, the present volume intends to bring together various viewpoints on the recent study of various ways of representing the fundamental idea of space, as it is worked with and modeled within differential geometry. Of particular interest for this project are the results focused on minimal submanifolds and their generalizations, and the related concept of harmonic mappings and their generalizations. In preparing this volume, the editors have been particularly interested in the study of curvature functionals in various contexts, from comparison geometry to new curvature invariants, relations between curvature and topology, as well as other related themes.

This volume published in the AMS *Contemporary Mathematics* series is based on contributions presented in two events:

- AMS Special Session on Recent Advances in Differential Geometry, 2023 Spring Eastern Sectional Meeting, Meeting virtually, hosted by the American Mathematical Society; Meeting # 1185, April 1–2, 2023;
- The Special Session on Recent Advances in the Geometry of Submanifolds, at the Tenth Congress of Romanian Mathematicians, Pitești, Romania, June 30–July 5, 2023.

The idea of having virtual AMS Special Sessions during the AMS Sectional Meetings was particularly useful during the pandemic period, as it allowed mathematicians from around the world to bypass isolation and discuss the mathematics they were working on at that time. As challenging as the period was, the present volume represents also a testament to the perseverance of mathematicians during that particular moment in the history of mathematics.

From among the speakers in the AMS Special Session on Recent Advances in Differential Geometry Spring Eastern Virtual Sectional Meeting, April 1–2, 2023, the following authors contributed to the present volume: Rareș Ambrosie, Mateo Anarella, Marie D'haene, Hsiao-Fan Liu, and Ramesh Sharma.

Additionally, from the speakers at the Tenth Congress of the Romanian Mathematicians, the following authors submitted their contributions to the present volume: Simona Decu, Rafael López, Ana Irina Nistor, and Magdalena Toda.

The Tenth Congress of the Romanian Mathematicians was organised by the Romanian Academy, the *Simion Stoilow* Institute of Mathematics of the Romanian Academy, Romanian Mathematical Society (a society in reciprocity with the American Mathematical Society), the Faculty of Mathematics and Computer Science at the University of Bucharest, and the University of Pitești.

It was a happy coincidence that Rafael López's paper included in the present volume was elaborated in the same period as his work (Amer. Math. Monthly, **130** (2023), no. 3, pp. 222–238), which was presented by the Mathematical Association of America with one of the 2023 Paul R. Halmos – Lester R. Ford Awards. From the MAA award citation, we retain: "This beautiful article considers the problem of determining the shape of a (suitably idealized) hanging surface or, equivalently, the shape of a cupola or dome that will stand under its own weight. The author analyzes the simplest case and treats readers to interesting connections to the work of architects, in particular Antonio Gaudí. The author closes with a new design for a roof whose rotational axis of symmetry is horizontal rather than vertical" ("Celebrating Excellence Across the MAA Community").

Last but not least, to better survey some intriguing themes of differential geometry today, the editors extended their invitation to several outstanding researchers, whose insight and valuable vision could only complement the general philosophy of our volume. This is how we extended our invitations to Yusei Aoki, Toshiaki Adachi, Volker Branding, Andreas Malmendier, Álvaro Pámpano, Stefano Montaldo and Andrea Ratto, and their contributions can be read in the present volume.

We would like to also express our thanks to the organisers of the Spring Eastern Virtual Sectional Meeting, 2023, in particular, to Steven H. Weintraub, who served as Associate Secretary for the AMS Scientific Program, and also to the technical support crew, as well as the organisers of the Tenth Congress of the Romanian Mathematicians. Additional thanks are due to the coauthors of the speakers whose papers are included in this volume. Our editorial work was not possible without the insight and the great expertise our referees had, and their care and thorough analysis made ultimately this book possible.

It is the editors' greatest hope that our volume will be useful to a reader who looks for new information and inspiration, in a fresh panorama of differential geometry elaborated today.

The Editors

Bibliography

[1] The Mathematical Association of America, "Celebrating Excellence Across the MAA Community", at `https://maa.org/community/awards/`, retrieved on August 18, 2024.

[2] F. C. Marques and A. Neves, *Min-max theory and the Willmore conjecture*, Ann. of Math. (2) **179** (2014), no. 2, 683–782, DOI 10.4007/annals.2014.179.2.6. MR3152944

[3] R. López, *What is the shape of a cupola?*, Amer. Math. Monthly **130** (2023), no. 3, 222–238, DOI 10.1080/00029890.2022.2154557. MR4550708

Contemporary Mathematics
Volume **821**, 2025
https://doi.org/10.1090/conm/821/16431

Harmonicity and biharmonicity of quadratic maps between spheres

Rareş Ambrosie

Abstract. In this paper, we present a short survey on homogeneous polynomial maps of degree 2 between spheres, which are quadratic forms, and their interconnection with the theory of harmonic and p-harmonic maps. We also discuss some new results for the biharmonic and λ-biharmonic quadratic forms, which prove their strong connection with the harmonic ones.

Introduction

One of the most famous minimal surfaces in the 4-dimensional Euclidean sphere is the Veronese surface. The immersion that defines it is a harmonic map and a quadratic form from $\mathbb{S}^2$ into $\mathbb{S}^4$ with appropriate radii. In the late 1920s and early 1930s, Borůvka generalized the Veronese surface to minimal surfaces in even dimensional Euclidean spheres given by harmonic k-forms, now referred as the Veronese–Borůvka spheres, and then he studied their structure of the second, third and higher order fundamental forms. Later on, Calabi [**8**] and Simons [**32**] revived the Veronese–Borůvka spheres. In particular, Calabi showed that these spheres were the only compact minimal surfaces with constant positive Gaussian curvature into $\mathbb{S}^n$, for any n.

In the late '60s, Wood [**36**] developed a great interest in quadratic forms. Among other interesting results, he discovered a class of harmonic quadratic forms (eigenmaps between spheres) as gradients of the harmonic eiconals among Cartan's isoparametric functions (see [**9**]). In his thesis [**33**], Smith proved interesting results concerning the existence of the harmonic representatives in the homotopy classes of the maps between spheres using the special harmonic maps obtained from k-forms. Afterward, many geometers, including Yiu [**38**], Toth [**30, 31**], Pettinati and Ratto [**28**], Chang [**10**], Gastel [**18, 19**], and Wu, Xiong, and Zhao [**37**], studied the k-forms—the quadratic ones, in particular.

Biharmonicity of the quadratic forms was recently studied in [**1**]. A characterization formula of the bitension field of such maps was found and some classification results for biharmonic quadratic forms were derived. Those results led to an open problem concerning the structure of the biharmonic quadratic forms. In [**2**], the authors gave a positive answer to that open problem. More precisely, they proved

2020 *Mathematics Subject Classification.* Primary 53C43, 35G20, 15A63.
Key words and phrases. Biharmonic maps, spherical maps, homogeneous polynomial maps.
Thanks are due to Cezar Oniciuc for useful comments and remarks.

that a quadratic form is nonharmonic biharmonic if and only if its energy density is constant, and then they gave a rigidity theorem which states that all nonharmonic biharmonic quadratic forms are obtained from the harmonic quadratic forms that take values in a certain small hypersphere of the target sphere.

In the present paper, we survey some of the most important results regarding the harmonicity and biharmonicity of the quadratic forms. We start with general facts regarding quadratic forms, then we present the harmonic, p-harmonic, (intrinsic) biharmonic, (intrinsic) λ-biharmonic, and extrinsic biharmonic quadratic forms.

For the p-harmonic quadratic forms, we first derive the characterization formula and then we prove that for $p \neq 2$ there is no nonharmonic p-harmonic quadratic form. Motivated by the rigidity result found for the biharmonic quadratic forms, one could ask whether there is a similar result for the λ-biharmonic case. The answer is positive and we present the result in Section 6. First, for the λ-biharmonic quadratic forms, we derive the characterization formula and then we prove that a quadratic form is nonharmonic λ-biharmonic map if and only if its energy density is a specific constant. Using this result, we give a rigidity theorem that correlates the nonharmonic λ-biharmonic quadratic forms to the harmonic ones. We end the paper by studying extrinsic biharmonic quadratic forms and we prove a nonexistence result. As in the previous sections, we derive a characterization formula for the extrinsic biharmonic quadratic forms, then we obtain that a quadratic form is extrinsic biharmonic if and only if it is harmonic.

Conventions. The following sign conventions for the rough Laplacian, which acts on the set $C\left(\phi^{-1}TN\right)$ of all sections of the pull-back bundle $\phi^{-1}TN$, and for the curvature tensor field are used

$$\Delta\sigma = -\mathrm{trace}_g \nabla^2\sigma, \quad R(X,Y)Z = \nabla_X\nabla_Y Z - \nabla_Y\nabla_X Z - \nabla_{[X,Y]}Z.$$

Also, by $\mathbb{S}^m(r)$ we indicate the m-dimensional Euclidean sphere of radius r. When $r = 1$, we write $\mathbb{S}^m$ instead of $\mathbb{S}^m(1)$.

1. Quadratic forms between spheres

In this section we consider homogeneous polynomial maps of degree 2 between spheres, which are quadratic forms. The commutative diagram below fixes the notations for this paper

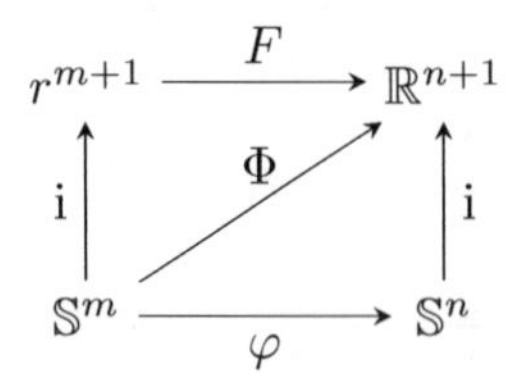

where $F : \mathbb{R}^{m+1} \to \mathbb{R}^{n+1}$ is a quadratic form, $|F(\overline{x})|^2 = |\overline{x}|^4$ for any $\overline{x} \in \mathbb{R}^{m+1}$, and i denotes the canonical inclusion. We note that we preserve the same terminology for the induced map φ. The quadratic form F can be written as

$$F(\overline{x}) = \left(X^t A_1 X, X^t A_2 X, \ldots, X^t A_{n+1}\ X\right),$$

where the vector $\overline{x} = \left(x^1, x^2, \ldots, x^{m+1}\right)$ is identified with the line matrix

$$X^t = \left[x^1\ x^2\ \ldots\ x^{m+1}\right]$$

and $A_1, \ldots, A_{n+1}$ are square matrices of order $m+1$. By a standard symmetrization process, we can always assume that A_i is symmetric, for any $i \in \{1, 2, \ldots, n+1\}$.

Assume that φ is not a constant map. Thus, there exists $i_0 \in \{1, 2, \ldots, n+1\}$ such that A_{i_0} is not the identity matrix I_{m+1} multiplied by a nonzero real constant.

In the following, we will use the Frobenius norm for the matrices, which is the square root of the sum of the squares of the matrix elements.

With $\overset{o}{\mathrm{d}}$, $\overset{o}{\nabla}$, $\overset{o}{\Delta}$ and $\overset{o}{\mathrm{grad}}$, we denote operators that act on $\mathbb{R}^{m+1}$. We observe that, since F is a quadratic form, then $\overset{o}{\Delta}\overset{o}{\Delta} F = \overline{0}$ on $\mathbb{R}^{m+1}$. Then

$$\overset{o}{\mathrm{d}}F(\overline{x}) = \begin{bmatrix} 2X^t A_1 \\ \vdots \\ 2X^t A_{n+1} \end{bmatrix}.$$

Thus, we obtain on $\mathbb{R}^{m+1}$

$$\left|\overset{o}{\mathrm{d}}F(\overline{x})\right|^2 = 4X^t\left(A_1^2 + \cdots + A_{n+1}^2\right)X = 4X^t S X, \tag{1.1}$$

where we denoted $S = A_1^2 + \cdots + A_{n+1}^2$. The following formulas hold on $\mathbb{R}^{m+1}$

$$\begin{aligned}
&\overset{o}{\Delta} F = -2\left(\mathrm{tr}A_1, \ldots, \mathrm{tr}A_{n+1}\right),\\
&\overset{o}{\Delta}\left(\left|\overset{o}{\mathrm{d}}F\right|^2\right) = -8\mathrm{tr}\left(A_1^2 + \cdots + A_{n+1}^2\right) = -8\mathrm{tr}S,\\
&\left|\overset{o}{\nabla}\overset{o}{\mathrm{d}}F\right|^2 = 4\left(|A_1|^2 + \cdots + |A_{n+1}|^2\right),\\
&\overset{o}{\mathrm{grad}}\left(\left|\overset{o}{\mathrm{d}}F\right|^2\right) = 8X^t\left(A_1^2 + \cdots + A_{n+1}^2\right) = 8X^t S,\\
&\overset{o}{\mathrm{d}}F\left(\overset{o}{\mathrm{grad}}\left(\left|\overset{o}{\mathrm{d}}F\right|^2\right)\right) = 16\left(X^t A_1 S X, \ldots, X^t A_{n+1} S X\right).
\end{aligned} \tag{1.2}$$

We notice that, since the matrices $A_1, \ldots, A_{n+1}$ are symmetric, then

$$|A_1|^2 + \cdots + |A_{n+1}|^2 = \mathrm{tr}S. \tag{1.3}$$

The quadratic forms have the following general property.

PROPOSITION 1.1 ([**1**]). *Let $F : \mathbb{R}^{m+1} \to \mathbb{R}^{n+1}$ be a quadratic form. Then,*

$$8\mathrm{tr}S + \left|\overset{o}{\Delta}F\right|^2 = 4(m+1)(m+3).$$

According to Toth in [**31**], by using the standard coordinates, any quadratic form $F : \mathbb{R}^{m+1} \to \mathbb{R}^{n+1}$ can be written as

$$F(\overline{x}) = \sum_{i=1}^{m+1} \overline{a}_i \left(x^i\right)^2 + \sum_{1 \le i < j \le m+1} \overline{a}_{ij} x^i x^j,$$

where $\overline{a}_i \in \mathbb{R}^{n+1}$, for $i = 1, \ldots, m+1$, and $\overline{a}_{ij} \in \mathbb{R}^{n+1}$, for $1 \le i < j \le m+1$. To simplify the notation, we set $\overline{a}_{ij} = \overline{a}_{ji}$, in order to have the vectors $\overline{a}_{ij}$ defined for all $1 \le i, j \le m+1$. The condition that F takes $\mathbb{S}^m$ into $\mathbb{S}^n$ is equivalent to the following (see [**31**] and [**36**]):

$$
\begin{aligned}
|\overline{a}_i| &= 1, && \text{for } 1 \le i \le m+1,\\
\langle \overline{a}_i, \overline{a}_{ij}\rangle &= 0, && \text{for } i \neq j,\\
|\overline{a}_{ij}|^2 + 2\langle \overline{a}_i, \overline{a}_j\rangle &= 2, && \text{for } i \neq j,\\
\langle \overline{a}_i, \overline{a}_{jk}\rangle + \langle \overline{a}_{ij}, \overline{a}_{ik}\rangle &= 0, && \text{for } i,j,k \text{ distinct},\\
\langle \overline{a}_{ij}, \overline{a}_{kl}\rangle + \langle \overline{a}_{ik}, \overline{a}_{jl}\rangle + \langle \overline{a}_{il}, \overline{a}_{jk}\rangle &= 0, && \text{for } i,j,k,l \text{ distinct}.
\end{aligned}
\tag{1.4}
$$

Further, we recall six classification results that we will use in Section 4. They are quite rigid since, under certain constraints on the dimension of the domain or codomain spheres, we obtain that the quadratic forms are either constant or the Hopf construction.

THEOREM 1.2 ([**36**]). *If m is a power of 2, then all polynomial maps from $\mathbb{S}^m$ to $\mathbb{S}^{m-1}$ are constant.*

THEOREM 1.3 ([**36**]). *All polynomial maps from $\mathbb{S}^m$ to $\mathbb{S}^n$ are constant if $m \ge 2n$.*

THEOREM 1.4 ([**38**]). *Let $\varphi : \mathbb{S}^{2^{l+1}-1} \to \mathbb{S}^{2^l}$ be a quadratic form. If $l \ge 4$, then φ is constant. If $l \le 3$ and φ is not constant, then φ is isometric to the Hopf fibration.*

A slightly stronger result is the following theorem.

THEOREM 1.5 ([**38**]). *Let $\varphi : \mathbb{S}^{2^{l+1}-1} \to \mathbb{S}^{2^l+1}$ be a quadratic form. If $l \ge 4$, then φ is constant. If $l \le 3$ and φ is not constant, then φ is either in fact mapping into $\mathbb{S}^{2^l}$, or, in case $l = 1$, the Hopf construction.*

THEOREM 1.6 ([**10**]). *If $\varphi : \mathbb{S}^{2n-2} \to \mathbb{S}^n$ is a nonconstant quadratic form, then $n \in \{2,4,8\}$ and, up to isometries, φ is the restriction of a Hopf fibration to a great $(2n-2)$-sphere.*

THEOREM 1.7 ([**10**]). *If $\varphi : \mathbb{S}^{2n-1} \to \mathbb{S}^n$ is a nonconstant quadratic form, then $n \in \{1,2,4,8\}$. Moreover, up to isometries, φ is the Hopf fibration when $n \in \{2,4,8\}$ and, when $n = 1$, $\varphi\left(e^{i\theta}\right) = e^{2\theta}$.*

2. Harmonic quadratic forms

As suggested by Eells and Sampson [**14, 15**], or Eells and Lemaire [**12**], harmonic maps $\varphi : (M^m, g) \to (N^n, h)$ between two Riemannian manifolds are the critical points of the energy functional

$$E : C^\infty(M,N) \to \mathbb{R}, \qquad E(\varphi) = \frac{1}{2}\int_M |\mathrm{d}\varphi|^2 \, v_g,$$

where M is compact. They are characterized by the vanishing of their associated tension field, which is given by

$$\tau(\varphi) = \operatorname{trace}\nabla \mathrm{d}\varphi.$$

From now on, we consider the special case of quadratic forms.

THEOREM 2.1. *Let $\varphi : \mathbb{S}^m \to \mathbb{S}^n$ be a quadratic form. Then its associated tension field is given by*

$$\tau(\varphi) = -\overset{o}{\Delta} F + \left(\left|\overset{o}{\mathrm{d}} F\right|^2 - 2(m+3)\right)\Phi. \tag{2.1}$$

Further, we recall two known results (see, for example, [**4**] and [**13**]).

PROPOSITION 2.2. *Let φ be a nonconstant map. Then φ is harmonic if and only if* $\overset{o}{\Delta} F = \overline{0}$.

The next result relates the harmonicity of the map φ with its energy density $e(\varphi) = |d\varphi|^2/2$.

PROPOSITION 2.3. *Let $\varphi : \mathbb{S}^m \to \mathbb{S}^n$ be a nonconstant map. Then φ is harmonic if and only if $e(\varphi) = m + 1$.*

In [**30**], Toth classified all full harmonic maps from $\mathbb{S}^3$ to $\mathbb{S}^n$. In this paper, a map between spheres is said to be full if its image is not contained in any great hypersphere of the target sphere.

THEOREM 2.4 ([**30**]). *Full quadratic harmonic maps of $\mathbb{S}^3$ into $\mathbb{S}^n$ exist only if $2 \le n \le 8$ and $n \ne 3$. Moreover, if $\varphi : \mathbb{S}^3 \to \mathbb{S}^n$ is such a map, then there exist $U \in O(4)$, $V \in O(n+1)$ and a symmetric positive definite matrix $B \in S^2\left(\mathbb{R}^{n+1}\right)$ such that*

$$V \circ \varphi \circ U = B \circ \varphi_n,$$

where $\varphi_n : \mathbb{S}^3 \to \mathbb{S}^n$ is defined by

$$\varphi_n(\overline{x}) = \begin{cases} \left(\left(x^1\right)^2 + \left(x^2\right)^2 - \left(x^3\right)^2 - \left(x^4\right)^2, 2\left(x^1x^3 - x^2x^4\right), 2\left(x^1x^4 + x^2x^3\right)\right), & n = 2 \\ \left(\left(x^1\right)^2 + \left(x^2\right)^2 - \left(x^3\right)^2 - \left(x^4\right)^2, 2x^1x^3, 2x^1x^4, 2x^2x^3, 2x^2x^4\right), & n = 4 \\ \left(\left(x^1\right)^2 - \left(x^2\right)^2, \left(x^3\right)^2 - \left(x^4\right)^2, 2x^1x^2, \sqrt{2}\left(x^1x^3 + x^2x^4\right),\right. & \\ \quad \left.\sqrt{2}\left(x^2x^3 - x^1x^4\right), 2x^3x^4\right), & n = 5 \\ \left(\frac{1}{\sqrt{2}}\left(\left(x^1\right)^2 + \left(x^2\right)^2 - \left(x^3\right)^2 - \left(x^4\right)^2\right), \frac{1}{\sqrt{2}}\left(\left(x^1\right)^2 - \left(x^2\right)^2\right),\right. & \\ \quad \frac{1}{\sqrt{2}}\left(\left(x^3\right)^2 - \left(x^4\right)^2\right), \sqrt{2}x^1x^2, \sqrt{3}\left(x^1x^3 + x^2x^4\right), & \\ \quad \left.\sqrt{3}\left(x^2x^3 - x^1x^4\right), \sqrt{2}x^3x^4\right), & n = 6 \\ \left(\left(x^1\right)^2 - \left(x^2\right)^2, \left(x^3\right)^2 - \left(x^4\right)^2, 2x^1x^2, \sqrt{2}x^1x^3, \sqrt{2}x^1x^4,\right. & \\ \quad \left.\sqrt{2}x^2x^3, \sqrt{2}x^2x^4, 2x^3x^4\right), & n = 7 \\ \varphi_{\lambda_2}\left(x^1, x^2, x^3, x^4\right), (\varphi_{\lambda_2} = \text{ a standard minimal immersion}) & n = 8. \end{cases}$$

Later, Wu, Xiong, and Zhao continued working on the classification of harmonic quadratic forms and also obtained the following result.

THEOREM 2.5 ([**37**]). *Any full harmonic quadratic form $\varphi : \mathbb{S}^7 \to \mathbb{S}^7$ is equivalent to either the gradient of the cubic isoparametric polynomial of Cartan or a map depending only on one parameter.*

We end this section by presenting the Veronese–Borůvka spheres. The Veronese surface is induced by the quadratic form from $\mathbb{R}^3$ into $\mathbb{R}^5$ defined by

$$F\left(x^1, x^2, x^3\right) = \left(\frac{1}{\sqrt{3}}x^2x^3, \frac{1}{\sqrt{3}}x^1x^3, \frac{1}{\sqrt{3}}x^1x^2, \frac{1}{2\sqrt{3}}\left(\left(x^1\right)^2 - \left(x^2\right)^2\right),\right.$$
$$\left.\frac{1}{6}\left(\left(x^1\right)^2 + \left(x^2\right)^2 - 2\left(x^3\right)^2\right)\right).$$

This form gives an (isometric) minimal immersion from $\mathbb{S}^2\left(\sqrt{3}\right)$ into $\mathbb{S}^4$. Further, for higher dimensions of the codomain sphere, the set of harmonic homogeneous polynomial functions of degree k defined on $\mathbb{R}^3$ is a $(2k+1)$ dimensional vector space. Considering the unit sphere $\mathbb{S}^{2k}$ in this vector space with the standard inner product, Borůvka constructed a full minimal immersion

$$\mathbb{S}^2\left(\sqrt{\frac{k(k+1)}{2}}\right) \to \mathbb{S}^{2k},$$

for any positive integer k, $k \geq 2$, and studied the structures of the second, third, and higher order fundamental forms of this minimal surface. The surface is now called the Veronese–Borůvka sphere. After about 40 years following Borůvka's studies, Calabi and Simons revived the Veronese–Borůvka spheres. In fact, they reminded us that minimal submanifolds in spheres are important for studying the singularities of minimal varieties in a Euclidean space. In [**8**], Calabi showed that, up to rigid motion, these Veronese–Borůvka spheres were the only compact minimal surfaces with constant positive Gaussian curvature in $\mathbb{S}^n$, for any n.

THEOREM 2.6 ([**8**]). *Let $\varphi : \mathbb{S}^2(r) \to \mathbb{S}^n$, $n \geq 4$, be a full minimal immersion. Then $n = 2k$ and*

(i) *the radius r is given by*

$$r = \sqrt{\frac{k(k+1)}{2}};$$

(ii) *the immersion φ is uniquely determined up to a rigid motion of $\mathbb{S}^{2k}$, and the $2k+1$ components of $\mathrm{i} \circ \varphi$ are a suitably normalized basis for the harmonic homogeneous polynomials on $\mathbb{R}^3$.*

3. p-harmonic quadratic forms

A generalization of harmonic maps is given by p-harmonic maps, where $p \geq 2$ is a positive integer, and for $p = 2$ they are the classical harmonic maps. They were studied by Takeuchi [**29**], Baird and Gudmunsdon [**3**], etc., and the p-harmonic maps continue to be a contemporary subject of interest; see Branding and Siffert [**5**].

The p-harmonic maps $\varphi : (M^m, g) \to (N^n, h)$ between two Riemannian manifolds are the critical points of the p-energy functional

$$E_p : C^\infty(M, N) \to \mathbb{R}, \qquad E_p(\varphi) = \frac{1}{p}\int_M |\mathrm{d}\varphi|^p \ v_g,$$

where M is compact. Equivalently, they are characterized by the vanishing of their p-tension field that is

$$\tau_p(\varphi) = \operatorname{trace}\nabla\left(|\mathrm{d}\varphi|^{p-2}\,\mathrm{d}\varphi\right).$$

Baird and Gudmundson gave a more explicit form for the p-tension field.

THEOREM 3.1 ([**3**]). *Let $\varphi : M \to N$ be a nonconstant map. Then its p-tension field is given by*

$$\tau_p(\varphi) = |\mathrm{d}\varphi|^{p-2}\left\{\tau(\varphi) + \mathrm{d}\varphi\left(\operatorname{grad}\left(\log|\mathrm{d}\varphi|^{p-2}\right)\right)\right\}. \tag{3.1}$$

In the special case of quadratic forms we have the following characterization formula.

THEOREM 3.2. *Let $\varphi : \mathbb{S}^m \to \mathbb{S}^n$ be a nonconstant quadratic form. Then its p-tension field is given by*

$$\begin{aligned}\tau_p(\varphi) =& 2\left(4X^tSX-4\right)^{\frac{p-2}{2}}\Big\{(\mathrm{tr}A_1,\mathrm{tr}A_2,\ldots,\mathrm{tr}A_{n+1})\\ &+\left(-\frac{(p-2)X^tSX}{X^tSX-1}+2X^tSX-(m+3)\right)\Phi \\ &+\frac{p-2}{X^tSX-1}\left(X^tA_1SX,X^tA_2SX,\ldots,X^tA_{n+1}SX\right)\Big\}.\end{aligned} \tag{3.2}$$

PROOF. First, we recall from [**1**] the following relation

$$|\mathrm{d}\varphi|^2 = \left|\overset{o}{\mathrm{d}}F\right|^2 - 4 > 0. \tag{3.3}$$

Therefore, by direct calculation we obtain on $\mathbb{S}^m$

$$\begin{aligned}&\mathrm{d}\varphi\left(\mathrm{grad}\left(\log|\mathrm{d}\varphi|^{p-2}\right)\right)\\ &=\frac{p-2}{2}\left\{\overset{o}{\mathrm{d}}F\left(\overset{o}{\mathrm{grad}}\left(\log\left(\left|\overset{o}{\mathrm{d}}F\right|^2-4\right)\right)\right)-\frac{4\left|\overset{o}{\mathrm{d}}F\right|^2}{\left|\overset{o}{\mathrm{d}}F\right|^2-4}\Phi\right\}.\end{aligned} \tag{3.4}$$

Replacing Equations (2.1), (3.3), and (3.4) in Equation (3.1), we obtain

(3.5)

$$\begin{aligned}\tau_p(\varphi) =& \left(\left|\overset{o}{\mathrm{d}}F\right|^2-4\right)^{\frac{p-2}{2}}\left\{-\overset{o}{\Delta}F+\left(-\frac{2(p-2)\left|\overset{o}{\mathrm{d}}F\right|^2}{\left|\overset{o}{\mathrm{d}}F\right|^2-4}+\left|\overset{o}{\mathrm{d}}F\right|^2-2(m+3)\right)\Phi\right.\\ &\left.+\frac{(p-2)}{2}\overset{o}{\mathrm{d}}F\left(\overset{o}{\mathrm{grad}}\left(\log\left(\left|\overset{o}{\mathrm{d}}F\right|^2-4\right)\right)\right)\right\}.\end{aligned}$$

Using Equations (1.1) and (1.2), by direct calculation Equation (3.5) becomes

$$\begin{aligned}\tau_p(\varphi) =& 2\left(4X^tSX-4\right)^{\frac{p-2}{2}}\Big\{(\mathrm{tr}A_1,\mathrm{tr}A_2,\ldots,\mathrm{tr}A_{n+1})\\ &+\left(-\frac{(p-2)X^tSX}{X^tSX-1}+2X^tSX-(m+3)\right)\Phi \\ &+\frac{p-2}{X^tSX-1}\left(X^tA_1SX,X^tA_2SX,\ldots,X^tA_{n+1}SX\right)\Big\}.\end{aligned}$$

□

THEOREM 3.3. *For $p \neq 2$, there is no nonharmonic p-harmonic quadratic form $\varphi : \mathbb{S}^m \to \mathbb{S}^n$.*

The proof has two parts. In the first part, we assume that φ has constant energy density. Using Equation (3.2) it follows that φ is p-harmonic if and only if it is harmonic, that is $p = 2$. In the second part of the proof, we assume that φ is p-harmonic with nonconstant energy density $e(\varphi)$. From Equation (3.2), we derive a homogeneous polynomial equation because if it vanishes on $\mathbb{S}^m$, then it vanishes on $\mathbb{R}^{m+1}$. We study the coefficient list in each component of the new homogeneous equation and it will lead to contradiction, but we will not present it. We note that the technique is similar to that used in the proof of of Theorem 5.2.

4. Intrinsic biharmonic quadratic forms

Intrisic biharmonic maps, or simply biharmonic maps, represent a higher order generalization of the well-known harmonic maps. As suggested by Eells and Sampson [**14, 15**], or Eells and Lemaire [**12**], biharmonic maps $\varphi : (M^m, g) \to (N^n, h)$ between two Riemannian manifolds are the critical points of the bienergy functional

$$E^2 : C^\infty(M, N) \to \mathbb{R}, \qquad E^2(\phi) = \frac{1}{2}\int_M |\tau(\phi)|^2 \; v_g,$$

where M is compact. In 1986, Jiang proved in [**20, 21**] that the biharmonic maps are characterized by the vanishing of their bitension field, which is given by

$$\tau^2(\phi) = -\Delta\tau(\varphi) - \operatorname{trace} R^N\left(d\varphi(\cdot), \tau(\varphi)\right) d\varphi(\cdot).$$

The equation $\tau^2(\varphi) = 0$ is called the biharmonic equation and it is a forth order semilinear elliptic equation. For recent result, one can see [**17**], [**27**].

Further, we recall the following result.

THEOREM 4.1 ([**24**]). *Let M be a compact manifold and consider $\psi : M \to \mathbb{S}^n(r/\sqrt{2})$ a nonconstant map, where $\mathbb{S}^n(r/\sqrt{2})$ is a small hypersphere of radius $r/\sqrt{2}$ of $\mathbb{S}^{n+1}(r)$. The map $\phi = \mathrm{i} \circ \psi : M \to \mathbb{S}^{n+1}(r)$, where i is the canonical inclusion, is proper biharmonic if and only if ψ is harmonic and the energy density $e(\psi)$ is constant.*

REMARK 4.2. As we can see from the proof, the compactness is needed only for the direct implication.

REMARK 4.3. When the map ψ is an isometric immersion, i.e., M is a submanifold, or a Riemannian submersion, the result was obtained in [**7**] and [**25**], respectively.

As an example of the above result, we can consider the map $\psi : \mathbb{S}^3(\sqrt{2}) \to \mathbb{S}^2\left(1/\sqrt{2}\right)$ as being the classical Hopf fibration. This map is a Riemannian submersion and then, by composing with the inclusion of $\mathbb{S}^2\left(1/\sqrt{2}\right)$ into $\mathbb{S}^3$, we obtain that

$$\varphi : \mathbb{S}^3(\sqrt{2}) = \left\{(z^1, z^2) \in \mathbb{C}^2 : \left|z^1\right|^2 + \left|z^2\right|^2 = 2\right\} \to \mathbb{S}^3$$

given by

$$\begin{aligned} \varphi(z^1, z^2) &= \frac{1}{2\sqrt{2}}\left(2z^1\overline{z^2}, \left|z^1\right|^2 - \left|z^2\right|^2, 2\right) \\ &= \frac{1}{2\sqrt{2}}\left(2z^1\overline{z^2}, \left|z^1\right|^2 - \left|z^2\right|^2, \left|z^1\right|^2 + \left|z^2\right|^2\right) \end{aligned} \tag{4.1}$$

is a proper biharmonic map (see [**25**]). As homothetic changes of the domain or codomain metrics preserve the harmoncity and biharmonicity, we can assume that φ maps $\mathbb{S}^3$ into $\mathbb{S}^3$. We also note that the components of φ are (restrictions of) homogeneous polynomials of degree 2.

As another application to Theorem 4.1 from above, using Theorem 2.5 we get Theorem 4.4.

THEOREM 4.4 ([**1**]). *There exist full proper biharmonic maps from $\mathbb{S}^7$ to $\mathbb{S}^8$, whose components are homogeneous polynomials of degree 2.*

Many examples of proper biharmonic maps were obtained when the target manifold is the Euclidean sphere and the maps take values in a small hypersphere of it. Taking into account the expression of the second fundamental form of a small hypersphere of the target sphere, we obtain a slight generalization of a result in [**24**].

THEOREM 4.5. *Let $\psi : M \to \mathbb{S}^{n-1}(r)$ be a nonconstant map, and consider $\varphi = \mathrm{i} \circ \psi : M \to \mathbb{S}^n(R)$, where $0 < r < R$ and $\mathrm{i} : \mathbb{S}^{n-1}(r) \to \mathbb{S}^n(R)$ is the inclusion map. Then, the bitension field of φ is given by*

$$\begin{aligned}\tau^2(\varphi) =& \tau^2(\psi) + 2\left(\frac{1}{R^2} - \frac{1}{r^2}\right) \mathrm{d}\psi\left(\operatorname{grad}|\mathrm{d}\psi|^2\right) + 4\left(\frac{1}{R^2} - \frac{1}{r^2}\right) e(\psi)\tau(\psi) \\ &+ \frac{\sqrt{R^2 - r^2}}{Rr}\left(2\Delta\left(e(\psi)\right) - 2\operatorname{div}\theta^\sharp + |\tau(\psi)|^2 - 4\left(\frac{2}{R^2} - \frac{1}{r^2}\right)\left(e(\psi)\right)^2\right)\eta,\end{aligned}$$

where $\theta(X) := \langle \mathrm{d}\psi(X), \tau(\psi)\rangle$ and $\theta^\sharp$ is the vector field associated to the 1-form θ through the musical isomorphism.

We have the following consequences.

COROLLARY 4.6. *Assume that φ is biharmonic.*

(1) *If the map ψ is harmonic with constant energy density, then $r = R/\sqrt{2}$.*
(2) *If M is compact, then $r \geq R/\sqrt{2}$. Moreover, if $r = R/\sqrt{2}$, then ψ is harmonic and has constant energy density (see [**24**]).*
(3) *If $e(\varphi)$ is constant and $\operatorname{div}\theta^\sharp = 0$, then $r \geq R/\sqrt{2}$. Moreover, if $r = R/\sqrt{2}$, then ψ is harmonic.*

Now, let $\varphi : (M^m, g) \to \mathbb{S}^n$ be a map and let $\mathrm{i} : \mathbb{S}^n \to \mathbb{R}^{n+1}$ be the standard isometric embedding of the unit Euclidean sphere into Euclidean space. We consider the composition map

$$\Phi = \mathrm{i} \circ \varphi : (M^m, g) \to \mathbb{R}^{n+1}.$$

As usual, we identify locally $\mathrm{d}\varphi(X)$ with $\mathrm{d}\Phi(X)$, for any vector fields X tangent to M.

We denote $\theta = \langle \mathrm{d}\varphi, \tau(\varphi)\rangle = \langle \mathrm{d}\Phi, \tau(\Phi)\rangle$, i.e., θ is the 1-form on M given by

$$\theta(X) = \langle \mathrm{d}\varphi(X), \tau(\varphi)\rangle = \langle \mathrm{d}\Phi(X), \tau(\Phi)\rangle.$$

By $\theta^\sharp$ we denote the vector field associated to the 1-form θ through the musical isomorphism. With these notations, we can state Theorem 4.7.

THEOREM 4.7 ([**1**, **26**]). *Let $\varphi : (M^m, g) \to \mathbb{S}^n$ be a map and let $\mathrm{i} : \mathbb{S}^n \to \mathbb{R}^{n+1}$ be the standard isometric embedding. Then, φ is a biharmonic map if and only if the vector function $\Phi = \mathrm{i} \circ \varphi : (M^m, g) \to \mathbb{R}^{n+1}$ solves the following PDE:*

$$\tau^2(\Phi) + 2|\mathrm{d}\Phi|^2\tau(\Phi) + \left(-\Delta|\mathrm{d}\Phi|^2 + 2\operatorname{div}\theta^\sharp - |\tau(\Phi)|^2 + 2|\mathrm{d}\Phi|^4\right)\Phi + 2\mathrm{d}\Phi\left(\operatorname{grad}|\mathrm{d}\Phi|^2\right) = 0.$$

Next, we give an application of Theorem 4.7 for a particular class of maps. Consider the diagram below

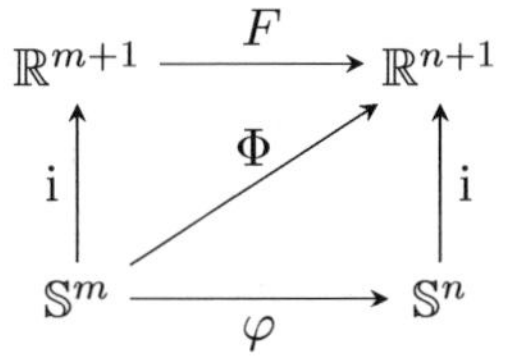

where $F : \mathbb{R}^{m+1} \to \mathbb{R}^{n+1}$ is a vector valued function such that each component is a homogeneous polynomial of degree k. Such a map F is called form of degree k.

THEOREM 4.8 ([**1**]). *The bitension field of the map φ is given by*

$$\begin{aligned}\tau^2(\varphi) = & \overset{o}{\Delta}\overset{o}{\Delta}F + 2\left(mk + 2k^2 - 3k - m + 3 - \left|\overset{o}{\mathrm{d}}F\right|^2\right)\overset{o}{\Delta}F \\ & + \left(-2\overset{o}{\Delta}\left(\left|\overset{o}{\mathrm{d}}F\right|^2\right) - 2\left|\overset{o}{\nabla}\overset{o}{\mathrm{d}}F\right|^2 + \left|\overset{o}{\Delta}F\right|^2\right. \\ & \left.-2\left(2mk + 6k^2 - 6k - m + 3\right)\left|\overset{o}{\mathrm{d}}F\right|^2 + 2\left|\overset{o}{\mathrm{d}}F\right|^4 + 4k^2(m + 2k - 1)\right)\Phi \\ & + 2\overset{o}{\mathrm{d}}F\left(\overset{o}{\mathrm{grad}}\left(\left|\overset{o}{\mathrm{d}}F\right|^2\right)\right),\end{aligned} \tag{4.2}$$

where $\overset{o}{\mathrm{d}}$, $\overset{o}{\nabla}$, $\overset{o}{\Delta}$ *and* $\overset{o}{\mathrm{grad}}$ *denote operators that act on* $\mathbb{R}^{m+1}$.

In the special case of quadratic forms, i.e., for $k = 2$, using Equations (1.1), (1.2), (1.3), and Proposition 1.1, from Equation (4.2) we obtain the following.

THEOREM 4.9 ([**1**]). *Let $F : \mathbb{R}^{m+1} \to \mathbb{R}^{n+1}$ be a quadratic form given by*

$$F(\overline{x}) = \left(X^t A_1 X, X^t A_2 X, \dots, X^t A_{n+1} X\right),$$

such that if $|\overline{x}| = 1$, then $|F(\overline{x})| = 1$. We consider $\varphi : \mathbb{S}^m \to \mathbb{S}^n$ defined by $\varphi(\overline{x}) = F(\overline{x})$ and $\Phi = \mathrm{i} \circ \varphi : \mathbb{S}^m \to \mathbb{R}^{n+1}$. If we denote $S = A_1^2 + A_2^2 + \cdots + A_{n+1}^2$, then at a point $\overline{x} \in \mathbb{S}^m$, the bitension field of φ has the following expression

$$\begin{aligned}\tau^2(\varphi)_{\overline{x}} = & -4\left(m + 5 - 4X^t S X\right)\left(\mathrm{tr}A_1, \mathrm{tr}A_2, \dots, \mathrm{tr}A_{n+1}\right) \\ & + 4\left((m+3)(m+5) - 6(m+5)X^t S X + 8\left(X^t S X\right)^2\right)\Phi(\overline{x}) \\ & + 32\left(X^t A_1 S X, X^t A_2 S X, \dots, X^t A_{n+1} S X\right).\end{aligned} \tag{4.3}$$

In general, when the map $\varphi : \mathbb{S}^m \to \mathbb{S}^n$ is the restriction of a quadratic form $F : \mathbb{R}^{m+1} \to \mathbb{R}^{n+1}$, because of the homogeneity, F has to bear certain conditions on $\mathbb{R}^{m+1}$ (see [**36**]). The next three classification results rely on those properties of F.

THEOREM 4.10 ([**1**]). *Up to orthogonal transformations of the domain and/or the codomain, the only proper biharmonic quadratic form $\varphi : \mathbb{S}^1 \to \mathbb{S}^n$, $n \geq 2$ is obtained from the restriction of the quadratic form $F : \mathbb{R}^2 \to \mathbb{R}^{n+1}$, given by*

$$F(x, y) = \left(x^2, c^1 y^2 + 2\gamma^1 xy, \dots, c^n y^2 + 2\gamma^n xy\right),$$

such that

$$\left(c^1\right)^2 + \cdots + \left(c^n\right)^2 = 1, \quad c^1\gamma^1 + \cdots + c^n\gamma^n = 0$$

and

$$\left(\gamma^1\right)^2 + \cdots + \left(\gamma^n\right)^2 = \frac{1}{2}.$$

Moreover, the image of φ is the circle of radius $1/\sqrt{2}$ of $\mathbb{S}^n$.

For the case when $\mathbb{S}^1$ is the target manifold, using Theorem 1.3, one can prove that any quadratic form $\varphi : \mathbb{S}^m \to \mathbb{S}^1$, for $m \geq 1$ is constant. Thus, we can state Proposition 4.11.

PROPOSITION 4.11 ([**1**]). *There is no proper biharmonic quadratic form from* $\mathbb{S}^m$ *to* $\mathbb{S}^1$, *for* $m \geq 1$.

For the proof of the next two theorems, one can use the results of Wood, Yiu, and Chang that were mentioned in Section 1.

THEOREM 4.12 ([**1**]). *There is no proper biharmonic quadratic form from* $\mathbb{S}^m$ *to* $\mathbb{S}^2$, *for* $m \geq 2$.

THEOREM 4.13 ([**1**]). *Up to homothetic transformations of the domain and/or codomain, the only proper biharmonic quadratic form from* $\mathbb{S}^m$ *to* $\mathbb{S}^3$, $m \geq 2$, *is the Hopf fibration* $\psi : \mathbb{S}^3 \to \mathbb{S}^2\left(1/\sqrt{2}\right)$ *followed by the inclusion, as described in example* (4.1).

All results obtained suggest the following.

OPEN PROBLEM. If $\varphi : \mathbb{S}^m \to \mathbb{S}^n$ is a proper biharmonic quadratic form then, up to an isometry of $\mathbb{S}^n$, the first n components of φ are harmonic polynomials on $\mathbb{R}^{m+1}$ and form a map $\psi : \mathbb{S}^m \to \mathbb{S}^{n-1}\left(1/\sqrt{2}\right)$.

In order to give an answer, we first give the following theorem.

THEOREM 4.14 ([**2**]). *Let* $\varphi : \mathbb{S}^m \to \mathbb{S}^n$ *be a quadratic form. Then,* φ *is proper-biharmonic if and only if its energy density* $e(\varphi)$ *is constant and equal to* $(m+1)/2$.

COROLLARY 4.15 ([**2**]). *Let* $\varphi : \mathbb{S}^m \to \mathbb{S}^n$ *be a proper biharmonic quadratic form. Then,* $\left|\overset{o}{\Delta} F\right|^2 = 2(m+1)^2$.

Now, using the results presented above, we can give a positive answer to this problem.

THEOREM 4.16 ([**2**]). *If* $\varphi : \mathbb{S}^m \to \mathbb{S}^n$ *is a proper biharmonic quadratic form, then up to an isometry of* $\mathbb{S}^n$, *the first* n *components of* φ *form a harmonic map* $\psi : \mathbb{S}^m \to \mathbb{S}^{n-1}\left(1/\sqrt{2}\right)$.

In the case of quadratic forms, Corollary 4.6 has a more rigid form, implying the reduction of the codimension for the map φ.

PROPOSITION 4.17 ([**2**]). *Let* $\varphi : \mathbb{S}^m \to \mathbb{S}^n$ *be a proper biharmonic quadratic form. Assume that* φ *lies in a small hypersphere* $\mathbb{S}^{n-1}(r)$ *of* $\mathbb{S}^n$. *Then,* $r \geq 1/\sqrt{2}$. *Moreover,*

(1) *if* $r = 1/\sqrt{2}$, *then the map* $\psi : \mathbb{S}^m \to \mathbb{S}^{n-1}\left(1/\sqrt{2}\right)$ *determined by* φ *is harmonic;*
(2) *if* $r > 1/\sqrt{2}$, *then* φ *lies in* $\mathbb{S}^{n-2}\left(1/\sqrt{2}\right) \subset \mathbb{S}^{n-1}(r)$ *and* $\psi : \mathbb{S}^m \to \mathbb{S}^{n-2}\left(1/\sqrt{2}\right)$ *is harmonic.*

As applications to Theorem 4.16, we give two classification results.

Using the result of Calabi concerning the uniqueness of compact minimal 2-dimensional round spheres in $\mathbb{S}^n$, i.e., the uniqueness of the Borůvka spheres (see [**8**] and also [**6**], [**22**]), we obtain Theorem 4.18.

THEOREM 4.18 ([**2**]). *Let* $\varphi : \mathbb{S}^2 \to \mathbb{S}^n$ *be a full quadratic form. Assume that* φ *is homothetic. Then,* φ *is proper biharmonic if and only if* $n = 5$, $\varphi\left(\mathbb{S}^2\right) \subset \mathbb{S}^4\left(1/\sqrt{2}\right)$, *and up to homothetic changes of domain and codomain,* $\psi : \mathbb{S}^2 \to \mathbb{S}^4\left(1/\sqrt{2}\right)$ *is the Veronese map.*

Next, using Theorem 2.4, i.e., the classification result of Toth concerning the harmonic maps from $\mathbb{S}^3$ to $\mathbb{S}^n$, we obtain the following theorem.

THEOREM 4.19 ([**2**]). *Full quadratic proper biharmonic maps of $\mathbb{S}^3$ into $\mathbb{S}^n$ exist only if $3 \leq n \leq 9$ and $n \neq 4$. Moreover, if $\varphi : \mathbb{S}^3 \to \mathbb{S}^n$ is such a map, then there exist $U \in O(4)$, $V \in O(n+1)$ and a symmetric positive definite matrix $B \in \mathbb{S}^2\left(\mathbb{R}^{n+1}\right)$ such that*

$$V \circ \varphi \circ U = \left(\frac{1}{\sqrt{2}} B \circ \varphi_n, \frac{1}{\sqrt{2}}\right).$$

5. λ-biharmonic quadratic forms

In 1974, Eliasson [**16**] started studying the λ-biharmonic maps, followed by Lemaire [**23**] in 1981, which are a generalization of the biharmonic maps. The λ-biharmonic maps $\varphi : (M^m, g) \to (N^n, h)$ between two Riemannian manifolds, where λ is a real number, are the critical points of the λ-bienergy functional

$$E^2_\lambda : C^\infty(M,N) \to \mathbb{R}, \qquad E^2_\lambda(\varphi) = E^2 + \lambda E,$$

where M is compact. Equivalently, a map ϕ is λ-biharmonic if its λ-bitension field vanishes, which is given by

$$\tau^2_\lambda(\varphi) = \tau^2(\varphi) - \lambda\tau(\varphi). \tag{5.1}$$

We note that for $\lambda = 0$ they are the classical biharmonic maps. A map that is nonharmonic λ-biharmonic is call proper λ-biharmonic.

Further in this section we consider the special case of quadratic forms. Using Theorems 2.1 and 4.4, from Equation (5.1) we obtain the following.

THEOREM 5.1. *Let $\varphi : \mathbb{S}^m \to \mathbb{S}^n$ be a quadratic form. Then its λ-bitension field is given by*

$$\begin{aligned}\tau^2_\lambda(\varphi)_{\overline{x}} = &-2\left(2m+10-8X^tSX+\lambda\right)\left(\mathrm{tr}A_1, \mathrm{tr}A_2, \ldots, \mathrm{tr}A_{n+1}\right)\\ &+\left(4(m+3)(m+5)+2\lambda(m+3)-\left(24(m+5)+4\lambda\right)X^tSX\right.\\ &\left.+32\left(X^tSX\right)^2\right)\Phi(\overline{x})+32\left(X^tA_1SX, X^tA_2SX, \ldots, X^tA_{n+1}SX\right).\end{aligned} \tag{5.2}$$

THEOREM 5.2. *Let $\varphi : \mathbb{S}^m \to \mathbb{S}^n$ be a quadratic form between spheres, $m \geq 2$. Then φ is proper λ-biharmonic if and only if it has constant energy density $e(\varphi) = (m+1)/2 + \lambda/4$ and $|\lambda| < 2(m+1)$.*

We note that in the proof, some computations for the coefficients of certain polynomials have been done using *Mathematica*® and the technique is similar to that used in the proof of Theorem 4.14.

PROOF. The proof is divided in two parts. In the first part we prove that if φ has constant energy density, then it is proper λ-biharmonic if and only if it has constant energy density $e(\varphi) = (m+1)/2 + \lambda/4$. Then we prove that $|\lambda| < 2(m+1)$. In the second part, we will assume that φ is proper λ-biharmonic with nonconstant energy density, but we will obtain a contradiction.

We suppose from the beginning that φ is not harmonic, thus

$$\overset{o}{\Delta} F \neq \overline{0} \quad \text{and} \quad e(\varphi) \neq m+1.$$

Assume that $e(\varphi) > 0$ is constant. Therefore, using Equations (1.1) and (3.3), this is equivalent to $S = cI_{m+1}$, where $c > 1$ is a real constant. Then the λ-bitension field from Equation (5.2) takes the form

(5.3)
$$\tau_\lambda^2(\varphi)_{\overline{x}} = \frac{1}{8}\left(\frac{2m+10+\lambda}{8} - c\right)\overset{o}{\Delta}F + 32\left(\frac{m+3}{2} - c\right)\left(\frac{2m+10+\lambda}{8} - c\right)\Phi.$$

Now, we take φ to be proper λ-biharmonic. The condition $e(\varphi) \neq m+1$ is equivalent to $c \neq (m+3)/2$. From Equation (5.3), since $\overset{o}{\Delta}F$ is a constant vector of $\mathbb{R}^{n+1}$, but Φ is not a constant map, it follows that $c = (2m+10+\lambda)/8$, which is equivalent to $e(\varphi) = (m+1)/2 + \lambda/4$.

Conversly, if $e(\varphi) = (m+1)/2 + \lambda/4$, then using Equation (5.3), it follows directly that φ is λ-biharmonic.

For a proper λ-biharmonic quadratic form, the condition $e(\varphi) > 0$ is equivalent to $\lambda > -2(m+1)$.

Using Proposition 1.1, since $e(\varphi) = (m+1)/2 + \lambda/4$, we get

$$\left|\overset{o}{\Delta}F\right|^2 = (m+1)(2m+2-\lambda). \tag{5.4}$$

The condition $\left|\overset{o}{\Delta}F\right|^2 > 0$ is equivalent to $\lambda < 2(m+1)$.

Further, we assume that φ is proper λ-biharmonic with nonconstant energy density. We transform the nonhomogeneous polynomial equation $\tau_\lambda^2 = \overline{0}$ into a homogeneous polynomial equation of degree 6 because if a homogeneous polynomial vanishes on $\mathbb{S}^m$, then it vanishes on $\mathbb{R}^{m+1}$. Thus, from Equation (5.2) we obtain

$$\begin{aligned}\overline{0} = &-2\left((2m+10+\lambda)\left|\overline{x}\right|^2 - 8X^tSX\right)\left|\overline{x}\right|^4(\operatorname{tr}A_1, \operatorname{tr}A_2, \ldots, \operatorname{tr}A_{n+1})\\ &+\left(4(m+3)(m+5)\left|\overline{x}\right|^4 + 2\lambda(m+3)\left|\overline{x}\right|^4 - (24(m+5)+4\lambda)\left|\overline{x}\right|^2X^tSX\right.\\ &\left.+32\left(X^tSX\right)^2\right)\Phi(\overline{x}) + 32\left|\overline{x}\right|^4\left(X^tA_1SX, X^tA_2SX, \ldots, X^tA_{n+1}SX\right).\end{aligned} \tag{5.5}$$

The matrix S defines a quadratic form. We perform an orthogonal change of the domain variables x^1, x^2, …, x^{m+1}, which brings S in diagonal form. Thus, we write S as

$$S = \begin{bmatrix} s_1 & 0 & \ldots & 0 \\ 0 & s_2 & \ldots & 0 \\ \vdots & \vdots & \ddots & \vdots \\ 0 & 0 & \ldots & s_{m+1} \end{bmatrix}.$$

We write the matrices A_α as

$$A_\alpha = \frac{1}{2}\begin{bmatrix} 2a_1^\alpha & a_{12}^\alpha & \ldots & a_{1(m+1)}^\alpha \\ a_{12}^\alpha & 2a_2^\alpha & \ldots & a_{2(m+1)}^\alpha \\ \vdots & \vdots & \ddots & \vdots \\ a_{1(m+1)}^\alpha & a_{2(m+1)}^\alpha & \ldots & 2a_{m+1}^\alpha \end{bmatrix}, \quad \forall \alpha \in \{1, 2, \ldots, n+1\}.$$

Now we study the coefficient list for each component of Equation (5.5). We see that in each αth component, for $\alpha \in \{1, 2, \ldots, n+1\}$, the coefficient of $\left(x^k\right)^6$, which has to vanish, is

(5.6)
$$2\left(10+2m+\lambda-8s_k\right)\left(a_k^\alpha\left(3+m-2s_k\right)-\mathrm{tr}A_\alpha\right)=0,\quad \forall k\in\{1,2,\dots,m+1\}.$$

Thus, for any k arbitrarily fixed we have of the following two situations: either

$$s_k=(10+2m+\lambda)/8,$$

or

$$\mathrm{tr}A_\alpha=a_k^\alpha\left(3+m-2s_k\right),\quad \forall\alpha\in\{1,2,\dots,n+1\}.$$

The last equation is equivalent to

$$-\frac{1}{2}\overset{o}{\Delta}F=\left(3+m-2s_k\right)\overline{a}_k,\quad \forall k\in\{1,2,\dots,m+1\}. \tag{5.7}$$

Since φ is not harmonic, from Equation (5.7) it follows that $s_k\neq(m+3)/2$, for any $k\in\{1,2,\dots,m+1\}$.

Further, we distinguish two cases:

CASE 1. There exists $k_0\in\{1,2,\dots,m+1\}$ such that $s_{k_0}=(10+2m+\lambda)/8$.

CASE 2. For any $k\in\{1,2,\dots,m+1\}$, we have $s_k\neq(10+2m+\lambda)/8$.

For *Case 1*, without loss of generality, we assume that $s_1=(10+2m+\lambda)/8$. From the coefficients of $\left(x^1\right)^5x_k$, for any $k\in\{2,\dots,m+1\}$, in each αth component of Equation (5.5), we obtain the following system of equations.

$$\begin{cases}-2a_{12}^\alpha\left(10+2m+\lambda-8s_2\right)=0,\\ -2a_{13}^\alpha\left(10+2m+\lambda-8s_3\right)=0,\\ \dots\\ -2a_{1(m+1)}^\alpha\left(10+2m+\lambda-8s_{m+1}\right)=0,\end{cases}\qquad \forall\alpha\in\{1,2,\dots,n+1\}. \tag{5.8}$$

By direct calculation of S and using Equations (1.4), for any $k\in\{1,2,\dots,m+1\}$ we obtain

$$\begin{aligned}s_k&=\left|\overline{a}_k\right|^2+\frac{1}{4}\sum_{i=1,\ i\neq k}^{m+1}\left|\overline{a}_{ik}\right|^2\\ &=1+\frac{1}{4}\sum_{i=1,\ i\neq k}^{m+1}\left|\overline{a}_{ik}\right|^2.\end{aligned}$$

Since $s_1>1$, it follows that there exists $k_0\in\{1,2,\dots,m+1\}$ such that $\overline{a}_{1k_0}\neq\overline{0}$. Without loss of generality, we assume that $\overline{a}_{12}\neq\overline{0}$. From Equations (5.8), it follows that also $s_2=(10+2m+\lambda)/8$.

If $m=1$, then $S=((10+2m+\lambda)/8)\,I_{m+1}$, which further implies that φ has constant energy density, which is a contradiction. Now assume $m\geq 2$. From the the coefficients of $\left(x^1\right)^3x^2\left(x^k\right)^2$, for any $k\in\{3,4,\dots,m+1\}$, in each α component of Equation (5.5), we obtain the following system of equations.

$$\begin{cases}\frac{1}{2}a_{12}^\alpha(10+2m-\lambda)\left(10+2m+\lambda-8s_3\right)=0,\\ \frac{1}{2}a_{12}^\alpha(10+2m-\lambda)\left(10+2m+\lambda-8s_4\right)=0,\\ \dots\\ \frac{1}{2}a_{12}^\alpha(10+2m-\lambda)\left(10+2m+\lambda-8s_{m+1}\right)=0,\end{cases}\qquad \forall\alpha\in\{1,2,\dots,n+1\}.$$

Since $\overline{a}_{12} \neq \overline{0}$, it follows that $\lambda = 10 + 2m$. Otherwise, $s_k = (10 + 2m + \lambda)/8$ for any $k \in \{1, 2, \dots, m+1\}$. But this implies that φ has constant energy density, which is in contradiction with our supposition. Since $\lambda = 10 + 2m$, we obtain $s_1 = s_2 = (m+5)/2$. There exists $k_0 \in \{3, 4, \dots, m+1\}$ such that $s_{k_0} \neq (m+5)/2$.

Without loss of generality, we assume that $k_0 = 3$. From Equation (5.6), for $k = 3$, we obtain

$$\mathrm{tr} A_\alpha = a_3^\alpha \left(3 + m - 2s_3\right), \quad \forall \alpha \in \{1, 2, \dots, n+1\}.$$

Then, by looking at the coefficient of $\left(x^1\right)^4 \left(x^3\right)^2$, we get

$$-8a_3^\alpha (5 + m - 2s_3)^2 = 0, \quad \forall \alpha \in \{1, 2, \dots, n+1\}.$$

which implies

$$\mathrm{tr} A_\alpha = -2a_3^\alpha, \quad \forall \alpha \in \{1, 2, \dots, n+1\}. \tag{5.9}$$

Since $|\overline{a}_3| = 1$, it follows that $s_3 = (m+5)/2$, which is a contradiction.

For *Case 2*, using Equation (5.7), we obtain that $\overline{a}_k \parallel \overline{a}_l$, for any $k, l \in \{1, 2, \dots, m+1\}$, and further, since they are unit vectors, it follows that for any $k, l \in \{1, 2, \dots, m+1\}$, $k \neq l$, either $\overline{a}_k = \overline{a}_l$, or $\overline{a}_k = -\overline{a}_l$. Thus, there exist k_0 and l_0 such that $\overline{a}_{k_0} = -\overline{a}_{l_0}$. Otherwise, $s_k = s_l$ for any $k, l \in \{1, 2, \dots, m+1\}$, and further we obtain that φ has constant energy density, which is a contradiction. Without loss of generality, we assume that $k_0 = 1$ and $l_0 = 2$. From Equation (5.7) we obtain

$$s_2 = 3 + m - s_1.$$

Using the third equation from System (1.4), we obtain

$$\left|\overline{a}_{12}\right|^2 = 2 - \langle \overline{a}_1, \overline{a}_2 \rangle = 4.$$

The coefficient of $\left(x^1\right)^5 x^2$ from each α component of Equation (5.5), gives us

$$2a_{12}^\alpha \left(18 + 2m + \lambda - 8s_1\right)\left(3 + m - 2s_1\right) = 0, \quad \forall \alpha \in \{1, 2, \dots, n+1\}.$$

Since $|\overline{a}_{12}| = 2$, and $s_1 \neq (m+3)/2$, it follows that $s_1 = (18 + 2m + \lambda)/8$. Further, from the coefficient of $\left(x^1\right)^3 \left(x^2\right)^3$ we obtain

$$a_{12}^\alpha (6 - 2m + \lambda)^2 = 0, \quad \forall \alpha \in \{1, 2, \dots, n+1\},$$

which implies that $\lambda = 2m - 6$, and further $s_1 = (m+3)/2$, which is a contradiction. Thus, we end *Case 2* and the proof is complete. □

THEOREM 5.3. *If $\varphi : \mathbb{S}^m \to \mathbb{S}^n$ is a proper λ-biharmonic quadratic form, $\lambda < 2(m+1)$, then and up to an isometry of $\mathbb{S}^n$, the first n components of φ form a harmonic map*

$$\psi : \mathbb{S}^m \to \mathbb{S}^{n-1}\left(\sqrt{\frac{1}{2} + \frac{\lambda}{4(m+1)}}\right).$$

PROOF. First, we recall that φ is proper biharmonic if and only if its energy density $e(\varphi)$ is constant and equal to $(m+1)/2 + \lambda/4$. We recall the following formula

$$\tau(\varphi) = -\overset{o}{\Delta} F + \left(\left|\overset{o}{\mathrm{d}} F\right|^2 - 2(m+3)\right)\Phi.$$

From the above equation, it follows that

$$\left\langle \overset{o}{\Delta} F, \Phi \right\rangle = -(m+1) + \frac{\lambda}{2}.$$

Thus, the image of φ also lies in the hyperplane

$$(\Pi) : \left\langle \overset{o}{\Delta} F, \Phi \right\rangle = -(m+1) + \lambda/2.$$

It follows that the image of φ lies in a smaller hypersphere $\mathbb{S}^{n-1}(r)$, where

$$\begin{aligned} r^2 &= 1 - d^2(O, (\Pi)) \\ &= 1 - \frac{\left((m+1) - \frac{\lambda}{2}\right)^2}{\left|\overset{o}{\Delta} F\right|^2}. \end{aligned} \tag{5.10}$$

Using Proposition 1.1, since $e(\varphi) = (m+1)/2 + \lambda/4$, we get

$$\left|\overset{o}{\Delta} F\right|^2 = (m+1)(2m+2-\lambda). \tag{5.11}$$

From Equations (5.10) and (5.11), it follows that

$$r = \sqrt{\frac{1}{2} + \frac{\lambda}{4(m+1)}}.$$

We note that the condition $0 < r < 1$ is equivalent to $|\lambda| < 2(m+1)$. Now, we act with an isometry on $\mathbb{S}^n$ such that $\overset{o}{\Delta} F$ becomes

$$\overset{o}{\Delta} F = \left(0, \ldots, 0, -\sqrt{(m+1)(2m+2-\lambda)}\right).$$

The last result implies that the first n components of φ form a harmonic map

$$\psi : \mathbb{S}^m \to \mathbb{S}^{n-1}\left(\sqrt{\frac{1}{2} + \frac{\lambda}{4(m+1)}}\right). \qquad \square$$

PROPOSITION 5.4. *Let $\varphi : \mathbb{S}^m \to \mathbb{S}^n$ be a proper λ-biharmonic quadratic form, $|\lambda| < 2(m+1)$. Assume that φ lies in a small hypersphere $\mathbb{S}^{n-1}(r)$ of $\mathbb{S}^n$. Then, $r \geq \sqrt{1/2 + \lambda/(4(m+1))}$. Moreover,*

1. *if $r = \sqrt{1/2 + \lambda/(4(m+1))}$, then the map*
$$\psi : \mathbb{S}^m \to \mathbb{S}^{n-1}\left(\sqrt{\frac{1}{2} + \frac{\lambda}{4(m+1)}}\right)$$
determined by φ is harmonic;
2. *if $r > \sqrt{1/2+\lambda/(4(m+1))}$, then φ lies in $\mathbb{S}^{n-2}\left(\sqrt{1/2 + \lambda/(4(m+1))}\right) \subset \mathbb{S}^{n-1}(r)$ and the map*
$$\psi : \mathbb{S}^m \to \mathbb{S}^{n-2}\left(\sqrt{\frac{1}{2} + \frac{\lambda}{4(m+1)}}\right)$$
determined by φ is harmonic.

PROOF. First, we make the following identification $S^{n-1}(r) \equiv S^{n-1}(r) \times \{\sqrt{1-r^2}\}$. It follows that the last component of F is $A_{n+1} = \sqrt{1-r^2} I_{m+1}$.

In Case (1), we obtain

$$A_{n+1} = \sqrt{\frac{1}{2} - \frac{\lambda}{4(m+1)}} I_{m+1}. \tag{5.12}$$

Since φ is proper λ-biharmonic, we get

$$\left|\overset{o}{\Delta}F\right|^2 = (m+1)(2m+2-\lambda), \tag{5.13}$$

and from Equations (1.2), (5.12) and (5.13) it follows directly that

$$\overset{o}{\Delta}F = \left(0, \ldots, 0, -\sqrt{(m+1)(2m+2-\lambda)}\right).$$

This implies that the first n components of φ form a harmonic map

$$\psi : \mathbb{S}^m \to \mathbb{S}^{n-1}\left(\sqrt{\frac{1}{2}+\frac{\lambda}{4(m+1)}}\right).$$

In Case (2), we observe the image of φ lies in the following two hyperplanes:

$$(\Pi) : y^{n+1} = \sqrt{1-r^2},$$
$$(\Pi_1) : \left\langle \overset{o}{\Delta}F, \overline{y}\right\rangle = -(m+1) + \lambda/2.$$

Thus, φ also lies in a hypersphere

$$\mathbb{S}^{n-1}\left(\sqrt{\frac{1}{2}+\frac{\lambda}{4(m+1)}}\right)$$

of (Π_1), centered in a point A. In order to find A we consider the line

$$d : \overline{y} = t\overset{o}{\Delta}F.$$

The intersection between the line d and the hyperplane (Π_1) is given by $t_0 = -1/(2(m+1))$. Therefore, the position vector of A is

$$t_0\overset{o}{\Delta}F = \left(\frac{\mathrm{tr}A_1}{m+1}, \frac{\mathrm{tr}A_2}{m+1}, \ldots, \frac{\mathrm{tr}A_n}{m+1}, \sqrt{1-r^2}\right).$$

We see that A is also contained in (Π), thus the intersection between (Π) and (Π_1) contains the center of the hypersphere

$$\mathbb{S}^{n-1}\left(\sqrt{\frac{1}{2}+\frac{\lambda}{4(m+1)}}\right).$$

It follows that φ lies in a totally geodesic sphere of

$$\mathbb{S}^{n-1}\left(\sqrt{\frac{1}{2}+\frac{\lambda}{4(m+1)}}\right),$$

and further, the map

$$\psi : \mathbb{S}^m \to \mathbb{S}^{n-2}\left(\sqrt{\frac{1}{2}+\frac{\lambda}{4(m+1)}}\right)$$

determined by φ is harmonic. It is obvious that

$$\mathbb{S}^{n-2}\left(\sqrt{\frac{1}{2}+\frac{\lambda}{4(m+1)}}\right)$$

is a smaller sphere of $\mathbb{S}^{n-1}(r)$ and it lies in a totally geodesic sphere $\mathbb{S}^{n-1}$ of $\mathbb{S}^n$. $\square$

6. Extrinsic biharmonic quadratic forms

Extrinsic biharmonic maps are another type of higher order generalization of the harmonic maps and they were studied by Chang, Wang, and Yang [**11**], Wang [**34,35**], etc. In order to define these maps, let (M^m, g) and (N^n, h) be two smooth and compact Riemannian manifolds without boundary. We consider N isometrically embedded into $\mathbb{R}^k$. For a smooth map $\varphi : M \to N$ we define the extrinsic bienergy functional

$$E^{2\text{ext}} : C^\infty(M,N) \to \mathbb{R}, \quad E^{2\text{ext}}(\varphi) = \frac{1}{2}\int_M |\Delta\varphi|^2 v_g,$$

where Δ is the Laplacian for a map from M to $\mathbb{R}^k$. Extrinsic biharmonic maps are defined as the critical points of the extrinsic bienergy functional.

Further, let $\varphi : (M^m, g) \to \mathbb{S}^n$ be a smooth map and $\Phi = \mathrm{i} \circ \varphi : M \to \mathbb{R}^{n+1}$ the composition map, where $\mathrm{i} : \mathbb{S}^n \to \mathbb{R}^{n+1}$ is the canonical inclusion. Then,

$$\tau(\Phi) = \tau(\varphi) - |\mathrm{d}\varphi|^2 \Phi \quad \text{and} \quad |\tau(\Phi)|^2 = |\tau(\varphi)|^2 + |\mathrm{d}\varphi|^4,$$

thus

$$E^{2\text{ext}}(\varphi) = E^2(\Phi) = E^2(\varphi) + 2E_4(\varphi),$$

where

$$E^2(\varphi) = \frac{1}{2}\int_M |\tau(\varphi)|^2 v_g \quad \text{and} \quad E_4(\varphi) = \frac{1}{4}\int_M |\mathrm{d}\varphi|^4 v_g.$$

Extrinsic biharmonic maps are characterized by the vanishing of their extrinsic bitension field, which, in this case, is given by

$$\begin{aligned}\tau^{2\text{ext}}(\varphi) = &- \Delta\tau(\varphi) - |\mathrm{d}\varphi|^2 \tau(\varphi) - \operatorname{trace}\langle \mathrm{d}\varphi, \tau(\varphi)\rangle \mathrm{d}\varphi \\ &- 2\mathrm{d}\varphi\left(\operatorname{grad}\left(|\mathrm{d}\varphi|^2\right)\right).\end{aligned} \tag{6.1}$$

THEOREM 6.1. *There is no nonharmonic extrinsic biharmonic quadratic form.*

PROOF. Let $\varphi : \mathbb{S}^m \to \mathbb{S}^n$ be a quadratic form. With the above notations, we recall from [**1**] the following results

$$\begin{aligned}-\Delta^\varphi\tau(\varphi) &= \tau^2(\Phi) + |\mathrm{d}\Phi|^2 \tau(\Phi) + \{-\Delta\left(|\mathrm{d}\Phi|^2\right) + 2\left(\operatorname{div}\theta^\sharp\right) - |\tau(\Phi)|^2 + |\mathrm{d}\Phi|^4\}\,\Phi \\ &\quad + 2\mathrm{d}\Phi\left(\operatorname{grad}\left(|\mathrm{d}\Phi|^2\right)\right) + \mathrm{d}\Phi(\theta^\sharp), \\ |\mathrm{d}\varphi|^2 \tau(\varphi) &= |\mathrm{d}\Phi|^2\left(\tau(\Phi) + |\mathrm{d}\Phi|^2 \Phi\right), \\ \operatorname{trace}\langle \mathrm{d}\varphi, \tau(\varphi)\rangle &= \mathrm{d}\varphi\left(\theta^\sharp\right) = \mathrm{d}\Phi\left(\theta^\sharp\right),\end{aligned}$$

where $\theta = \langle \mathrm{d}\varphi, \tau(\varphi)\rangle = \langle \mathrm{d}\Phi, \tau(\Phi)\rangle$.

We replace the above results in Equation (6.1) and we obtain

$$\tau^{2\text{ext}}(\varphi) = \tau^2(\Phi) + \left\{-\Delta\left(|\mathrm{d}\Phi|^2\right) + 2\operatorname{div}\theta^\sharp - |\tau(\Phi)|^2\right\}\Phi. \tag{6.2}$$

We also recall from [**1**] that for a k-form we have the following 4 formulas.

$$\begin{aligned}
\tau^2(\Phi) =& \overset{o}{\Delta}\overset{o}{\Delta}F + 2(mk + k^2 - 3k - m + 3)\overset{o}{\Delta}F + k^2(m + k - 1)^2\Phi, \\
\Delta\left(|\mathrm{d}\Phi|^2\right) =& \overset{o}{\Delta}\left(\left|\overset{o}{\mathrm{d}}F\right|^2\right) + 2(k-1)(m+2k-3)\left|\overset{o}{\mathrm{d}}F\right|^2, \\
\operatorname{div}\theta^\sharp =& \left|\overset{o}{\Delta}F\right|^2 + k(m+2k-3)\left|\overset{o}{\mathrm{d}}F\right|^2 - k^2(m+2k-3)(m+2k-1) \\
& - \frac{1}{2}\overset{o}{\Delta}\left(\left|\overset{o}{\mathrm{d}}F\right|^2\right) - \left|\overset{o}{\nabla}\overset{o}{\mathrm{d}}F\right|^2, \\
|\tau(\Phi)|^2 =& \left|\overset{o}{\tau}(F)\right|^2 + 2k(m+k-1)\left|\overset{o}{\mathrm{d}}F\right|^2 - k^2(m+k-1)(m+3k-1).
\end{aligned}$$

For $k = 2$, we replace them in Equation (6.2) and we obtain

(6.3)

$$\tau^{2\mathrm{ext}}(\varphi) = 2(m+1)\overset{o}{\Delta}F + \left\{-2\overset{o}{\Delta}\left(\left|\overset{o}{\mathrm{d}}F\right|^2\right) - 2\left|\overset{o}{\nabla}\overset{o}{\mathrm{d}}F\right|^2 + \left|\overset{o}{\Delta}F\right|^2 - 2(m+1)\left|\overset{o}{\mathrm{d}}F\right|^2\right\}\Phi.$$

Using Equations (1.1) and (1.2), Proposition 1.1, and Theorem 2.1, from Equation (6.3) we obtain

$$\begin{aligned}
\tau^{2\mathrm{ext}}(\varphi) &= -2(m+1)\left(-\overset{o}{\Delta}F + \left(\left|\overset{o}{\mathrm{d}}F\right|^2 - 2(m+3)\right)\Phi\right) \\
&= -2(m+1)\tau(\varphi).
\end{aligned}$$

Therefore, φ is extrinsic biharmonic if and only if it is harmonic. □

References

[1] R. Ambrosie, C. Oniciuc, and Y.-L. Ou, *Biharmonic homogeneous polynomial maps between spheres*, Results Math. **78** (2023), no. 4, Paper No. 159, 40, DOI 10.1007/s00025-023-01935-1. MR4599666

[2] R. Ambrosie and C. Oniciuc, *The energy density of biharmonic quadratic maps between spheres*, Differential Geom. Appl. **93** (2024), Paper No. 102096, 12, DOI 10.1016/j.difgeo.2023.102096. MR4682222

[3] P. Baird and S. Gudmundsson, *p-harmonic maps and minimal submanifolds*, Math. Ann. **294** (1992), no. 4, 611–624, DOI 10.1007/BF01934344. MR1190447

[4] P. Baird and J. C. Wood, *Harmonic morphisms between Riemannian manifolds*, London Mathematical Society Monographs. New Series, vol. 29, The Clarendon Press, Oxford University Press, Oxford, 2003, DOI 10.1093/acprof:oso/9780198503620.001.0001. MR2044031

[5] V. Branding and A. Siffert, *On p-harmonic self-maps of spheres*, Calc. Var. Partial Differential Equations **62** (2023), no. 4, Paper No. 139, 15, DOI 10.1007/s00526-023-02481-y. MR4579395

[6] R. L. Bryant, *Minimal surfaces of constant curvature in S^n*, Trans. Amer. Math. Soc. **290** (1985), no. 1, 259–271, DOI 10.2307/1999793. MR787964

[7] R. Caddeo, S. Montaldo, and C. Oniciuc, *Biharmonic submanifolds in spheres*, Israel J. Math. **130** (2002), 109–123, DOI 10.1007/BF02764073. MR1919374

[8] E. Calabi, *Minimal immersions of surfaces in Euclidean spheres*, J. Differential Geometry **1** (1967), 111–125. MR233294

[9] E. Cartan, *Sur des familles remarquables d'hypersurfaces isoparamétriques dans les espaces sphériques* (French), Math. Z. **45** (1939), 335–367, DOI 10.1007/BF01580289. MR169

[10] S. Chang, *On quadratic forms between spheres*, Geom. Dedicata **70** (1998), no. 2, 111–124, DOI 10.1023/A:1004976419110. MR1620738

[11] S.-Y. A. Chang, L. Wang, and P. C. Yang, *A regularity theory of biharmonic maps*, Comm. Pure Appl. Math. **52** (1999), no. 9, 1113–1137, DOI 10.1002/(SICI)1097-0312(199909)52:9¡1113::AID-CPA4¿3.0.CO;2-7. MR1692148

[12] J. Eells and L. Lemaire, *Selected topics in harmonic maps*, CBMS Regional Conference Series in Mathematics, vol. 50, Published for the Conference Board of the Mathematical Sciences, Washington, DC; by the American Mathematical Society, Providence, RI, 1983, DOI 10.1090/cbms/050. MR703510

[13] J. Eells and A. Ratto, *Harmonic maps and minimal immersions with symmetries: Methods of ordinary differential equations applied to elliptic variational problems*, Annals of Mathematics Studies, vol. 130, Princeton University Press, Princeton, NJ, 1993, DOI 10.1515/9781400882502. MR1242555

[14] J. Eells Jr. and J. H. Sampson, *Énergie et déformations en géométrie différentielle* (French), Ann. Inst. Fourier (Grenoble) **14** (1964), no. fasc. 1, 61–69. MR172310

[15] J. Eells Jr. and J. H. Sampson, *Variational theory in fibre bundles*, Proc. U.S.-Japan Seminar in Differential Geometry (Kyoto, 1965), Nippon Hyoronsha Co., Ltd., Tokyo, 1966, pp. 22–33. MR216519

[16] H. I. Eliasson, *Introduction to global calculus of variations*, Global analysis and its applications (Lectures, Internat. Sem. Course, Internat. Centre Theoret. Phys., Trieste, 1972), Internat. Atomic Energy Agency, Vienna, 1974, pp. 113–135. MR440602

[17] D. Fetcu and C. Oniciuc, *Biharmonic and biconservative hypersurfaces in space forms*, Differential geometry and global analysis—in honor of Tadashi Nagano, Contemp. Math., vol. 777, Amer. Math. Soc., [Providence], RI, [2022] ©2022, pp. 65–90, DOI 10.1090/conm/777/15628. MR4410183

[18] A. Gastel, *Construction of harmonic maps between spheres by joining three eigenmaps*, Internat. J. Math. **9** (1998), no. 7, 821–844, DOI 10.1142/S0129167X98000348. MR1651045

[19] A. Gastel, *On the harmonic Hopf construction*, Proc. Amer. Math. Soc. **132** (2004), no. 2, 607–615, DOI 10.1090/S0002-9939-03-07062-X. MR2022387

[20] G. Y. Jiang, *2-harmonic isometric immersions between Riemannian manifolds* (Chinese), Chinese Ann. Math. Ser. A **7** (1986), no. 2, 130–144. An English summary appears in Chinese Ann. Math. Ser. B **7** (1986), no. 2, 255. MR858581

[21] G. Y. Jiang, *2-harmonic maps and their first and second variational formulas* (Chinese), Chinese Ann. Math. Ser. A **7** (1986), no. 4, 389–402. An English summary appears in Chinese Ann. Math. Ser. B **7** (1986), no. 4, 523. MR886529

[22] K. Kenmotsu, *Minimal surfaces with constant curvature in 4-dimensional space forms*, Proc. Amer. Math. Soc. **89** (1983), no. 1, 133–138, DOI 10.2307/2045079. MR706526

[23] L. Lemaire, *Minima and critical points of the energy in dimension two*, Global differential geometry and global analysis (Berlin, 1979), Lecture Notes in Math., vol. 838, Springer, Berlin, 1981, pp. 187–193. MR636281

[24] E. Loubeau and C. Oniciuc, *On the biharmonic and harmonic indices of the Hopf map*, Trans. Amer. Math. Soc. **359** (2007), no. 11, 5239–5256, DOI 10.1090/S0002-9947-07-03934-7. MR2327029

[25] C. Oniciuc, *New examples of biharmonic maps in spheres*, Colloq. Math. **97** (2003), no. 1, 131–139, DOI 10.4064/cm97-1-12. MR2010548

[26] C. Oniciuc, Y.-L. Ou, *Unpublished note*, (2018).

[27] Y.-L. Ou and B.-Y. Chen, *Biharmonic submanifolds and biharmonic maps in Riemannian geometry*, World Scientific Publishing Co. Pte. Ltd., Hackensack, NJ, [2020] ©2020. MR4265170

[28] V. Pettinati and A. Ratto, *Existence and nonexistence results for harmonic maps between spheres*, Ann. Scuola Norm. Sup. Pisa Cl. Sci. (4) **17** (1990), no. 2, 273–282. MR1076254

[29] H. Takeuchi, *Stability and Liouville theorems of p-harmonic maps*, Japan. J. Math. (N.S.) **17** (1991), no. 2, 317–332, DOI 10.4099/math1924.17.317. MR1145657

[30] G. Tóth, *Classification of quadratic harmonic maps of S^3 into spheres*, Indiana Univ. Math. J. **36** (1987), no. 2, 231–239, DOI 10.1512/iumj.1987.36.36013. MR891772

[31] G. Tóth, *Quadratic eigenmaps between spheres*, Geom. Dedicata **56** (1995), no. 1, 35–52, DOI 10.1007/BF01263612. MR1338316

[32] J. Simons, *Minimal varieties in Riemannian manifolds*, Ann. of Math. (2) **88** (1968), 62–105, DOI 10.2307/1970556. MR233295

[33] R. T. Smith, *Harmonic mappings of spheres*, Bull. Amer. Math. Soc. **78** (1972), 593–596, DOI 10.1090/S0002-9904-1972-13018-0. MR298590

[34] C. Wang, *Remarks on biharmonic maps into spheres*, Calc. Var. Partial Differential Equations **21** (2004), no. 3, 221–242, DOI 10.1007/s00526-003-0252-7. MR2094320

[35] C. Wang, *Stationary biharmonic maps from $\mathbb{R}^m$ into a Riemannian manifold*, Comm. Pure Appl. Math. **57** (2004), no. 4, 419–444, DOI 10.1002/cpa.3045. MR2026177

[36] R. Wood, *Polynomial maps from spheres to spheres*, Invent. Math. **5** (1968), 163–168, DOI 10.1007/BF01425547. MR227999

[37] F. Wu, Y. Xiong, and X. Zhao, *Classification of quadratic harmonic maps of S^7 into S^7*, J. Geom. Anal. **25** (2015), no. 3, 1992–2010, DOI 10.1007/s12220-014-9501-6. MR3358081

[38] P. Y. H. Yiu, *Quadratic forms between spheres and the nonexistence of sums of squares formulae*, Math. Proc. Cambridge Philos. Soc. **100** (1986), no. 3, 493–504, DOI 10.1017/S0305004100066226. MR857724

Faculty of Mathematics, Alexandru Ioan Cuza University of Iaşi, Blvd. Carol I, 11, 700506 Iaşi, Romania

Email address: rares_ambrosie@yahoo.com

Contemporary Mathematics
Volume **821**, 2025
https://doi.org/10.1090/conm/821/16432

A survey on submanifolds of nearly Kähler spaces

Mateo Anarella

ABSTRACT. In this survey, we give an introduction to nearly Kähler manifolds and list some results on submanifolds of these spaces. This survey tries by no means to be complete.

1. Introduction

We understand by "submanifold theory" the study of submanifolds from a Riemannian point of view. Namely, the study of Riemannian immersions, their intrinsic and extrinsic geometry and the local invariants. Nonetheless, submanifold theory borrows some concepts from other areas in geometry. For instance, a Lagrangian submanifold is a submanifold for which the symplectic form vanishes everywhere. Although not every Riemannian manifold is a symplectic manifold, for almost Hermitian manifolds there is a canonical 2-form: the Kähler form.

Gray and Hervella [**35**] divided almost Hermitian manifolds into 16 different classes. Among them we can find Kähler manifolds (Riemannian, symplectic, and complex), almost Kähler (Riemannian and symplectic but not complex) and nearly Kähler (only Riemannian). In principle, this last kind does not have any integrable structure (e.g., symplectic, complex, etc.). However, six-dimensional nearly Kähler manifolds possess other interesting properties, such as being Einstein, admitting Killing spinors, etc. Hence, they received great interest in the last years, as we will see in the next sections.

In this survey, I talk about submanifolds of the different nearly Kähler manifolds. Due to works of Nagy and Butruille, mentioned in Section 2, we will focus mostly on the four homogeneous ones: $\mathbb{S}^6$, $\mathbb{S}^3 \times \mathbb{S}^3$, $\mathbb{C}P^3$, and $F(\mathbb{C}^3)$.

2. Nearly Kähler manifolds and splitting theorems

Let (N, g, J) be an almost Hermitian manifold. That is, (N, g) is a Riemannian manifold and J is an almost complex structure on N compatible with g:

$$J^2 = -\mathrm{Id}, \qquad g(JX, JY) = g(X, Y),$$

for all $X, Y \in \Gamma(TN)$.

2020 *Mathematics Subject Classification.* Primary 53C42.

The author was supported by CERAMATHS, University Polytechnique Hauts-de-France / Methusalem grant METH/21/03–long term structural funding of the Flemish Goverment.

DEFINITION 2.1. A Kähler manifold is an almost Hermitian manifold (N, g, J) such that J is integrable ($\mathcal{N}_J \equiv 0$, where $\mathcal{N}$ is the Nijenhuis tensor) and the Kähler form $\omega = g(J\cdot, \cdot)$ is closed ($d\omega = 0$). Equivalently, N is Kähler if $\nabla_X J = 0$ for all $X \in \Gamma(TN)$.

A **nearly Kähler** manifold is an almost Hermitian manifold (N, g, J) such that $\tilde{\nabla} J$ is skew-symmetric. Here, $\tilde{\nabla}$ refers to the Levi-Civita connection associated to g. We say that N is a **strict nearly Kähler manifold** if $\tilde{\nabla}_X J \neq 0$ for every $X \in \Gamma(TN)$ different from zero.

Nearly Kähler manifolds are even dimensional, yet we only find interesting examples from dimension six onwards.

PROPOSITION 1. *Two-dimensional and four-dimensional nearly Kähler manifolds are Kähler.*

Another interesting property of nearly Kähler manifolds is that they are nicely decomposable. Gray and Nagy did significant work in this direction.

PROPOSITION 2 (Gray '76 [**34**], see also [**51**]). *Let (N, g, J) be a complete and simply connected nearly Kähler manifold. Then N is a Riemannian product $M_1 \times M_2$ where M_1 is a Kähler manifold and M_2 is a strict nearly Kähler manifold.*

THEOREM 2.2 (Gray '76 [**34**]). *There do not exist eight-dimensional strict nearly Kähler manifolds.*

In fact, this implies that all eight-dimensional nearly Kähler manifolds are either Kähler or a product of a six-dimensional strict nearly Kähler manifold and a two-dimensional Kähler manifold. In the book of Schäfer [**53**] we see that this is true also for pseudo-Riemannian nearly Kähler manifolds (or pseudo-nearly Kähler manifolds for short), as long as the three-form $g(\tilde{\nabla}_. J\cdot, \cdot)$ has nonzero length.

THEOREM 2.3 (Nagy '02 [**51**]). *A 10-dimensional complete nearly Kähler manifold is either the universal cover of a Riemannian product of a six-dimensional nearly Kähler manifold with a Kähler surface, or the twistor space over a positive, eight-dimensional quaternionic manifold.*

As quaternionic Kähler manifolds are always Einstein, "positive" means positive Einstein constant. All eight-dimensional positive quaternionic manifolds are classified, so the 10-dimensional strict nearly Kähler manifolds are as well (see [**44**]).

To conclude the discussion on splitting theorems, I present Nagy's theorem, the culmination of these results.

THEOREM 2.4 (Nagy '02 [**50**]). *Let (N, g, J) be a complete, simply connected strict nearly Kähler manifold. Then N is a Riemannian product whose factors belong to one of the following three classes:*

(1) *Six-dimensional nearly Kähler manifold,*
(2) *Homogeneous nearly Kähler manifolds (with some conditions),*
(3) *Twistor spaces over quaternionic Kähler manifolds with positive scalar curvature.*

The conditions for homogeneous nearly Kähler manifolds in the theorem above are quite involved; for more information see Nagy's original work.

We see in Nagy's theorem that six-dimensional and homogeneous nearly Kähler are among the most important classes. Butruille classified all nearly Kähler manifolds satisfying both properties.

THEOREM 2.5 (Butruille '05 [**17**]). *The only simply connected homogeneous strict nearly Kähler spaces of dimension six are isometric to G/H with the metric induced from the Killing form, where G and H are the Lie groups given in the list.*

(1) $G = \mathbb{S}^3 \times \mathbb{S}^3 \times \mathbb{S}^3$, $H = \Delta\mathbb{S}^3$.
(2) $G = G_2$, $H = \mathrm{SU}(3)$.
(3) $G = \mathrm{Sp}(2)$, $H = \mathbb{S}^1 \times \mathrm{SU}(2)$.
(4) $G = \mathrm{SU}(3)$, $H = \mathbb{S}^1 \times \mathbb{S}^1$.

The manifolds above are, respectively, $\mathbb{S}^3 \times \mathbb{S}^3$ with a different metric than the product metric, the six-sphere $\mathbb{S}^6$ with the canonical metric, the complex projective space $\mathbb{C}P^3$ with a metric different than the Fubini–Study metric, and the manifold of full flags in $\mathbb{C}^3$.

Since Tachibana introduced the concept in [**56**], the question if there exist six-dimensional nearly Kähler manifolds which are not homogeneous remained open until Foscolo and Haskins [**32**] gave two examples of complete six-dimensional nearly Kähler manifolds which are not homogeneous, but cohomogeneity-one instead.

Butruille's classification is still open for the pseudo-Riemannian case. Using construction by T-duality, Kath [**42**] gave six analogues of the six-dimensional homogeneous examples:

- $\mathrm{SL}(2,\mathbb{R}) \times \mathrm{SL}(2,\mathbb{R})$,
- $\mathbb{S}^6_4$,
- $\frac{\mathrm{SO}^+(2,3)}{\mathrm{U}(1,1)}$,
- $\frac{\mathrm{SO}^+(4,1)}{\mathrm{U}(2)}$,
- $\frac{\mathrm{SU}(2,1)}{\mathrm{U}(1)\times\mathrm{U}(1)}$,
- $\frac{\mathrm{SL}(3,\mathbb{R})}{\mathbb{R}^*\times\mathrm{SO}(2)}$.

A natural assumption would be that every six-dimensional homogeneous pseudo-nearly Kähler manifold is a T-dual of a Riemannian nearly Kähler, but Alekseevsky, Kurglikov, and Winther constructed in [**3**] an example of a pseudo-nearly Kähler six-manifold which is not a T-dual of a Riemannian one.

3. Submanifold theory

In this section, we present the basics of submanifold theory, and we provide the definitions of the submanifolds we will discuss about in the next sections.

Let $f : M \to N$ be an isometric immersion of a (pseudo-)Riemannian manifold M into a (pseudo-)Riemannian manifold N. In submanifold theory, we consider two submanifolds to be "equal" if there exists an isometry of the ambient space that maps one into the other. We denote by $\mathrm{Iso}(N)$ the isometry group of N, and by $\mathrm{Iso}_o(N)$ the connected component of the identity.

Denote by ∇ and $\tilde{\nabla}$ the Levi-Civita connections of M and N, respectively. The connection $\tilde{\nabla}$ splits into a tangent and normal part given by the Gauss and Weingarten formulas below. For simplicity we omit the immersion f. For $X, Y \in \Gamma(TM)$ we have that

$$\text{(Gauss formula)} \qquad \tilde{\nabla}_X Y = \nabla_X Y + h(X,Y),$$

where h is a symmetric $TM^\perp$-valued tensor, called the **second fundamental form**. For $X \in \Gamma(TM)$ and $\xi \in \Gamma(TM^\perp)$ we have

$$\text{(Weingarten formula)} \qquad \tilde{\nabla}_X \xi = -A_\xi X + \nabla^\perp_X \xi,$$

where A_ξ is symmetric tensor linear in ξ, known as **the shape operator**. The connection $\nabla^\perp\colon \Gamma(TM)\times\Gamma(TM^\perp)\to\Gamma(TM^\perp)$ is known as the **normal connection**.

We could say that the simplest kind of submanifolds are those with no extrinsic geometry. These are know as totally geodesic submanifolds.

DEFINITION 3.1. A submanifold is said to be **totally geodesic** if its geodesics are also geodesics of the ambient space. Equivalently, a submanifold is totally geodesic if and only if the second fundamental form vanishes everywhere.

An important vector field in submanifold theory that encodes extrinsic curvature is the one known as the mean curvature vector field.

DEFINITION 3.2. Let M^m be a (pseudo-)Riemannian immersed manifold of a (pseudo-)Riemannian manifold (N^n, g), and let $\{e_i\}_{i=1,\dots,n}$ be a (pseudo-)orthonormal frame with signatures $g(e_i, e_i) = \kappa_i = \pm 1$. The **mean curvature vector field** is defined as

$$H = \frac{1}{n}\sum_{i=1}^{n} \kappa_i h(e_i, e_i).$$

DEFINITION 3.3. A submanifold is said to be **minimal** if the mean curvature vector field H vanishes everywhere.

The name minimal comes from the fact that these submanifolds are critical points of the energy functional.

DEFINITION 3.4. A submanifold is said to be **totally umbilical** if for every normal vector field ξ the shape operator A_ξ is a multiple of the identity. Equivalently, there exist a normal vector field η such that $h(X,Y) = g(X,Y)\eta$. Moreover, $\eta = H$ in this case.

As nearly Kähler manifolds are in particular almost Hermitian manifolds, we give an important class of submanifolds involving the complex structure.

DEFINITION 3.5. Let (N, g, J) be an almost Hermitian manifold. A **CR-submanifold** M of N is a submanifold such that there exists a holomorphic smooth distribution D on M, i.e., $JD \subset D$, and the orthonormal distribution $D^\perp$ is a totally real distribution, i.e., $J(D^\perp) \subset TM^\perp$.

If $D = TM$, we say that M is an **almost complex submanifold** (if it is a surface it is also known as pseudo-holomorphic curve or J-holomorphic curve). If instead D is the trivial distribution then we say that M is a **totally real submanifold**.

DEFINITION 3.6. A **Lagrangian submanifold** M of an almost Hermitian manifold N is a totally real submanifold such that $2\dim M = \dim N$. In the pseudo-Riemannian setting, we add the condition of being nondegenerate.

Lagrangian submanifolds of six-dimensional nearly Kähler manifolds are particularly interesting, since they are automatically minimal and orientable (see [**53**]).

DEFINITION 3.7. Given a (pseudo-)Riemannian manifold N, a submanifold is **extrinsically homogeneous** (also known as equivariant) if it is the orbit of a Lie subgroup of $\mathrm{Iso}_o(N)$.

DEFINITION 3.8. An action of a Lie group H on a (pseudo-)Riemannian manifold N has **cohomogeneity one** if H is a Lie subgroup of $\mathrm{Iso}_o(N)$ and the codimension of the principal orbits is one. Note that these principal orbits are extrinsically homogeneous hypersurfaces.

Given a hypersurface M in a nearly Kähler manifold (N, g, J) with unit normal ξ, the tangent component of the vector field JX for $X \in \Gamma(TM)$ defines an almost contact structure $(M, J\xi, \varphi, \eta)$ on M, where φ is the tensor given by

$$\varphi(X) = JX - g(JX, \xi)\xi.$$

DEFINITION 3.9. Let M be a hypersurface in a nearly Kähler manifold (N, g, J), and let ξ be the unit normal vector field. The vector field $J\xi$ is the **structure vector field** (also known as Reeb vector field).

DEFINITION 3.10. A Hopf hypersurface is a hypersurface whose structure vector field is a principal vector field, i.e. $A_\xi J\xi = \alpha J\xi$, where α is a smooth function. Equivalently, the integral curves of $J\xi$ are geodesics.

DEFINITION 3.11. A hypersurface M of a nearly Kähler manifold (N, g, J) is said to be **nearly cosymplectic** if $\tilde{\nabla}\varphi$ is skew symmetric.

A result by Blair [**15**] says that nearly cosymplectic hypersurfaces satisfy

$$A_\xi \circ \varphi = \varphi \circ A_\xi. \tag{3.1}$$

Moreover, all hypersurfaces satisfying (3.1) are Hopf hypersurfaces.

4. State-of-the-art

Because of the splitting theorems by Nagy and Gray, there are not many works on submanifolds of nearly Kähler manifolds of dimension higher than six. However, following a similar construction as for $\mathbb{S}^3 \times \mathbb{S}^3$, we can define a nearly Kähler structure (not necessarily strict) on $G \times G$ (see [**1**]), where G is a Lie group.

Aguilar-Suárez and Ruiz-Hernández [**2**] gave a characterization of minimal Lagrangian submanifolds of $G \times G$, since in higher dimensions Lagrangian submanifolds are not necessarily minimal. Lubbe and Schäfer [**46**] studied pseudo-holomorphic curves in $G \times G$. They prove that any totally geodesic pseudo-holomorphic curve is locally determined by a Lie subalgebra $\mathfrak{a}$ of $\mathfrak{g}$, with $0 < \dim \mathfrak{a} \leq 2$.

In the rest of the survey we only consider submanifolds of six-dimensional nearly Kähler manifolds.

4.1. In six-dimensional nearly Kähler manifolds.

For general six-dimensional nearly Kähler manifolds, Lê and Schwachhöfer [**43**] studied the variation of the volume functional for Lagrangian submanifolds, coupled with an analysis of Lagrangian deformations.

Podestà and Spiro [**52**] studied cohomogeneity-one actions on six-dimensional nearly Kähler manifolds. They found that the group acting with cohomogeneity one is either SU(3), SO(4) and the manifold is $\mathbb{S}^6$, or SU(2) $\times$ SU(2) and the manifolds is $\mathbb{C}P^3$ or $\mathbb{S}^3 \times \mathbb{S}^3$.

Storm [**55**] constructed examples of Lagrangian submanifolds of $\mathbb{C}P^3$ and $F(\mathbb{C}^3)$ from minimal submanifolds of their underlying spaces from the twistor fibration.

In [**23**], Deschamps and Loubeau studied hypersurfaces of both $\mathbb{C}P^3$ and $F(\mathbb{C}^3)$. They proved that there are no hypersurfaces satisfying (3.1). Consequently, the only nearly cosymplectic hypersurface of six-dimensional homogeneous nearly Kähler manifolds is $\mathbb{S}^5 \hookrightarrow \mathbb{S}^6$. Moreover, there are no totally geodesic nor totally umbilical hypersurfaces in $\mathbb{C}P^3$ or $F(\mathbb{C}^3)$.

Hu, Yao, and Zhang showed in [**40**] that there does not exists a hypersurface in $\mathbb{S}^3 \times \mathbb{S}^3$ for which the shape operator and the almost contact structure tensor φ anticommute. On the other hand, they showed that the only hypersurfaces satisfying such equation in $\mathbb{S}^6$, are totally geodesic.

4.2. In the nearly Kähler $\mathbb{S}^6$. The sphere $\mathbb{S}^6$ is the most traditional example of a six-dimensional nearly Kähler manifold. It carries the metric inherited from the immersion in $\mathbb{R}^7$ and the almost complex structure is determined by the cross product in the imaginary octonions.

Due to the extensive research in this manifold, to list all the results on submanifolds would be a hard task. Luckily, there are already some surveys that focus particularly on $\mathbb{S}^6$. For instance, Sekigawa and Hashimoto [**36**] (2004) wrote about CR-submanifolds, J-holomorphic curves and totally real submanifolds. Dillen also wrote about Lagrangian submanifolds in [**24**] (1996). Two decades later (2016), Antić and Vrancken also listed works on CR-submanifolds [**9**]. Thus, we are going to list only works after this date.

In [**47**] Madnick studied holomorphic curves inside geodesic balls. He proved that if the boundary of the holomorphic curve meets the boundary of the geodesic ball, then the curve is totally geodesic. Along the same lines, Tsouri and Vlachos [**57**] studied relations between minimal surfaces and holomorphic curves inside $\mathbb{S}^5 \hookrightarrow \mathbb{S}^6$. Also, Eschenburg and Vlachos [**31**] gave a characterization of holomorphic curves in $\mathbb{S}^6$ and $\mathbb{S}^5$.

Many results regarding Lagrangian submanifolds of $\mathbb{S}^6$ have been obtained. Hu, Yin, and Yao [**39**] studied restrictions on the Ricci curvature of Lagrangian submanifolds. They conclude that, under some pinching conditions, the only Lagrangian submanifolds are $\mathbb{S}^3$ or Berger spheres. Also, Lê and Schwachhöfer [**43**] studied deformations in Lagrangian submanifolds.

Moruz [**48**] studied warped product Lagrangian immersions, finding a relation of such submanifolds with minimal surfaces in $\mathbb{S}^5$. Moreover, Ali et al. [**4**] studied the homology of warped product Lagrangian submanifolds. In another work by Şahin [**19**] the topology of submanifolds is also studied.

Enoyoshi and Tsuakada [**30**] studied Gauss maps and Lagrangian submanifolds: Gauss maps $M \to \widetilde{\mathrm{Gr}}_{\mathrm{ass}}(\mathrm{Im}\,\mathbb{O})$ associated to Lagrangian immersions $M \to \mathbb{S}^6$ are harmonic.

Bae, Park, and Sekigawa [**12**] studied quasi contact metric hypersurfaces, and gave a one-parameter family of totally umbilical hypersurfaces.

4.3. In the nearly Kähler $\mathbb{S}^3 \times \mathbb{S}^3$. This space does not possess the traditional product metric. Instead, the metric on $\mathbb{S}^3 \times \mathbb{S}^3$ is the one that makes the map $\pi : \mathbb{S}^3 \times \mathbb{S}^3 \times \mathbb{S}^3 \to \mathbb{S}^3 \times \mathbb{S}^3$ given by

$$\pi(a, b, c) = (ac^{-1}, bc^{-1}),$$

into a Riemannian submersion. This space also carries an almost product structure P, which is defined by swapping the tangent vectors at each factor. Most of the authors working on $\mathbb{S}^3 \times \mathbb{S}^3$ make use of this tensor in their classifications.

Djorić, Djorić, and Moruz [**29**] gave a parametrization of the geodesics of $\mathbb{S}^3 \times \mathbb{S}^3$ and they provided conditions for when the geodesics (with respect to the nearly Kähler metric) coincide with the geodesics associated to the product metric.

Almost complex surfaces were studied by Bolton et al. in [**16**]. In particular, they classified all totally geodesic almost complex surfaces in $\mathbb{S}^3 \times \mathbb{S}^3$: a flat torus T^2 and $\mathbb{S}^2(\frac{2}{3})$. A Bonnet-type theorem for existence an uniqueness is given by Dioos et al. in [**26**]. Moreover, they proved that all flat almost complex surfaces are part of a family of immersions of T^2 into $\mathbb{S}^3 \times \mathbb{S}^3$. Dioos in his PhD thesis [**25**] studied submanifolds of $\mathbb{S}^3 \times \mathbb{S}^3$ in great extent. There, he provided a classification of totally geodesic totally real surfaces. He found that the only one is $\mathbb{S}^2(1)$ immersed in one of the factors of $\mathbb{S}^3 \times \mathbb{S}^3$

Antić et al. [**8**] classified all the surfaces that are invariant under the almost product structure P. They found that they are either the torus T^2 of the classifications in [**16**] and [**26**], any surface in $p : N^2 \to \mathbb{S}^3$ immersed by $p \mapsto (p, p)$, or solutions of a specific differential equation. Furthermore, they classified all slant P-invariant surfaces (surfaces with constant Kähler angle). This last condition facilitates the solution of the aforementioned differential equation.

Dioos et al. [**61**] proved that any Lagrangian submanifold of a six-dimensional nearly Kähler manifold with parallel second fundamental form is totally geodesic (this is also true for the pseudo-Riemannian case [**5**]). Moreover, they provided a full classification of totally geodesic Lagrangian submanifolds of $\mathbb{S}^3 \times \mathbb{S}^3$. Although they listed six examples, because of [**49**] these are only two. They are either a round sphere $\mathbb{S}^3(\frac{3}{4})$ or a Berger sphere.

In [**27**], Dioos, Vrancken, and Wang classified all Lagrangian submanifolds with constant sectional curvature: the already mentioned totally geodesic sphere $\mathbb{S}^3(\frac{3}{4})$, a flat torus T^3 and $\mathbb{R}P^3(\frac{3}{16})$.

Bektaş et al. [**13**] studied Lagrangian submanifolds where the angles between PE_i and E_i are constant, for a specific frame $\{E_i\}_i$. They found that they must be either totally geodesic submanifolds or submanifolds with constant sectional curvature. As a consequence, the authors classified all extrinsically homogeneous Lagrangian submanifolds, since these have constant angle functions.

We can find examples of nonextrinsically homogeneous Lagrangian submanifolds in [**14**]. There, the same authors as in [**13**] showed a way to construct Lagrangian submanifolds from a minimal submanifold immersed in $\mathbb{S}^3$.

Three-dimensional CR-submanifolds have been studied by Antić et al in [**7**]. They considered those CR-submanifolds whose associated distributions are totally geodesic. As a continuation, the authors studied in [**6**] how the almost product structure P acts on the distributions of three-dimensional CR-submanifolds.

Hu, Yao, and Zhang [**41**] showed that there are no totally umbilical hypersurfaces nor hypersurfaces with parallel second fundamental form in $\mathbb{S}^3 \times \mathbb{S}^3$. Furthermore, they classified all hypersurfaces that satisfy (3.1) and have three different principal curvatures. In [**38**], Hu and Yao proved that there does not exist a Hopf hypersurface with only two different principal curvatures. This motivated the authors [**38, 59**] to study Hopf hypersurfaces with at least three different principal curvatures. Additionally, they require that P preserves the orthogonal space to $J\xi$, where ξ is the unit normal. They found a family of immersions from $\mathbb{S}^3 \times \mathbb{S}^2$ into $\mathbb{S}^3 \times \mathbb{S}^3$ which are Hopf hypersurfaces.

Hu et al. [**37**] proved that there are no locally conformally flat hypersurfaces nor Einstein Hopf hypersurfaces. Moreover, they provided a Simons type integral inequality for minimal hypersurfaces.

In the same line as the previous articles, Chen and Yang [**18**] studied hypersurfaces satisfying a similar equation to (3.1), replacing the shape operator by the Ricci tensor. Adding the condition of being Hopf, they proved that the mean curvature is 2α, where α is the principal curvature associated to the structure vector field. They also show that the mean curvature is constant if P preserves the normal space to the structure vector field. Hu, Yao and Zhang [**60**] also proved that the following conditions are equivalent:

- P preserving the normal space to the structure vector field,
- the associated principal curvature being constant,
- the hypersurface being congruent to a product $\mathbb{S}^3 \times \Sigma$ with $\Sigma \hookrightarrow \mathbb{S}^3$.

This generalized a previous result obtained by Djorić [**28**].

4.4. In the nearly Kähler $\mathbb{C}P^3$. The nearly Kähler complex projective space does not carry the traditional Fubini–Study metric. Instead, the nearly Kähler metric arises from the fact that $\mathbb{C}P^3$ is the twistor space over the quaternionic Kähler $\mathbb{H}P^1 \cong \mathbb{S}^4$. Namely, $\mathbb{C}P^3$ is a fiber bundle by 2-spheres over $\mathbb{S}^4$. Then, there exist two distinguished distributions $\mathcal{D}_1^2$ and $\mathcal{D}_2^4$. By modifying the Kähler complex structure on $\mathcal{D}_1$ and the Kähler metric on $\mathcal{D}_2$, we get a nearly Kähler manifold.

It is clear that the flag manifold and $\mathbb{C}P^3$ are the least studied of the homogeneous six-dimensional nearly Kähler spaces. Xu studied J-holomorphic curves [**58**]. Also, Aslan studied them in [**10**].

Aslan [**11**] and Liefsoens [**45**] independently studied extrinsically homogeneous and totally geodesic Lagrangian submanifolds. Among them, we find $\mathbb{S}^1 \times \mathbb{S}^2$, a Berger sphere, and $\mathbb{R}P^3$, the last one being the only totally geodesic Lagrangian submanifold.

For hypersurfaces see the work by Deschamps and Loubeau in Section 4.1.

4.5. In the nearly Kähler $F(\mathbb{C}^3)$. The manifold of full flags in $\mathbb{C}^3$ carries a nearly Kähler structure similar to $\mathbb{C}P^3$. It is also a twistor space, but over $\mathbb{C}P^2$. So in the same way, from a Kähler structure we construct a nearly Kähler structure.

Vrancken and Cwiklinski classified in [**20**] all the totally geodesic almost complex surfaces. There, they found that there are 3 such surfaces, all spaces forms: a sphere of radius 1, a sphere of radius $\frac{1}{2}$ and the flat torus.

Storm classified in [**54**] all totally geodesic Lagrangian submanifolds. These are either the manifold of full flags in $\mathbb{R}^3$ or a Berger sphere. Moreover, he classified all extrinsically homogeneous Lagrangian submanifolds. He showed that in addition to the previous examples we have $\mathbb{R}P^3$, with a Berger-like metric.

For hypersurfaces see the work by Deschamps and Loubeau in Section 4.1.

4.6. In pseudo-nearly Kähler manifolds. The study of submanifolds of six-dimensional homogeneous pseudo-nearly Kähler manifolds has just started. In his book, Schäfer [**53**] did great work on the general theory of Lagrangian submanifolds.

Similar to the classification in $\mathbb{S}^3 \times \mathbb{S}^3$ in [**16**], Ghandour and Vrancken [**33**] classified almost complex totally geodesic surfaces in $\mathrm{SL}(2,\mathbb{R}) \times \mathrm{SL}(2,\mathbb{R})$. These are a flat torus T^2, $\mathbb{R}^2$ or $H^2(-\frac{4}{3})$. Anarella and Dekimpe [**21**] proved instead that the only two totally real totally geodesic surfaces are $H_1^2(-\frac{3}{2})$ and $H^2(-\frac{3}{2})$. Dekimpe also studied degenerate almost complex surfaces in [**22**].

When it comes to Lagrangian submanifolds of $\mathrm{SL}(2,\mathbb{R}) \times \mathrm{SL}(2,\mathbb{R})$, Anarella and Van der Veken [**5**] found a full classification up to congruence of totally geodesic Lagrangian submanifolds. There are three noncongruent totally geodesic Lagrangian submanifolds, $H_1^3(-\frac{3}{2})$, and two anti-de Sitter spaces with different Berger-like metrics.

In the same paper [**20**] as for $F(\mathbb{C}^3)$, Cwiklinski and Vrancken classified all totally geodesic surfaces of $\frac{\mathrm{SU}(2,1)}{\mathrm{U}(1)\times\mathrm{U}(1)}$. Analogous to the Riemannian case they found three surfaces: a sphere $\mathbb{S}^2(\frac{1}{2})$, $H^2(-4)$ and $H^2(-1)$.

5. Personal comments

To finish, I would like to point out the similarity of the works for $\mathbb{S}^3 \times \mathbb{S}^3$ [**13**], $\mathbb{C}P^3$ and $F(\mathbb{C}^3)$ [**55**], and $\mathbb{S}^6$ [**48**]. In all of these works, the authors construct Lagrangian submanifolds from minimal surfaces immersed in a smaller space. These are the only examples of non-extrinsically homogeneous Lagrangian submanifolds I am aware of.

It would be interesting to know if we can construct this kind of Lagrangian submanifolds for any nearly Kähler space, and if they are the only homogeneous nonextrinsically homogeneous Lagrangian submanifolds.

References

[1] E. Abbena and S. Garbiero, *Almost Hermitian homogeneous manifolds and Lie groups*, Nihonkai Math. J. **4** (1993), no. 1, 1–15. MR1232120

[2] R. Aguilar-Suárez and G. Ruiz-Hernández, *A characterization of minimal Lagrangian submanifolds of the nearly Kähler $G \times G$*, Bull. Belg. Math. Soc. Simon Stevin **30** (2023), no. 2, 140–162, DOI 10.36045/j.bbms.220331. MR4644946

[3] D. V. Alekseevsky, B. S. Kruglikov, and H. Winther, *Homogeneous almost complex structures in dimension 6 with semi-simple isotropy*, Ann. Global Anal. Geom. **46** (2014), no. 4, 361–387, DOI 10.1007/s10455-014-9428-y. MR3276441

[4] A. Ali, F. Mofarreh, N. Alluhaibi, and P. Laurian-Ioan, *Null homology in warped product Lagrangian submanifolds of the nearly Kaehler $\mathbb{S}^6$ and its applications*, J. Geom. Phys. **158** (2020), 103859, 13, DOI 10.1016/j.geomphys.2020.103859. MR4145611

[5] M. Anarella and J. Van der Veken, *Totally geodesic Lagrangian submanifolds of the pseudo-nearly Kähler* $\mathrm{SL}(2,\mathbb{R}) \times \mathrm{SL}(2,\mathbb{R})$, Math. Nachr. **297** (2024), no. 5, 1922–1944, DOI 10.1002/mana.202300351. MR4755743

[6] M. Antić, N. Djurdjević, and M. Moruz, *CR submanifolds of the nearly Kähler $\mathbb{S}^3 \times \mathbb{S}^3$ characterised by properties of the almost product structure*, Mediterr. J. Math. **15** (2018), no. 3, Paper No. 111, 28, DOI 10.1007/s00009-018-1152-6. MR3800244

[7] M. Antić, N. Djurdjević, M. Moruz, and L. Vrancken, *Three-dimensional CR submanifolds of the nearly Kähler $\mathbb{S}^3 \times \mathbb{S}^3$*, Ann. Mat. Pura Appl. (4) **198** (2019), no. 1, 227–242, DOI 10.1007/s10231-018-0770-8. MR3918627

[8] M. Antić, Z. Hu, M. Moruz, and L. Vrancken, *Surfaces of the nearly Kähler $\mathbb{S}^3 \times \mathbb{S}^3$ preserved by the almost product structure*, Math. Nachr. **294** (2021), no. 12, 2286–2301, DOI 10.1002/mana.201900376. MR4371298

[9] M. Antić and L. Vrancken, *CR-submanifolds of the nearly Kähler 6-sphere*, Geometry of Cauchy-Riemann submanifolds, Springer, Singapore, 2016, pp. 57–90. MR3525852

[10] B. Aslan, *Transverse J-holomorphic curves in nearly Kähler $\mathbb{CP}^3$*, Ann. Global Anal. Geom. **61** (2022), no. 1, 115–157, DOI 10.1007/s10455-021-09806-0. MR4367903

[11] B. Aslan, *Special Lagrangians in nearly Kähler $\mathbb{CP}^3$*, J. Geom. Phys. **184** (2023), Paper No. 104713, 18, DOI 10.1016/j.geomphys.2022.104713. MR4521332

[12] J. Bae, J. Park, and K. Sekigawa, *A one-parameter family of totally umbilical hyperspheres in the nearly Kähler 6-sphere*, J. Korean Math. Soc. **55** (2018), no. 4, 963–974, DOI 10.4134/JKMS.j170569. MR3820144

[13] B. Bektaş, M. Moruz, J. Van der Veken, and L. Vrancken, *Lagrangian submanifolds with constant angle functions of the nearly Kähler* $\mathbb{S}^3 \times \mathbb{S}^3$, J. Geom. Phys. **127** (2018), 1–13, DOI 10.1016/j.geomphys.2018.01.011. MR3774331

[14] B. Bektaş, M. Moruz, J. Van der Veken, and L. Vrancken, *Lagrangian submanifolds of the nearly Kähler* $\mathbb{S}^3 \times \mathbb{S}^3$ *from minimal surfaces in* $\mathbb{S}^3$, Proc. Roy. Soc. Edinburgh Sect. A **149** (2019), no. 3, 655–689, DOI 10.1017/prm.2018.43. MR3961687

[15] D. E. Blair, *Almost contact manifolds with Killing structure tensors*, Pacific J. Math. **39** (1971), 285–292. MR305299

[16] J. Bolton, F. Dillen, B. Dioos, and L. Vrancken, *Almost complex surfaces in the nearly Kähler* $S^3 \times S^3$, Tohoku Math. J. (2) **67** (2015), no. 1, 1–17, DOI 10.2748/tmj/1429549576. MR3337960

[17] J.-B. Butruille, *Classification des variétés approximativement kähleriennes homogènes* (French, with English summary), Ann. Global Anal. Geom. **27** (2005), no. 3, 201–225, DOI 10.1007/s10455-005-1581-x. MR2158165

[18] X. Chen and Y. Yang, *Hopf hypersurfaces of the homogeneous nearly Kähler* $\mathbb{S}^3\times\mathbb{S}^3$ *satisfying certain commuting conditions*, Bull. Korean Math. Soc. **59** (2022), no. 6, 1567–1594, DOI 10.4134/BKMS.b210904. MR4515849

[19] F. Şahin, *Homology of submanifolds of six dimensional sphere*, J. Geom. Phys. **145** (2019), 103471, 6, DOI 10.1016/j.geomphys.2019.07.002. MR3982036

[20] K. Cwiklinski and L. Vrancken, *Almost complex surfaces in the nearly Kähler flag manifold*, Results Math. **77** (2022), no. 3, Paper No. 134, 17, DOI 10.1007/s00025-022-01670-z. MR4417452

[21] K. Dekimpe, *Contributions to the theory of surfaces in almost complex spaces*, PhD Thesis, KU Leuven, 2023 (eng).

[22] K. Dekimpe, *Degenerate almost complex surfaces in the nearly Kähler* $\mathrm{SL}_2\mathbb{R} \times \mathrm{SL}_2\mathbb{R}$, 2023.

[23] G. Deschamps and E. Loubeau, *Hypersurfaces of the nearly Kähler twistor spaces* $\mathbb{C}\mathrm{P}^3$ *and* $\mathbb{F}_{1,2}$, Tohoku Math. J. (2) **73** (2021), no. 4, 627–642, DOI 10.2748/tmj.20200930. MR4355062

[24] F. Dillen, *Lagrangian submanifolds of* $S^6(1)$, Geometry and topology of submanifolds, VIII (Brussels, 1995/Nordfjordeid, 1995), World Sci. Publ., River Edge, NJ, 1996, pp. 138–144. MR1434562

[25] B. Dioos, *Submanifolds of the nearly Kähler manifold* $\mathbb{S}^3 \times \mathbb{S}^3$, PhD thesis, KU Leuven, 2015 (eng).

[26] B. Dioos, H. Li, H. Ma, and L. Vrancken, *Flat almost complex surfaces in the homogeneous nearly Kähler* $S^3\times S^3$, Results Math. **73** (2018), no. 1, Paper No. 38, 24, DOI 10.1007/s00025-018-0784-y. MR3764538

[27] B. Dioos, L. Vrancken, and X. Wang, *Lagrangian submanifolds in the homogeneous nearly Kähler* $\mathbb{S}^3 \times \mathbb{S}^3$, Ann. Global Anal. Geom. **53** (2018), no. 1, 39–66, DOI 10.1007/s10455-017-9567-z. MR3746515

[28] M. Djorić, *Hypersurfaces of the homogeneous nearly Kähler* $\mathbb{S}^3\times\mathbb{S}^3$ *whose normal vector field is* $\mathcal{P}$*-principal*, Mediterr. J. Math. **18** (2021), no. 6, Paper No. 251, 20, DOI 10.1007/s00009-021-01916-0. MR4330448

[29] M. Djorić, M. Djorić, and M. Moruz, *Geodesic lines on nearly Kähler* $\mathbb{S}^3\times\mathbb{S}^3$, J. Math. Anal. Appl. **466** (2018), no. 1, 1099–1108, DOI 10.1016/j.jmaa.2018.06.040. MR3818161

[30] K. Enoyoshi and K. Tsukada, *Lagrangian submanifolds of* S^6 *and the associative Grassmann manifold*, Kodai Math. J. **43** (2020), no. 1, 170–192, DOI 10.2996/kmj/1584345693. MR4077210

[31] J.-H. Eschenburg and T. Vlachos, *Pseudoholomorphic curves in* $\mathbb{S}^6$ *and* $\mathbb{S}^5$, Rev. Un. Mat. Argentina **60** (2019), no. 2, 517–537, DOI 10.33044/revuma.v60n2a16. MR4049801

[32] L. Foscolo and M. Haskins, *New* G_2*-holonomy cones and exotic nearly Kähler structures on* S^6 *and* $S^3 \times S^3$, Ann. of Math. (2) **185** (2017), no. 1, 59–130, DOI 10.4007/annals.2017.185.1.2. MR3583352

[33] E. Ghandour and L. Vrancken, *Almost complex surfaces in the nearly KÄhler* $\mathrm{SL}(2,\mathbb{R}) \times \mathrm{SL}(2,\mathbb{R})$, Mathematics **8** (2020), no. 7, 1160.

[34] A. Gray, *The structure of nearly Kähler manifolds*, Math. Ann. **223** (1976), no. 3, 233–248, DOI 10.1007/BF01360955. MR417965

[35] A. Gray and L. M. Hervella, *The sixteen classes of almost Hermitian manifolds and their linear invariants*, Ann. Mat. Pura Appl. (4) **123** (1980), 35–58, DOI 10.1007/BF01796539. MR581924

[36] H. Hashimoto and K. Sekigawa, *Submanifolds of a nearly Kähler 6-dimensional sphere*, Proceedings of the Eighth International Workshop on Differential Geometry, Kyungpook Nat. Univ., Taegu, 2004, pp. 23–45. MR2061867

[37] Z. Hu, M. Moruz, L. Vrancken, and Z. Yao, *On the nonexistence and rigidity for hypersurfaces of the homogeneous nearly Kähler* $\mathbb{S}^3 \times \mathbb{S}^3$, Differential Geom. Appl. **75** (2021), Paper No. 101717, 22, DOI 10.1016/j.difgeo.2021.101717. MR4205149

[38] Z. Hu and Z. Yao, *On Hopf hypersurfaces of the homogeneous nearly Kähler* $S^3 \times S^3$, Ann. Mat. Pura Appl. (4) **199** (2020), no. 3, 1147–1170, DOI 10.1007/s10231-019-00915-z. MR4102805

[39] Z. Hu, Z. Yao, and J. Yin, *On Ricci curvature pinching of Lagrangian submanifolds in the homogeneous nearly Kähler* $\mathbb{S}^6(1)$, Results Math. **75** (2020), no. 2, Paper No. 52, 7, DOI 10.1007/s00025-020-1179-4. MR4075306

[40] Z. Hu, Z. Yao, and X. Zhang, *Hypersurfaces of the homogeneous nearly Kähler* $\mathbb{S}^6$ *and* $\mathbb{S}^3\times\mathbb{S}^3$ *with anticommutative structure tensors*, Bull. Belg. Math. Soc. Simon Stevin **26** (2019), no. 4, 535–549, DOI 10.36045/bbms/1576206356. MR4042400

[41] Z. Hu, Z. Yao, and Y. Zhang, *On some hypersurfaces of the homogeneous nearly Kähler* $\mathbf{S}^3 \times \mathbf{S}^3$, Math. Nachr. **291** (2018), no. 2-3, 343–373, DOI 10.1002/mana.201600398. MR3767142

[42] I. Kath, *Pseudo-Riemannian T-duals of compact Riemannian homogeneous spaces*, Transform. Groups **5** (2000), no. 2, 157–179, DOI 10.1007/BF01236467. MR1762116

[43] H. V. Lê and L. Schwachhöfer, *Lagrangian submanifolds in strict nearly Kähler 6-manifolds*, Osaka J. Math. **56** (2019), no. 3, 601–629. MR3981303

[44] C. LeBrun and S. Salamon, *Strong rigidity of positive quaternion-Kähler manifolds*, Invent. Math. **118** (1994), no. 1, 109–132, DOI 10.1007/BF01231528. MR1288469

[45] M. Liefsoens, *Nearly KÄhler* $\mathbb{C}P^3$*: From the Hopf fibration to Lagrangian submanifolds*, Msc thesis, KU Leuven. Faculteit Wetenschappen, Leuven, 2022 (eng).

[46] F. Lubbe and L. Schäfer, *Pseudo-holomorphic curves in nearly Kähler manifolds*, Differential Geom. Appl. **36** (2014), 24–43, DOI 10.1016/j.difgeo.2014.07.002. MR3262895

[47] J. Madnick, *Free-boundary problems for holomorphic curves in the 6-sphere*, Math. Z. **303** (2023), no. 4, Paper No. 85, 17, DOI 10.1007/s00209-023-03234-5. MR4559050

[48] M. Moruz, *Lagrangian warped product immersions in* $\mathbb{S}^6$, Arch. Math. (Basel) **113** (2019), no. 3, 325–336, DOI 10.1007/s00013-019-01325-6. MR3988827

[49] M. Moruz and L. Vrancken, *Properties of the nearly Kähler* $\mathbb{S}^3 \times \mathbb{S}^3$, Publ. Inst. Math. (Beograd) (N.S.) **103(117)** (2018), 147–158, DOI 10.2298/pim1817147m. MR3812054

[50] P.-A. Nagy, *Nearly Kähler geometry and Riemannian foliations*, Asian J. Math. **6** (2002), no. 3, 481–504, DOI 10.4310/AJM.2002.v6.n3.a5. MR1946344

[51] P.-A. Nagy, *On nearly-Kähler geometry*, Ann. Global Anal. Geom. **22** (2002), no. 2, 167–178, DOI 10.1023/A:1019506730571. MR1923275

[52] F. Podestà and A. Spiro, *Six-dimensional nearly Kähler manifolds of cohomogeneity one*, J. Geom. Phys. **60** (2010), no. 2, 156–164, DOI 10.1016/j.geomphys.2009.09.008. MR2587385

[53] L. Schäfer, *Nearly pseudo-Kähler manifolds and related special holonomies*, Lecture Notes in Mathematics, vol. 2201, Springer, Cham, 2017, DOI 10.1007/978-3-319-65807-0. MR3645443

[54] R. Storm, *Lagrangian submanifolds of the nearly Kähler full flag manifold* $F_{1,2}(\mathbb{C}^3)$, J. Geom. Phys. **158** (2020), 103844, 16, DOI 10.1016/j.geomphys.2020.103844. MR4139114

[55] R. Storm, *A note on Lagrangian submanifolds of twistor spaces and their relation to superminimal surfaces*, Differential Geom. Appl. **73** (2020), 101669, 10, DOI 10.1016/j.difgeo.2020.101669. MR4140592

[56] S.-i. Tachibana, *On almost-analytic vectors in almost-Kählerian manifolds*, Tohoku Math. J. (2) **11** (1959), 247–265, DOI 10.2748/tmj/1178244584. MR109329

[57] A.-S. Tsouri and T. Vlachos, *Minimal surfaces in spheres and a Ricci-like condition*, Manuscripta Math. **166** (2021), no. 3-4, 561–588, DOI 10.1007/s00229-020-01254-7. MR4325963

[58] F. Xu, *Pseudo-holomorphic curves in nearly Kähler* $\mathbf{CP}^3$, Differential Geom. Appl. **28** (2010), no. 1, 107–120, DOI 10.1016/j.difgeo.2009.05.005. MR2579386

[59] Z. Yao and Z. Hu, *On Hopf hypersurfaces of the homogeneous nearly Kähler* $\mathbf{S}^3 \times \mathbf{S}^3$*, II*, Manuscripta Math. **168** (2022), no. 3-4, 371–402, DOI 10.1007/s00229-021-01308-4. MR4440708

[60] Z. Yao, X. Zhang, and Z. Hu, *Hypersurfaces of the homogeneous nearly Kähler* $\mathbf{S}^3 \times \mathbf{S}^3$ *with* P*-invariant holomorphic distributions*, J. Geom. Anal. **32** (2022), no. 7, Paper No. 209, 16, DOI 10.1007/s12220-022-00931-1. MR4425371

[61] Y. Zhang, B. Dioos, Z. Hu, L. Vrancken, and X. Wang, *Lagrangian submanifolds in the 6-dimensional nearly Kähler manifolds with parallel second fundamental form*, J. Geom. Phys. **108** (2016), 21–37, DOI 10.1016/j.geomphys.2016.06.004. MR3539487

Department of Mathematics, KU Leuven, Celestijnenlaan 200 B - Box 2400, 3001 Leuven, Belgium

Current address: Université Polytechnique Hauts-de-France, Campus Mont Houy 59313 Valenciennes Cedex 9, France

Email address: mateo.anarella@kuleuven.be

Contemporary Mathematics
Volume **821**, 2025
https://doi.org/10.1090/conm/821/16433

Expressions of circles in a complex hyperbolic space by trajectories on tubes around complex hyperplanes

Yusei Aoki and Toshiaki Adachi

ABSTRACT. In this paper, we study how circles on a complex hyperbolic space are expressed by geodesics and by trajectories on tubes around totally geodesic and totally complex hyperplanes. For bounded circles we essentially have two kinds of expressions up to congruency, and for unbounded circles whose velocity and acceleration vectors do not form complex lines we have a unique expression by trajectories on tubes around totally geodesic complex hyperplanes.

1. Introduction

A circle on a Riemannian manifold is a Frenet curve of osculating order 2 whose geodesic curvature is constant. Circles and their extensions play an important role in the study of submanifolds. For example, Nomizu and Yano [**20**] characterized extrinsic spheres in terms of circles, Naitoh [**18**] studied a parallel immersion of a torus into a complex projective space which maps geodesics to circles and applied to investigate isotropic submanifolds in a complex projective space, and Sakamoto [**22**] and Maeda and Sato [**15**] characterized isotropic submanifolds with parallel second fundamental forms in real and complex space forms, respectively (see also [**12, 16, 17, 21, 23**]). In these works, properties of circles are used to investigate submanifolds.

By changing a viewing angle, we focus our mind to circles themselves on complex space forms, and study how they are related to homogeneous submanifolds. We recall the elementary fact that every circle in a Euclidean 3-space is expressed uniquely as a geodesic on an embedded sphere. We are interested in extensions of this result to circles in a complex projective space and in a complex hyperbolic space. This paper is a sequel of the preceding papers [**2, 3**]. Though circles in a Euclidean space are classified with respect to isometries by their geodesic curvatures, those in a nonflat complex space form are classified by their geodesic curvatures and complex torsions. Here, the complex torsion of a circle is defined by the angle formed by its acceleration vector and the complex line spanned by its velocity vector. In the case of a complex projective space, circles are expressed by geodesics

2020 *Mathematics Subject Classification.* Primary 53C22, 53C30; Secondary 53D10, 53C40.
Key words and phrases. Circles, trajectories for Sasakian magnetic fields, tubes around complex hyperplanes, complex hyperbolic space.
The second author was partially supported by Grant-in-Aid for Scientific Research (C) (No. 20K03581), Japan Society for the Promotion of Science.

on geodesic spheres if and only if either their velocity and acceleration vectors span complex lines or acceleration vectors are orthogonal to complex lines spanned by velocity vectors, that is, their complex torsions are either ± 1 or null.

In order to study expressions of other circles on a complex projective space, we extend our family of curves on geodesic spheres. Since each geodesic sphere admits an almost contact metric structure induced by the complex structure on the ambient space, we take a family of curves which are closely associated with this structure, which is a family of trajectories for Sasakian magnetic fields (see §3). Since trajectories for the trivial magnetic field are geodesics, we may say that trajectories are natural generalizations associated with the structure. Using these trajectories, we can show that every circle on a complex projective space is expressed by a trajectory on some geodesic sphere. If this circle has complex torsion $\tau \neq 0, \pm 1$, there are three kinds of expressions by nongeodesic trajectories (see [**2**]).

In the case of complex hyperbolic space, we have both bounded and unbounded circles. Since geodesic spheres are compact, expressions of circles by trajectories on geodesic spheres stand only for bounded circles. In [**3**], we showed similar properties on expressions by geodesics and by non-geodesic trajectories on geodesic spheres. The difference from the case of a complex projective space is that there are only one kind of expressions for each bounded circle of complex torsion $\tau \neq 0, \pm 1$. In order to study expressions of unbounded circles in a complex hyperbolic space, in this paper we treat tubes around totally geodesic and totally complex hyperplanes. If we recall the case of complex projective space $\mathbb{C}P^n$, a tube around $\mathbb{C}P^{n-1}$ coincides with a geodesic sphere. It is therefore expected that we can get the same results by considering both geodesic spheres and tubes around complex hypersurfaces. In [**5**], Bao and the second author studied extrinsic shapes of trajectories on tubes around complex hypersurfaces in a complex hyperbolic space. But as they do not consider congruency of expressions, their statement corresponding to this point is vague. In this paper, we make clear this point and study the number of congruence classes of expressions for circles. When they are bounded we have two kinds of expressions by nongeodesic trajectories on tubes, hence have a corresponding property as for circles in a complex projective space. But when they are unbounded they are uniquely expressed by a trajectory on a tube up to congruency.

2. Circles on a complex hyperbolic space

Let $\mathbb{C}H^n(c)$ be a complex hyperbolic space of constant holomorphic sectional curvature c. A smooth curve σ parameterized by its arclength is said to be a *circle* on $\mathbb{C}H^n(c)$ if there exist a nonnegative constant k_σ and a field Y_σ of unit tangent vectors along σ satisfying the system of differential equations

$$\begin{cases} \widetilde{\nabla}_{\dot\sigma}\dot\sigma &= k_\sigma Y_\sigma, \\ \widetilde{\nabla}_{\dot\sigma}Y_\sigma &= -k_\sigma \dot\sigma, \end{cases}$$

which is equivalent to the equation $\widetilde{\nabla}_{\dot\sigma}\widetilde{\nabla}_{\dot\sigma}\dot\sigma = -k_\sigma^{\,2}\dot\sigma$ (cf. [**20**]). Here, $\widetilde{\nabla}$ denotes the Riemannian connection on $\mathbb{C}H^n(c)$. The constant k_σ and the frame $\{\dot\sigma, Y_\sigma\}$ along σ are called the geodesic curvature and the Frenet frame of σ, respectively. When $k_\sigma = 0$, it is a geodesic with an arbitrary parallel vector field along σ.

We here classify circles on $\mathbb{C}H^n$. We say two smooth curves σ_1, σ_2 on a Riemannian manifold N which are parameterized by their arclengths to be *congruent* to each other if there is an isometry φ of N satisfying $\sigma_2 = \varphi\circ\sigma_1$. For a circle σ of positive geodesic curvature on $\mathbb{C}H^n$, we set $\tau_\sigma = \langle \dot\sigma, JY_\sigma\rangle$ by using its Frenet frame and complex structure J on $\mathbb{C}H^n$. It is called the *complex torsion* of σ. Clearly, it is constant along σ. Geodesic curvatures and complex torsions of circles on $\mathbb{C}H^n$ are important invariants. It is known that two circles σ_1, σ_2 on $\mathbb{C}H^n$ are congruent to each other if and only if either $k_{\sigma_1} = k_{\sigma_2} = 0$ or $k_{\sigma_1} = k_{\sigma_2} > 0$ and $|\tau_{\sigma_1}| = |\tau_{\sigma_2}|$ (see [**14**]). Therefore, the moduli space $\mathcal{C}(\mathbb{C}H^n)$ of circles, the set of all congruence classes of circles, on $\mathbb{C}H^n$ is set theoretically identified with a band $[0,\infty)\times[0,1]/\sim$. Here, for $(k_1,\tau_1), (k_2,\tau_2) \in [0,\infty)\times[0,1]$, we define $(k_1,\tau_1)\sim(k_2,\tau_2)$ if and only if either $k_1 = k_2 = 0$ or $k_1 = k_2 > 0$ and $\tau_1 = \tau_2$.

On $\mathbb{C}H^n$ we have both bounded and unbounded circles. Since $\mathbb{C}H^n$ is a typical example of Hadamard manifolds, we can compactify it by using its ideal boundary $\partial\mathbb{C}H^n$ formed by asymptotic classes of geodesics (see [**10**]). We say a smooth curve σ on $\mathbb{C}H^n$ which is parameterized by its arclength to be unbounded in both directions if both of the sets $\sigma\big((-\infty,0]\big)$, $\sigma\big([0,\infty)\big)$ are unbounded. In this case, we set $\sigma(-\infty) = \lim_{t\to-\infty}\sigma(t)$ and $\sigma(\infty) = \lim_{t\to\infty}\sigma(t)$ if they exist in $\partial\mathbb{C}H^n$. They are called points at infinity of σ. We say a curve σ which is parametrized by its arclength and is unbounded in both directions to be *horocyclic* if it satisfies the following two conditions: It has single point at infinity (i.e., $\sigma(-\infty) = \sigma(\infty)$), and if it crosses to a geodesic α with $\alpha(\infty) = \sigma(\infty)$ then they cross orthogonally. If we set a function $\nu : [0,\infty) \to \mathbb{R}$ by

$$\nu(k) = \begin{cases} 0, & \text{if } k < \sqrt{|c|}/2, \\ (4k^2 + c)^{3/2}/(3\sqrt{3}|c|k), & \text{if } \sqrt{|c|}/2 \le k \le \sqrt{|c|}, \\ 1, & \text{if } k > \sqrt{|c|}, \end{cases}$$

then a circle γ on $\mathbb{C}H^n(c)$ of geodesic curvature k_γ and complex torsion τ_γ is as follows (see [**4**]):

(1) If $k_\gamma > \sqrt{|c|}$ or $\tau_\gamma < \nu(k_\gamma)$, it is bounded;
(2) If it does not satisfies the condition in 1), it is unbounded in both directions and has points at infinity;
(3) It is horocyclic if and only if $\sqrt{|c|}/2 \le k_\gamma \le \sqrt{|c|}$ and $|\tau_\gamma| = \nu(k_\gamma)$ hold;
(4) When $\tau_\gamma = \pm 1$, it lies on a totally geodesic $\mathbb{C}H^1$, and when $\tau_\gamma = 0$, it lies on a totally geodesic $\mathbb{R}H^2$.

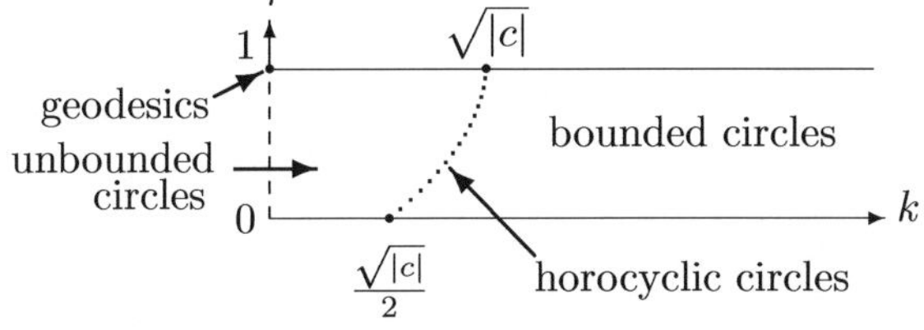

FIGURE 1. The moduli space $\mathcal{C}(\mathbb{C}H^n)$ of circles on $\mathbb{C}H^n(c)$

3. Sasakian magnetic fields

In a complex hyperbolic space $\mathbb{C}H^n(c)$, we take a real hypersurface $M = T(r)$ which is a tube of radius r around totally geodesic complex hyperplane $\mathbb{C}H^{n-1}$. On this real hypersurface we have an almost contact metric structure $(\xi, \eta, \phi, \langle\ ,\ \rangle)$ induced by the complex structure J on $\mathbb{C}H^n$. If we take a local unit normal $\mathcal{N}$ on M in $\mathbb{C}H^n$, then the vector field ξ is defined by $\xi = -J\mathcal{N}$, the 1-form η by $\eta(v) = \langle v, \xi\rangle$, the $(1,1)$-tensor field ϕ by $\phi(v) = Jv - \eta(v)\mathcal{N}$, and $\langle\ ,\ \rangle$ is the induced metric. We denote by ∇ and $\widetilde{\nabla}$ the Riemannian connections on M and $\mathbb{C}H^n$, respectively. Since the complex structure J is parallel, by use of the formula of Gauss which states $\widetilde{\nabla}_X Y = \nabla_X Y + \langle A_M X, Y\rangle\mathcal{N}$ for arbitrary vector fields X, Y tangent to M, we have

$$(\nabla_X \phi)Y = \eta(Y)A_M X - \langle A_M X, Y\rangle\xi. \tag{3.1}$$

Here, A_M denotes the shape operator of M with respect to $\mathcal{N}$. This equality guarantees that the 2-form $\mathbb{F}_\phi$ defined by $\mathbb{F}_\phi(v, w) = \langle v, \phi(w)\rangle$ is closed. We say a constant multiple $\mathbb{F}_\kappa = \kappa\mathbb{F}_\phi$ $(\kappa \in \mathbb{R})$ of this fundamental form to be a *Sasakian magnetic field* (see [**1**]). It is called also a contact magnetic field ([**8**, **9**, **11**, **13**]). Generally, a closed 2-form on a Riemannian manifold is said to be a magnetic field because it can be regarded as a generalization of a static magnetic field on a Euclidean 3-space (see [**7**, **24**], for example).

A smooth curve γ on M parameterized by its arclength is said to be a *trajectory* for $\mathbb{F}_\kappa$ if it satisfies the differential equation $\nabla_{\dot\gamma}\dot\gamma = \kappa\phi\dot\gamma$. When $\kappa = 0$, it is a geodesic on M. Hence, we may say that trajectories for Sasakian magnetic fields on M are natural generalizations of geodesics which correspond to the almost contact metric structure (cf. [**6**]). We adopt these curves to represent circles on $\mathbb{C}H^n$. For a trajectory γ for $\mathbb{F}_\kappa$, we set $\rho_\gamma = \langle\dot\gamma, \xi\rangle$, and call it its *structure torsion*. By using Weingarten formula which states $\widetilde{\nabla}_X\mathcal{N} = -A_M X$ for every vector field X tangent to M, we have $\nabla_X\xi = \phi A_M X$. Therefore, we find

$$\frac{d}{dt}\rho_\gamma = \langle\kappa\phi\dot\gamma, \xi\rangle + \langle\dot\gamma, \phi A_M\dot\gamma\rangle = \langle\dot\gamma, \phi A_M\dot\gamma\rangle = -\langle A_M\phi\dot\gamma, \dot\gamma\rangle,$$

because A_M is symmetric and ϕ is skew-symmetric. Hence, we have

$$\frac{d}{dt}\rho_\gamma = \frac{1}{2}\langle\dot\gamma, (\phi A_M - A_M\phi)\dot\gamma\rangle.$$

The shape operator A_M of a tube $M = T(r)$ satisfies $A_M\xi = \delta_M\xi$ and $A_M v = \lambda_M v$ for every tangent vector $v \in TM$ orthogonal to ξ with

$$\lambda_M = \big(\sqrt{|c|}/2\big)\tanh\big(\sqrt{|c|}\,r/2\big), \quad \delta_M = \sqrt{|c|}\coth\big(\sqrt{|c|}\,r\big)$$

(see [**19**], for example). In particular, the shape operator A_M and the structure tensor field ϕ are simultaneously diagonalizable, i.e., $A_M\phi = \phi A_M$. Therefore, we obtain that the structure torsion ρ_γ is constant along γ for each trajectory γ. The structure torsion is an important invariant for a trajectory. Since every isometry φ of M is equivariant, that is, there is an isometry $\tilde\varphi$ of $\mathbb{C}H^n(c)$ satisfying $\iota\circ\varphi = \tilde\varphi\circ\iota$ with an isometric immersion $\iota : M \to \mathbb{C}H^n(c)$, and since every isometry of $\mathbb{C}H^n(c)$ is $\pm$-holomorphic, that is, it is either holomorphic or anti-holomorphic, we find that two trajectories γ_1, γ_2 for $\mathbb{F}_{\kappa_1}, \mathbb{F}_{\kappa_2}$, respectively, are congruent to each other if and only if one of the following conditions holds (see [**1**]):

(1) $|\rho_{\gamma_1}| = |\rho_{\gamma_2}| = 1$,
(2) $|\rho_{\gamma_1}| = |\rho_{\gamma_2}| < 1$, $|\kappa_1| = |\kappa_2|$ and $\kappa_1\rho_{\gamma_1} = \kappa_2\rho_{\gamma_2}$.

Therefore, the moduli space $\mathcal{T}\big(T(r)\big)$ of all trajectories for Sasakian magnetic fields on $T(r)$ is set theoretically identified with $[0,\infty) \times [-1,1]/\approx$. Here, for $(\kappa_1,\rho_1), (\kappa_2,\rho_2) \in [0,\infty)\times[-1,1]$, we define $(\kappa_1,\rho_1) \approx (\kappa_2,\rho_2)$ if and only if either $(\kappa_1,\rho_1) = (\kappa_2,\rho_2)$ or $|\rho_1| = |\rho_2| = 1$. When the structure torsion of a trajectory γ is ±1, as we have $\nabla_{\dot\gamma}\dot\gamma = \kappa\phi(\pm\xi) = 0$, it is a geodesic and does not depend on κ.

We now study the extrinsic shape of a trajectory for $\mathbb{F}_\kappa$ on M by following to [**5**]. Generally, for a curve γ on M, considering the curve $\iota\circ\gamma$ on $\mathbb{C}H^n$, we call it its *extrinsic shape.* Usually, we denote $\iota\circ\gamma$ also by γ for the sake of simplicity.

LEMMA 3.1 ([**5**]). *Let γ be a trajectory for $\mathbb{F}_\kappa$ on a tube $T(r)$ in $\mathbb{C}H^n(c)$. Its extrinsic shape is a circle on $\mathbb{C}H^n$ if and only if it satisfies one of the following conditions:*

(i) $\rho_\gamma = \pm1$,
(ii) $|\rho_\gamma| < 1$ *and* $\lambda_M - \kappa\rho_\gamma + (\delta_M - \lambda_M)\rho_\gamma^2 = 0$,
(iii) $\kappa + (\delta_M - \lambda_M)\rho_\gamma = 0$.

Corresponding to these cases, the geodesic curvature k_γ and the complex torsion τ_γ of the extrinsic shape are as follows:

(i) $k_\gamma = |\delta_M|$ *and* $\tau_\gamma = \mp1$,
(ii) $k_\gamma = |\kappa|$ *and* $\tau_\gamma = -\mathrm{sgn}(\kappa)1$,
(iii) $k_\gamma = \sqrt{\kappa^2 - 2\lambda_M\kappa\rho_\gamma + \lambda_M^2}$ *and* $\tau_\gamma = (2\kappa\rho_\gamma^2 - \kappa - \lambda_M\rho_\gamma)/k_\gamma$.

Here, $\mathrm{sgn}(\kappa)$ *denotes the signature of* κ.

PROOF. By use of Gauss and Weingarten formulas, we have

$$\widetilde\nabla_{\dot\gamma}\dot\gamma = \kappa\phi\dot\gamma + \langle A_M\dot\gamma,\dot\gamma\rangle\mathcal{N} = \kappa J\dot\gamma + \big(\langle A_M\dot\gamma,\dot\gamma\rangle - \kappa\rho_\gamma\big)\mathcal{N}$$
$$\widetilde\nabla_{\dot\gamma}\big(\kappa\phi\dot\gamma + \langle A_M\dot\gamma,\dot\gamma\rangle\mathcal{N}\big) = -\kappa^2\dot\gamma - \big(\langle A_M\dot\gamma,\dot\gamma\rangle - \kappa\rho_\gamma\big)(\kappa\xi + A_M\dot\gamma).$$

Since we have $\|\widetilde\nabla_{\dot\gamma}\dot\gamma\|^2 = \kappa^2(1-\rho_\gamma^2) + \langle A_M\dot\gamma,\dot\gamma\rangle^2 > 0$ and since the second equality turns to

$$\begin{aligned}\widetilde\nabla_{\dot\gamma}&\big(\kappa\phi\dot\gamma + \langle A_M\dot\gamma,\dot\gamma\rangle\mathcal{N}\big)\\ &= -\big\{\kappa^2(1-\rho_\gamma^2) + \langle A_M\dot\gamma,\dot\gamma\rangle^2\big\}\dot\gamma\\ &\qquad + \big(\langle A_M\dot\gamma,\dot\gamma\rangle - \kappa\rho_\gamma\big)\big\{\big(\langle A_M\dot\gamma,\dot\gamma\rangle + \kappa\rho_\gamma\big)\dot\gamma - \kappa\xi - A_M\dot\gamma\big\}\\ &= -\big\{\kappa^2(1-\rho_\gamma^2) + \langle A_M\dot\gamma,\dot\gamma\rangle^2\big\}\dot\gamma\\ &\qquad + \big(\langle A_M\dot\gamma,\dot\gamma\rangle - \kappa\rho_\gamma\big)\big\{\kappa + (\delta_M - \lambda_M)\rho_\gamma\big\}(\xi - \rho_\gamma\dot\gamma),\end{aligned}$$

we find that the extrinsic shape is a circle if and only if one of the three conditions in our assertion holds. When $\rho_\gamma = \pm1$, we have $k_\gamma^2 = \delta_M^2$ and $Y_\gamma = \mathcal{N}$ because $\delta_M > 0$. We therefore have $\tau_\gamma = \mp1$. When $\langle A_M\dot\gamma,\dot\gamma\rangle = \kappa\rho_\gamma$, which is the second case, we have $k_\gamma^2 = \kappa^2$ and $Y_\gamma = \mathrm{sgn}(\kappa)J\dot\gamma$, hence have $\tau_\gamma = -\mathrm{sgn}(\kappa)1$. In the last case, as we have $k_\gamma Y_\gamma = \kappa J\dot\gamma + \big(\langle A_M\dot\gamma,\dot\gamma\rangle - \kappa\rho_\gamma\big)\mathcal{N}$, we get the geodesic curvature and the complex torsion of the extrinsic shape by direct computation. □

If we denote by $\mathcal{E}\big(T(r)\big)$ $\big(\subset \mathcal{T}\big(T(r)\big)\big)$ the moduli space of trajectories whose extrinsic shapes are circles, Lemma 3.1 shows that we have a map Φ_M of $\mathcal{E}\big(T(r)\big)$ into $\mathcal{C}(\mathbb{C}H^n)$. To study this map, the third case in Lemma 3.1 is important. Since we have $\delta_M - \lambda_M = |c|/(4\lambda_M)$, the condition in the third case turns to $4\kappa\lambda_M + |c|\rho_\gamma = 0$. Therefore, we have

$$k_\gamma = \sqrt{\lambda_M^2 + \frac{|c|\rho_\gamma^2}{2} + \frac{c^2\rho_\gamma^2}{16\lambda_M^2}}, \quad \tau_\gamma = \frac{\rho_\gamma(|c| - 2|c|\rho_\gamma^2 - 4\lambda_M^2)}{4k_\gamma\lambda_M}.$$

As $|\rho_\gamma| \le 1$, we find $\lambda_M \le k_\gamma \le \delta_M$. By using these two equalities, if we set a function g_M on the interval $[\lambda_M, \delta_M]$ by

$$g_M(k) = \frac{(k^2 - \lambda_M^2)\{32\lambda_M^2 k^2 + 4c\lambda_M^2 - c^2\}^2}{|c|(8\lambda_M^2 - c)^3 k^2},$$

we can express the complex torsion as a function of the geodesic curvature as $\tau_\gamma^2 = g(k_\gamma)$. The function g is monotone decreasing on the interval $[\alpha_M, \beta_M]$ with

$$\alpha_M = \sqrt{2(4\lambda_M^2 - c)}/4, \quad \beta_M = \sqrt{2(c^2 - 4c\lambda_M^2)}/(8\lambda_M)$$

and is monotone increasing on other part of the domain, and takes the following values:

$$g_M(\lambda_M) = g_M(\beta_M) = 0, \;\; g_M(\delta_M) = 1, \;\; g_M(\alpha_M) = \frac{-(4\lambda_M^2 + c)^3(4\lambda_M^2 - c)}{|c|(8\lambda_M^2 - c)^3} \; (< 1).$$

We note that the second condition shows that $\rho_\gamma \neq 0$ and $\kappa = \frac{\lambda_M}{\rho_\gamma} + \frac{|c|}{4\lambda_M}\rho_\gamma$. Thus, in this case we have $|\kappa| \ge \sqrt{|c|}$. Since $\sqrt{2(4\lambda_M^2 - c)}/4 < \sqrt{|c|}/2$, we find that the image of the map Φ_M is like Figure 2.

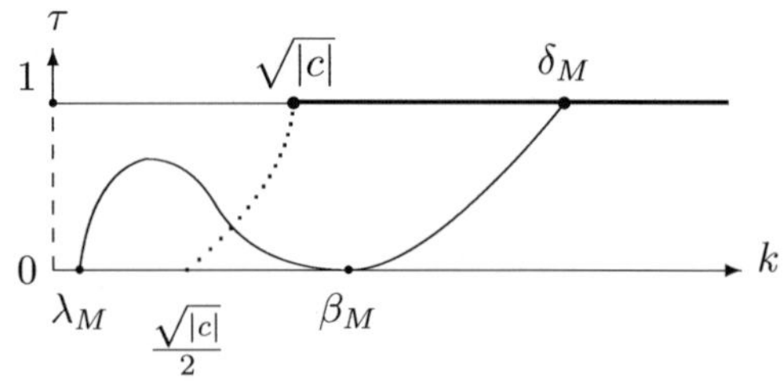

FIGURE 2. The image of Φ_M in $\mathcal{C}(\mathbb{C}H^n)$

Here we check the case when extrinsic shapes of trajectories on M are horocyclic. Trajectories for $\mathbb{F}_{\pm\kappa_H}$ on M of structure torsion $\mp\rho_H$ with

$$\kappa_H = \kappa_H(M) = \frac{\sqrt{|c|}}{2\sqrt{9\sinh^2(\sqrt{|c|}\,r/2) + 1}}, \; \rho_H = \rho_H(M) = \frac{2\tanh(\sqrt{|c|}\,r/2)}{\sqrt{9\sinh^2(\sqrt{|c|}\,r/2) + 1}}$$

satisfy the third condition, and their extrinsic shapes are horocyclic circles of geodesic curvature ϵ_M and of structure torsion $\pm\tau_H$ with

$$\epsilon_M^2 = \frac{|c|(5\tanh^2(\sqrt{|c|}\,r/2) + 4)}{4(8\tanh^2(\sqrt{|c|}\,r/2) + 1)} = \frac{|c|(9\sinh^2(\sqrt{|c|}\,r/2) + 4)}{4(9\sinh^2(\sqrt{|c|}\,r/2) + 1)},$$

$$\tau_H = \tau_H(M) = \frac{2}{\left(9\sinh^2(\sqrt{|c|}\,r/2) + 1\right)\sqrt{9\sinh^2(\sqrt{|c|}\,r/2) + 4}}.$$

Extrinsic shapes of other trajectories on M are not horocyclic circles.

REMARK 3.2. Through the map Φ_M, we have the following correspondences:

(1) The congruence class $[0,0] \in \mathcal{E}(M)$ of geodesics with null structure torsion corresponds to the congruence class $[\lambda_M, 0] \in \mathcal{C}(\mathbb{C}H^n)$ of circles with geodesic curvature λ_M and null complex torsion;

(2) The congruence class

$$\left[\pm\frac{\sqrt{|c|}}{2\sqrt{2}\sinh(\sqrt{|c|}\,r/2)}, \mp\frac{1}{\sqrt{2}\cosh(\sqrt{|c|}\,r/2)}\right] \in \mathcal{E}(M)$$

of trajectories corresponds to the congruence class $[\beta_M, 0] \in \mathcal{C}(\mathbb{C}H^n)$ of circles.

4. Expressions of circles by trajectories

For a smooth curve σ on $\mathbb{C}H^n(c)$, we say a pair (M,γ) of a real hypersurface M and a smooth curve γ on M to be its *representation* if σ is an extrinsic shape of γ. The purpose of this paper is to study whether every circle on $\mathbb{C}H^n$ is expressed by a trajectory on a tube around complex hyperplane $\mathbb{C}H^{n-1}$, and also to study how many kinds of expressions we have for this circle. We shall say that two expressions (M_1,γ_1) and (M_2,γ_2) of a curve σ on $\mathbb{C}H^n(c)$ are congruent to each other if there is an isometry $\tilde{\varphi}$ of $\mathbb{C}H^n(c)$ with $\tilde{\varphi}(M_1) = M_2$ by regarding M_1, M_2 as subsets of $\mathbb{C}H^n$ through isometric embeddings which either preserves σ (i.e., $\tilde{\varphi}\circ\sigma = \sigma$) or reverses σ (i.e., $\tilde{\varphi}\circ\sigma(t) = \sigma(-t)$ for all t).

First we study expressions of circles by geodesics.

THEOREM 4.1. *Let σ be a circle of geodesic curvature k_σ and of complex torsion τ_σ on a complex hyperbolic space $\mathbb{C}H^n(c)$ of constant holomorphic sectional curvature c.*

(1) *When $k_\sigma > \sqrt{|c|}$ and $\tau_\sigma = \pm 1$, there exists a unique expression by a geodesic on a tube around complex hyperplane $\mathbb{C}H^{n-1}$ up to congruency. The radius of this tube is $\frac{1}{2\sqrt{|c|}}\log\frac{k_\sigma+\sqrt{|c|}}{k_\sigma-\sqrt{|c|}}$.*

(2) *When $k_\gamma < \sqrt{|c|}/2$ and $\tau_\gamma = 0$, there exists a unique expression by a geodesic on a tube around complex hyperplane $\mathbb{C}H^{n-1}$ up to congruency. The radius of this tube is $\frac{1}{\sqrt{|c|}}\log\frac{\sqrt{|c|}+2k_\sigma}{\sqrt{|c|}-2k_\sigma}$.*

(3) *Otherwise, there are no such expressions by geodesics on a tube around complex hyperplane $\mathbb{C}H^{n-1}$.*

To show this, by Lemma 3.1, we only need to study the first case and the third case with $\rho_\gamma = 0$. In these cases, complex torsions of extrinsic shapes are ∓ 1 and 0, respectively. Since their geodesic curvatures are δ_M and λ_M, respectively, they vary in the interval $(\sqrt{|c|},\infty)$ and $(0,\sqrt{|c|}/2)$, respectively, corresponding to varying of radii of tubes. Since $\mathbb{C}H^n$ is a symmetric space of rank 1, for two circles σ_1,σ_2, we have an isometry $\tilde{\varphi}$ of $\mathbb{C}H^n$ satisfying $\sigma_2 = \tilde{\varphi}\circ\sigma_1$. Therefore, we get the third assertion and the existence parts of the first and the second assertion. We shall discuss congruency of representations later.

Next we study representations of circles by nongeodesic trajectories on tubes around $\mathbb{C}H^{n-1}$. The following is our main result in this paper.

THEOREM 4.2. *Let γ be a circle of geodesic curvature k_γ and of complex torsion τ_γ on a complex hyperbolic space $\mathbb{C}H^n(c)$ of constant holomorphic sectional curvature c.*

(1) *When $k_\gamma \geq \sqrt{|c|}$ and $\tau_\gamma = \pm 1$, that is, when γ is a bounded circle of complex torsion ± 1 or is a horocyclic circle of complex torsion ± 1, there are infinitely many kinds of expressions by nongeodesic trajectories on a tube around complex hyperplane $\mathbb{C}H^{n-1}$ up to congruency.*
(2) *When it is a bounded circle with $0 < |\tau_\gamma| < 1$, there are two kinds of expression by nongeodesic trajectories on a tube around complex hyperplane $\mathbb{C}H^{n-1}$ up to congruency.*
(3) *When it is a unbounded circle with $0 < |\tau_\gamma| < 1$ or is a bounded circle of null complex torsion, there exists a unique expression by a nongeodesic trajectory on a tube around complex hyperplane $\mathbb{C}H^{n-1}$ up to congruency.*
(4) *When it is a unbounded nonhorocyclic circle with $\tau_\gamma = 0, \pm 1$ or is a horocyclic circle of null structure torsion, there are no expressions by nongeodesic trajectories on a tube around complex hyperplane $\mathbb{C}H^{n-1}$.*

REMARK 4.3. For unbounded circle γ with $k_\gamma \leq \sqrt{|c|}/2$, the third assertion of Theorem 4.2 was stated by Bao and Adachi [**5**, Theorem 3]. But they missed making a clear congruency of expressions.

In Figures 3 and 4, we give images on how circles on a complex hyperbolic space are expressed by geodesics and by nongeodesic trajectories on tubes around totally geodesic complex hyperplanes. Since our proofs of theorems are a bit long, we give a consequence of these theorems before we show them.

COROLLARY 4.4. *For non-horocyclic unbounded circles of complex torsion ± 1 and horocyclic circles of null structure torsion on $\mathbb{C}H^n$, there are no expressions by trajectories on a tube around a totally geodesic complex hyperplane.*

We may say that nonhorocyclic unbounded circles of complex torsion ± 1 have different properties compared with other circles on a complex hyperbolic space.

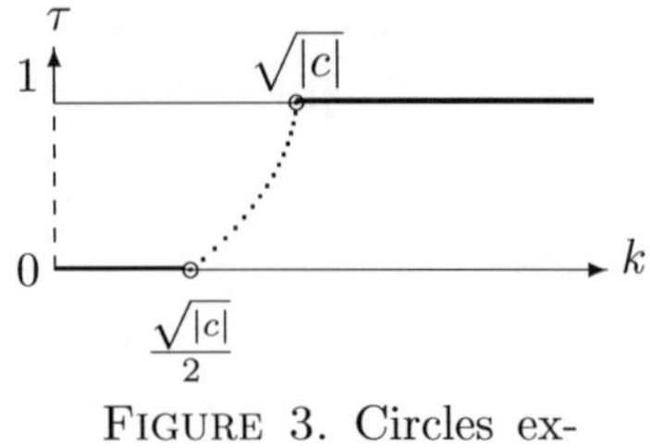

FIGURE 3. Circles expressed by geodesics

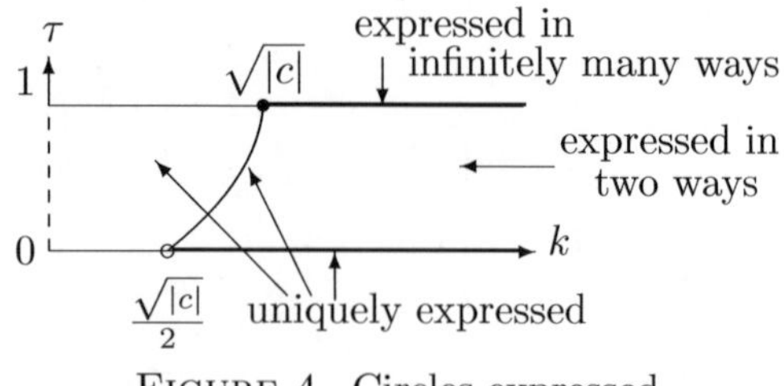

FIGURE 4. Circles expressed by nongeodesic trajectories

PROOF OF THEOREM 4.2. To show the first assertion, we study the second case of Lemma 3.1. In this case, the extrinsic shape of a trajectory has geodesic curvature $|\kappa|$ and complex torsion ± 1. Since we have $\kappa = \frac{\lambda_M}{\rho_\gamma} + \frac{|c|}{4\lambda_M}\rho_\gamma$, the strength $|\kappa|$ runs in the interval $\big[\sqrt{|c|}, \infty\big)$. It takes twice for values in the interval $\big(\sqrt{|c|}, \sqrt{|c|}\coth(\sqrt{|c|}\,r)\big)$. We hence get the first assertion, because two expressions are not congruent to each other if their underlying manifolds are not isometric.

In order to study other assertions, we study the third case of Lemma 3.1 with nonzero structure torsions. For a tube M around $\mathbb{C}H^{n-1}$, we set three subsets $\mathcal{CT}_i(M)$ $(i = 1, 2, 3)$ of $\mathcal{C}(\mathbb{C}H^n)$ by

$$\mathcal{CT}_1(M) = \{[k,\tau] \in \mathcal{C}(\mathbb{C}H^n) \mid \tau^2 = g_M(k),\ \lambda_M < k \le \epsilon_M\},$$
$$\mathcal{CT}_2(M) = \{[k,\tau] \in \mathcal{C}(\mathbb{C}H^n) \mid \tau^2 = g_M(k),\ \epsilon_M < k < \beta_M\},$$
$$\mathcal{CT}_3(M) = \{[k,\tau] \in \mathcal{C}(\mathbb{C}H^n) \mid \tau^2 = g_M(k),\ \beta_M \le k < \delta_M\}.$$

These subsets depend only on radii of tubes. We therefore identify $\mathcal{CT}_i(M)$ and $\mathcal{CT}_i(M')$ if the radii of tubes M and M' are the same. To study their dependence on radii, we consider g_M as a function of λ_M. That is, for each positive k, we set a function f_k on a suitable domain by

$$f_k(x) = \frac{(k^2 - x)\{4(8k^2 + c)x - c^2\}^2}{|c|k^2(8x - c)^3}.$$

Its differential is given as

$$\frac{df_k}{dx} = \frac{(8k^2 - c)^2(4x + c)\{c^2 - 4(c + 8k^2)x\}}{|c|k^2(8x - c)^4}.$$

We take a point $[k_0, \tau_0] \in \mathcal{CT}_3(M)$ with $k_0 > \beta_M$. Since $4\lambda_M^2 + c < 0$, we find that $\dfrac{df_{k_0}}{dx}(\lambda_M^2) > 0$. This means that if we increase the radius of M then the image $\mathcal{CT}_3(M)$ moves up in $\mathcal{C}(\mathbb{C}H^n) \cong [0, \infty) \times [0, 1]/\sim$. In Figure 5, the radius of M' is larger than that of M. Moreover, if we vary radii of tubes in the interval $(0, \infty)$, we have

$$\lim_{r\downarrow 0} \lambda_M = 0, \quad \lim_{r\to\infty} \lambda_M = \frac{\sqrt{|c|}}{2}, \quad \lim_{r\downarrow 0} \delta_M = \infty, \quad \lim_{r\to\infty} \delta_M = \sqrt{|c|},$$

hence we obtain

$$\lim_{r\downarrow 0} \beta_M = \infty, \quad \lim_{r\to\infty} \beta_M = \frac{\sqrt{|c|}}{2}, \quad \lim_{r\to\infty} g_M(k) = \nu(k).$$

Therefore, we find that $\{\mathcal{CT}_3(M) \mid M\}$ forms a foliation of the moduli space of bounded circles of complex torsion τ with $|\tau| < 1$ on $\mathbb{C}H^n$.

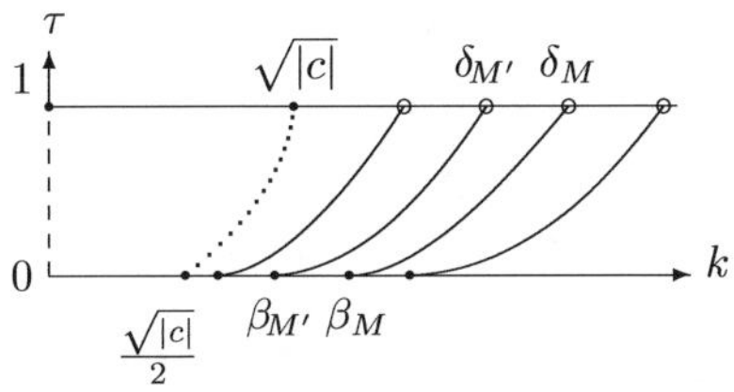

FIGURE 5. The foliation $\{\mathcal{CT}_3(M) \mid M\}$

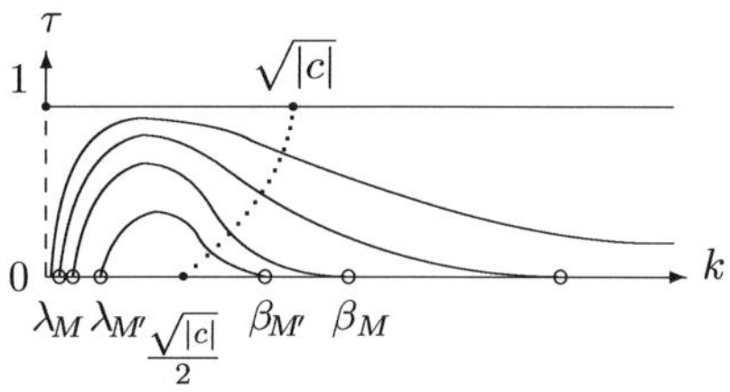

FIGURE 6. The foliation $\{\mathcal{CT}_1(M) \cup \mathcal{CT}_2(M) \mid M\}$

Next we study $\{\mathcal{CT}_1(M)\}$ and $\{\mathcal{CT}_2(M)\}$. We take a point $[k_0, \tau_0] \in \mathcal{CT}_1(M) \cup \mathcal{CT}_2(M)$ with $k_0 < \beta_M$. We then have

$$\frac{df_{k_0}}{dx}(\lambda_M^2) < 0,$$

which means that if we increase the radius of M then the image $\mathcal{CT}_1(M) \cup \mathcal{CT}_2(M)$ moves down in $\mathcal{C}(\mathbb{C}H^n) \cong [0, \infty) \times [0, 1]/\sim$. In Figure 6, the radius of M' is larger than that of M. Since we have

$$\lim_{r\downarrow 0} \epsilon_M = \sqrt{|c|}, \quad \lim_{r\to\infty} \epsilon_M = \lim_{r\to\infty} \beta_M = \frac{\sqrt{|c|}}{2}, \quad \lim_{r\downarrow 0} \beta_M = \infty,$$
$$\lim_{r\downarrow 0} f_k(\lambda_M^2) = 1 \quad (k \geq \sqrt{|c|}),$$

we find that $\{\mathcal{CT}_2(M) \mid M\}$ forms a foliation of the moduli space of bounded circles of complex torsion τ with $0 < |\tau| < 1$. Also, since we have

$$\lim_{r\downarrow 0} \epsilon_M = \sqrt{|c|}, \quad \lim_{r\to\infty} \epsilon_M = \lim_{r\to\infty} \lambda_M = \frac{\sqrt{|c|}}{2}, \quad \lim_{r\downarrow 0} \lambda_M = 0,$$
$$\lim_{r\downarrow 0} f_k(\lambda_M^2) = 1 \quad (k < \sqrt{|c|}),$$

we find that $\{\mathcal{CT}_1(M) \mid M\}$ forms a foliation of the moduli space of unbounded nongeodesic circles of complex torsion τ with $0 < |\tau| < 1$. Thus, we get the fourth assertion and the existence parts of the second and the third assertions. $\square$

In order to complete proofs of our theorems, we need to study congruency of two expressions of circles by trajectories on tubes. Our discussion on this study is the same as for expressions of circles by trajectories on geodesic spheres (see [**2**, **3**]). But for the sake of readers' convenience, we show in below.

LEMMA 4.5. *Let σ be a circle on $\mathbb{C}H^n(c)$, and let (M_1, γ_1) and (M_2, γ_2) be two expressions of σ by trajectories γ_1, γ_2 for $\mathbb{F}_{\kappa_1}$ on a tube M_1 around $\mathbb{C}H^{n-1}$ and for $\mathbb{F}_{\kappa_2}$ on a tube M_2 around $\mathbb{C}H^{n-1}$, respectively. They are congruent to each other if and only if these tubes have the same radius and trajectories satisfy either the condition $|\rho_{\gamma_1}| = |\rho_{\gamma_2}| = 1$ or the condition $|\rho_{\gamma_1}| = |\rho_{\gamma_2}| < 1$, $|\kappa_1| = |\kappa_2|$ and $\kappa_1\rho_{\gamma_1} = \kappa_2\rho_{\gamma_2}$.*

PROOF. First, we suppose that two expressions $(M_1, \gamma_1), (M_2, \gamma_2)$ are congruent to each other. Considering these two tubes as subsets through isometric immersions, we have an isometry $\tilde{\varphi}$ of $\mathbb{C}H^n$ satisfying $\tilde{\varphi}(M_1) = M_2$ which preserves or reverses σ. We set

$$\varepsilon = \begin{cases} 1, & \text{when } \tilde{\varphi} \circ \sigma(t) = \sigma(t) \text{ for all } t, \\ -1, & \text{when } \tilde{\varphi} \circ \sigma(t) = \sigma(-t) \text{ for all } t. \end{cases}$$

We then have $(d\varphi \circ \dot{\sigma})(t) = \varepsilon\dot{\sigma}(\varepsilon t)$. We may suppose $d\tilde{\varphi}(\mathcal{N}_{M_1}) = \mathcal{N}_{M_2}$. Since $\tilde{\varphi}$ is $\pm$-holomorphic, we have $d\tilde{\varphi} \circ J = \pm J \circ d\tilde{\varphi}$. As the extrinsic shapes of γ_1 and γ_2 coincide with σ, we have

$$\begin{aligned} \rho_{\gamma_2} &= \langle \dot{\gamma}_2, -J\mathcal{N}_{M_2} \rangle = \langle \varepsilon d\tilde{\varphi}(\dot{\gamma}_1), -Jd\tilde{\varphi}(\mathcal{N}_{M_1}) \rangle \\ &= \pm\varepsilon \langle d\tilde{\varphi}(\dot{\gamma}_1), d\tilde{\varphi}(-J\mathcal{N}_{M_1}) \rangle = \pm\varepsilon \langle \dot{\gamma}_1, -J\mathcal{N}_{M_1} \rangle = \pm\varepsilon\rho_{\gamma_1}. \end{aligned}$$

In particular, we have $|\rho_{\gamma_1}| = |\rho_{\gamma_2}|$. Also, since we have

$$\langle A_{M_1}\dot{\gamma}_1, \dot{\gamma}_1 \rangle = \langle d\tilde{\varphi}(A_{M_1}\dot{\gamma}_1), d\tilde{\varphi}(\dot{\gamma}_1) \rangle = \langle A_{M_2} d\tilde{\varphi}(\dot{\gamma}_1), d\tilde{\varphi}(\dot{\gamma}_1) \rangle = \langle A_{M_2}\dot{\gamma}_2, \dot{\gamma}_2 \rangle,$$

we find

$$\begin{aligned}\kappa_2\phi\dot\gamma_2 &= \nabla_{\dot\gamma_2}\dot\gamma_2 = \widetilde\nabla_{\dot\gamma_2}\dot\gamma_2 - \langle A_{M_2}\dot\gamma_2,\dot\gamma_2\rangle\mathcal{N}_{M_2}\\ &= \widetilde\nabla_{d\tilde\varphi\circ\dot\gamma_1} d\tilde\varphi\circ\dot\gamma_1 - \langle A_{M_1}\dot\gamma_1,\dot\gamma_1\rangle d\tilde\varphi(\mathcal{N}_{M_1})\\ &= d\tilde\varphi(\widetilde\nabla_{\dot\gamma_1}\dot\gamma_1) - d\tilde\varphi(\langle A_{M_1}\dot\gamma_1,\dot\gamma_1\rangle\mathcal{N}_{M_1})\\ &= d\tilde\varphi(\nabla_{\dot\gamma_1}\dot\gamma_1) = d\tilde\varphi(\kappa_1\phi\dot\gamma_1) = \kappa_1 d\tilde\varphi(J\dot\gamma_1 - \rho_{\gamma_1}\mathcal{N}_{M_1})\\ &= \pm\kappa_1 J d\tilde\varphi(\dot\gamma_1) - \kappa_1\rho_{\gamma_1}\mathcal{N}_{M_2} = \pm\varepsilon\kappa_1 J\dot\gamma_2 \mp \kappa_1\varepsilon\rho_{\gamma_2}\mathcal{N}_{M_2} = \pm\varepsilon\kappa_1\phi\dot\gamma_2.\end{aligned}$$

Here, through an isometric immersion we regard the tangent bundle TM_1 as a subbundle $T\mathbb{C}H^n$. When $|\rho_{\gamma_1}| = |\rho_{\gamma_2}| = 1$, the above tells nothing. When $|\rho_{\gamma_1}| = |\rho_{\gamma_2}| < 1$, we obtain $\kappa_2\rho_{\gamma_2} = \kappa_1\rho_{\gamma_1}$. Thus, the conditions hold. If $d\tilde\varphi(\mathcal{N}_{M_1}) = -\mathcal{N}_{M_2}$, we have $d\tilde\varphi(A_{M_1}v) = -A_{M_2}(d\tilde\varphi v)$. Hence, our computation goes through in the same way.

Next, we suppose that $(M_1,\gamma_1),(M_2,\gamma_2)$ satisfy the condition. Since these tubes are of the same radius, we have an isometry $\tilde\varphi$ of $\mathbb{C}H^n$ with $\tilde\varphi(M_1) = M_2$. Since $\tilde\varphi$ is $\pm$-holomorphic, by the same computation as above, we find $\rho_{\varphi\circ\gamma_1} = \pm\rho_{\gamma_1}$ and $\nabla_{d\tilde\varphi\circ\dot\gamma_1} d\tilde\varphi\circ\dot\gamma_1 = \pm\kappa_1\phi(d\tilde\varphi\circ\dot\gamma_1)$. Therefore $\tilde\varphi\circ\gamma_1$ is a trajectory for $\mathbb{F}_{\pm\kappa_1}$ on M_2. Since either $|\rho_{\tilde\varphi\circ\gamma_1}| = |\rho_{\gamma_2}| = 1$ holds or $|\rho_{\tilde\varphi\circ\gamma_1}| = |\rho_{\gamma_2}| < 1$, $|\pm\kappa_1| = |\kappa_2|$ and $\pm\kappa_1\rho_{\tilde\varphi\circ\gamma_1} = \kappa_2\rho_{\gamma_2}$ hold, we find that $(\tilde\varphi|_{M_2})\circ\gamma_1$ and γ_2 are congruent to each other as curves on M_2. We take an isometry ψ on M_2 satisfying $\gamma_2 = \psi\circ(\tilde\varphi|_{M_2})\circ\gamma_1$. It is well known that isometries of M_2 are equivariant, that is, we have an isometry $\tilde\psi$ of $\mathbb{C}H^n$ satisfying $\tilde\psi|_{M_2} = \psi$. The isometry $\tilde\psi\circ\tilde\varphi$ shows the congruency of two expressions. □

REMARK 4.6. If γ is a trajectory for $\mathbb{F}_\kappa$ on a tube M, the reversed curve $\check\gamma$ defined by $\check\gamma(t) = \gamma(-t)$ is a trajectory for $\mathbb{F}_{-\kappa}$. Since we have $\rho_{\check\gamma} = \langle -\dot\gamma, \xi_M\rangle = -\rho_\gamma$, we find that γ and $\check\gamma$ are congruent to each other. Therefore, in the proof of Lemma 4.5, we have only the case that $\tilde\psi\circ\tilde\varphi$ preserves the direction of γ_1.

By the proof of the first assertion of Theorem 4.2, even if we fix the radii of tubes around totally geodesic complex hyperplane $\mathbb{C}H^{n-1}$, we can express bounded circles of complex torsion ±1.

PROPOSITION 4.7. *Let σ be a bounded circle or a horocyclic circle on $\mathbb{C}H^n(c)$. Suppose that its complex torsion is $\tau_\sigma = \pm1$.*

(1) *It is expressed by nongeodesic trajectory on a tube of fixed radius r around totally geodesic $\mathbb{C}H^{n-1}$.*
(2) *When its geodesic curvature k_σ satisfies $k_\sigma \geq \sqrt{|c|}\coth(\sqrt{|c|}\,r)$ or $k_\sigma = \sqrt{|c|}$, it is expressed uniquely by nongeodesic trajectory on a tube of radius r around $\mathbb{C}H^{n-1}$ up to congruency relation.*
(3) *When $\sqrt{|c|} < k_\sigma < \sqrt{|c|}\coth(\sqrt{|c|}\,r)$, it has two kinds of expressions by nongeodesic trajectory on a tube of radius r around $\mathbb{C}H^{n-1}$ up to congruency.*
(4) *When $k_\sigma = \sqrt{|c|}\coth(\sqrt{|c|}\,r)$, it is expressed by a geodesic and by a nongeodesic trajectory on tubes of radius r around $\mathbb{C}H^{n-1}$.*

We here compare our result to that on circles in a complex projective space. In [**3**] we showed the following:

(1) Every bounded circle on $\mathbb{C}H^n$ of complex torsion either ±1 or 0 is expressed uniquely by a geodesic on some geodesic sphere up to congruency;

(2) Every bounded circle on $\mathbb{C}H^n$ of complex torsion τ with $0 < |\tau| < 1$ is expressed uniquely by a non-geodesic trajectory on some geodesic sphere up to congruency;
(3) Bounded circles of null complex torsion on $\mathbb{C}H^n$ are not expressed by nongeodesic trajectories on geodesic spheres.

Combining Theorem 4.2 and the above, we find that bounded circles of complex torsion τ with $0 < |\tau| < 1$ on a complex hyperbolic space have the same property on expressions by nongeodesic trajectories as for the case of complex projective space by considering both geodesic spheres and tubes around totally geodesic complex hyperplanes ([**2**]).

COROLLARY 4.8. *For bounded circles on* $\mathbb{C}H^n$ *of complex torsion* $0 < |\tau| < 1$, *there are three kinds of expressions by non-geodesic trajectories on geodesic spheres or on tubes around* $\mathbb{C}H^{n-1}$ *up to congruency.*

REMARK 4.9. For each bounded circle of null complex torsion on $\mathbb{C}H^n$, even if we consider non-geodesic trajectories both on geodesic spheres and on tubes around totally geodesic complex hyperplanes, it is uniquely expressed up to congruency.

5. Expressions of horocyclic circles by trajectories on horospheres

In this section, we study expressions of circles by trajectories on horospheres in a complex hyperbolic space. The shape operator A_{HS} of a horosphere HS in $\mathbb{C}H^n(c)$ satisfies $A_{HS}\xi = \sqrt{|c|}\,\xi$ and $A_{HS}v = (\sqrt{|c|}/2)v$ for each tangent vector v orthogonal to ξ (see [**19**], for example). The condition that extrinsic shapes of trajectories on HS are circles is the same as in Lemma 3.1 corresponding to these principal curvatures. We take a trajectory γ for $\mathbb{F}_\kappa$ on HS.

(1) If $\rho_\gamma = \pm 1$, its extrinsic shape is a circle of geodesic curvature $\sqrt{|c|}$ and of complex torsion ∓ 1.
(2) If $2\kappa\rho_\gamma = \sqrt{|c|}(1+\rho_\gamma^2)$ holds, its extrinsic shape is a circle of geodesic curvature $|\kappa|$ and of complex torsion $-\mathrm{sgn}(\kappa)1$.
(3) If $2\kappa + \sqrt{|c|}\rho_\gamma = 0$ holds, its extrinsic shape is a circle of geodesic curvature $\sqrt{|c|(1+3\rho_\gamma^2)/2}$ and of complex torsion $-2\rho_\gamma^3/\sqrt{1+3\rho_\gamma^2}$, which is horocyclic.

Therefore, we have the following.

THEOREM 5.1. *Let* σ *be a circle on a complex hyperbolic space* $\mathbb{C}H^n(c)$.

(1) *If it is of geodesic curvature* $k_\sigma = \sqrt{|c|}$ *and of complex torsion* $\tau_\sigma = \pm 1$, *or if it is of* $k_\sigma = \sqrt{|c|}/2$ *and of* $\tau_\sigma = 0$, *then it is expressed by a geodesic on a horosphere HS.*
(2) *If it is of* $k_\sigma > \sqrt{|c|}$ *and of* $\tau_\sigma = \pm 1$, *or if it is a horocyclic circle of* $0 < |\tau_\sigma| < 1$, *then it is expressed by a non-geodesic trajectory on a horosphere HS.*
(3) *These expressions are unique as expressions by trajectories on HS up to congruence relation.*
(4) *Except the cases* (1) *and* (2), *it is not expressed by trajectories on horospheres.*

Combining the third assertion of Theorem 4.2 and the second assertion of Theorem 5.1, we have the following.

COROLLARY 5.2. *For each horocyclic circle σ of complex torsion τ_σ with $0 < |\tau_\sigma| < 1$, we have two kinds of expressions by non-geodesic trajectories on a horosphere and on a tube around totally geodesic complex hyperplane.*

The first and the third assertions of Theorem 4.2, Corollaries 4.8, and 5.2 show how circles on $\mathbb{C}H^n$ are expressed by nongeodesic trajectories on totally η-umbilic submanifolds in $\mathbb{C}H^n$, which are geodesic spheres, horospheres, and tubes around totally geodesic complex hyperplanes.

References

[1] T. Adachi, *Trajectories on geodesic spheres in a non-flat complex space form*, J. Geom. **90** (2008), no. 1-2, 1–29, DOI 10.1007/s00022-008-1941-3. MR2465783

[2] Y. Aoki and T. Adachi, *Expressions of circles on a complex projective space by geodesics and by non-geodesic trajectories on geodesic spheres*, To appear in Osaka J. Math., 2025.

[3] Y. Aoki and T. Adachi, *Bounded circles on a complex hyperbolic space are expressed by trajectories on geodesic spheres* (English, with English and French summaries), C. R. Math. Acad. Sci. Soc. R. Can. **46** (2024), no. 1, 1–10, DOI 10.1007/s00283-023-10299-x. MR4695958

[4] T. Adachi and S. Maeda, *Global behaviours of circles in a complex hyperbolic space*, Tsukuba J. Math. **21** (1997), no. 1, 29–42, DOI 10.21099/tkbjm/1496163159. MR1467219

[5] T. Bao and T. Adachi, *Extrinsic circular trajectories on totally η-umbilic real hypersurfaces in a complex hyperbolic space*, Kodai Math. J. **39** (2016), no. 3, 615–631, DOI 10.2996/kmj/1478073776. MR3567237

[6] C.-L. Bejan and S.-L. Druţă-Romaniuc, *F-geodesics on manifolds*, Filomat **29** (2015), no. 10, 2367–2379, DOI 10.2298/FIL1510367B. MR3434176

[7] A. Comtet, *On the Landau levels on the hyperbolic plane*, Ann. Physics **173** (1987), no. 1, 185–209, DOI 10.1016/0003-4916(87)90098-4. MR870891

[8] S. L. Druţă-Romaniuc, J.-i. Inoguchi, M. I. Munteanu, and A. I. Nistor, *Magnetic curves in Sasakian manifolds*, J. Nonlinear Math. Phys. **22** (2015), no. 3, 428–447, DOI 10.1080/14029251.2015.1079426. MR3378108

[9] S.-L. Druţă-Romaniuc, J.-i. Inoguchi, M. I. Munteanu, and A. I. Nistor, *Magnetic curves in cosymplectic manifolds*, Rep. Math. Phys. **78** (2016), no. 1, 33–48, DOI 10.1016/S0034-4877(16)30048-9. MR3542529

[10] P. Eberlein and B. O'Neill, *Visibility manifolds*, Pacific J. Math. **46** (1973), 45–109. MR336648

[11] M. Jleli, M. I. Munteanu, and A. I. Nistor, *Magnetic trajectories in an almost contact metric manifold* $\mathbb{R}^{2N+1}$, Results Math. **67** (2015), no. 1-2, 125–134, DOI 10.1007/s00025-014-0398-y. MR3304035

[12] S. Maeda, *A characterization of constant isotropic immersions by circles*, Arch. Math. (Basel) **81** (2003), no. 1, 90–95, DOI 10.1007/s00013-003-4677-1. MR2002720

[13] M. I. Munteanu and A. I. Nistor, *A note on magnetic curves on* $\mathbb{S}^{2n+1}$ (English, with English and French summaries), C. R. Math. Acad. Sci. Paris **352** (2014), no. 5, 447–449, DOI 10.1016/j.crma.2014.03.006. MR3194254

[14] S. Maeda and Y. Ohnita, *Helical geodesic immersions into complex space forms*, Geom. Dedicata **30** (1989), no. 1, 93–114, DOI 10.1007/BF02424315. MR995941

[15] S. Maeda and N. Sato, *On submanifolds all of whose geodesics are circles in a complex space form*, Kodai Math. J. **6** (1983), no. 2, 157–166. MR702937

[16] S. Maeda and R. Takagi, *Circular geodesic submanifolds with parallel mean curvature vector in a non-flat complex space form*, Proc. Japan Acad. Ser. A Math. Sci. **73** (1997), no. 4, 55–57. MR1453530

[17] S. Maeda and S. Udagawa, *Surfaces with constant Kaehler angle all of whose geodesics are circles in a complex space form*, Tokyo J. Math. **13** (1990), no. 2, 341–351, DOI 10.3836/tjm/1270132266. MR1088236

[18] H. Naitoh, *Isotropic submanifolds with parallel second fundamental form in $P^m(c)$*, Osaka Math. J. **18** (1981), no. 2, 427–464. MR628843

[19] R. Niebergall and P. J. Ryan, *Real hypersurfaces in complex space forms*, Tight and taut submanifolds (Berkeley, CA, 1994), Math. Sci. Res. Inst. Publ., vol. 32, Cambridge Univ. Press, Cambridge, 1997, pp. 233–305. MR1486875

[20] K. Nomizu and K. Yano, *On circles and spheres in Riemannian geometry*, Math. Ann. **210** (1974), 163–170, DOI 10.1007/BF01360038. MR348674

[21] J. S. Pak and K. Sakamoto, *Submanifolds with proper d-planar geodesics immersed in complex projective spaces*, Tohoku Math. J. (2) **38** (1986), no. 2, 297–311, DOI 10.2748/tmj/1178228495. MR843814

[22] K. Sakamoto, *Planar geodesic immersions*, Tohoku Math. J. (2) **29** (1977), no. 1, 25–56, DOI 10.2748/tmj/1178240693. MR470913

[23] K. Suizu, *Characterizations of parallel imbeddings of projective spaces into space forms by circles*, Sci. Math. Jpn. **55** (2002), no. 1, 43–52. MR1885775

[24] T. Sunada, *Magnetic flows on a Riemann surface*, Proc. KAIST Math. Workshop **8** (1993), 93–108.

DIVISION OF MATHEMATICS AND MATHEMATICAL SCIENCE, NAGOYA INSTITUTE OF TECHNOLOGY, NAGOYA 466-8555, JAPAN

Email address: clf14002@nitech.jp

OSAKA CENTRAL ADVANCED MATHEMATICAL INSTITUTE, OSAKA METROPOLITAN UNIVERSITY; and DEPARTMENT OF MATHEMATICS, PROFESSOR EMERITUS OF NAGOYA INSTITUTE OF TECHNOLOGY, NAGOYA 466-8555, JAPAN

Email address: adachi.nitech@gmail.com

Contemporary Mathematics
Volume **821**, 2025
https://doi.org/10.1090/conm/821/16434

On conservation laws for polyharmonic maps

Volker Branding

ABSTRACT. This article provides an overview on various conservation laws for polyharmonic maps between Riemannian manifolds. Besides recalling that the variation of the energy for polyharmonic maps with respect to the domain metric gives rise to the stress-energy tensor, we also show how the presence of a Killing vector field on the target manifold leads to a conservation law. For harmonic and biharmonic maps we also point out a number of applications of such conservation laws.

1. Introduction and results

One of the most famous variational problems for maps between two Riemannian manifolds (M, g) and (N, h) is the one of harmonic maps. For a map $\phi\colon M \to N$ one defines its energy by

$$E(\phi) := \frac{1}{2}\int_M |d\phi|^2 \, dv_g. \tag{1.1}$$

The critical points of (1.1) are characterized by the vanishing of the so-called *tension field*, that is

$$0 = \tau(\phi) := \operatorname{Tr}_g \bar{\nabla} d\phi, \tag{1.2}$$

where $\bar{\nabla}$ represents the connection on the pull-back bundle ϕ^*TN. Solutions of (1.2) are called *harmonic maps*. The harmonic map equation constitutes a second order semilinear elliptic partial differential equation for which many results have been established; for the current status of research, see [**HW**] and [**BW**].

A fourth order variational problem for maps between Riemannian manifolds that has received growing attention over the last decades is the one of biharmonic maps. To define the latter, one considers the *bienergy* of a map defined as follows

$$E_2(\phi) := \frac{1}{2}\int_M |\tau(\phi)|^2 \, dv_g. \tag{1.3}$$

The critical points of (1.3) are called *biharmonic maps* and are determined by the vanishing of the bitension field $\tau_2(\phi)$, that is,

$$0 = \tau_2(\phi) := -\bar{\Delta}\tau(\phi) - \operatorname{Tr}_g R^N(d\phi(\cdot), \tau(\phi))d\phi(\cdot). \tag{1.4}$$

2020 *Mathematics Subject Classification.* Primary 58E20; Secondary 53C43.

Key words and phrases. Polyharmonic maps, conservation laws, Killing vector field.

The author gratefully acknowledges the support of the Austrian Science Fund (FWF) through the project "Geometric Analysis of Biwave Maps" (DOI:10.55776/P34853).

Here, R^N represents the curvature tensor of the manifold N. The biharmonic map equation (1.4) comprises a semilinear elliptic partial differential equation of fourth order and, due to the higher number of derivatives, additional technical problems arise in its analysis.

A direct inspection of the biharmonic map equation (1.4) reveals that every harmonic map is automatically biharmonic. For this reason, one is interested in finding biharmonic maps which are nonharmonic and these are usually referred to as *proper biharmonic*.

In the case that M is compact and N has nonpositive sectional curvature, the maximum principle shows that all biharmonic maps are necessarily harmonic [**J**]. For this reason most of the research carried out on biharmonic maps considers the case of a target with positive curvature and many examples of proper biharmonic maps into the sphere have been obtained.

For the current status of research on biharmonic maps between Riemannian manifolds one may consult the recent book [**OC**] and the survey [**FO**].

Besides harmonic and biharmonic maps we will also focus on the following k-order versions of the energy functional.

If $k = 2s$, $s \geq 1$, we set

$$E_{2s}(\phi) := \frac{1}{2}\int_M |\bar{\Delta}^{s-1}\tau(\phi)|^2 \, dv_g. \tag{1.5}$$

If $k = 2s+1$, $s \geq 1$, we consider

$$E_{2s+1}(\phi) := \frac{1}{2}\int_M |\bar{\nabla}\bar{\Delta}^{s-1}\tau(\phi)|^2 \, dv_g. \tag{1.6}$$

A *polyharmonic map of order k* (briefly, a *k-harmonic map*) is a critical point of the k-energy functional E_k.

The Euler–Lagrange equations of (1.5), (1.6) are as follows (with $\bar{\Delta}^{-1} = 0$):

(1) The critical points of (1.5) are those which satisfy

$$\begin{aligned} 0 = \tau_{2s}(\phi) :=& \bar{\Delta}^{2s-1}\tau(\phi) - R^N(\bar{\Delta}^{2s-2}\tau(\phi), d\phi(e_j))d\phi(e_j) \\ & - \sum_{\ell=1}^{s-1}\Big(R^N(\bar{\nabla}_{e_j}\bar{\Delta}^{s+\ell-2}\tau(\phi), \bar{\Delta}^{s-\ell-1}\tau(\phi))d\phi(e_j) \\ & \qquad - R^N(\bar{\Delta}^{s+\ell-2}\tau(\phi), \bar{\nabla}_{e_j}\bar{\Delta}^{s-\ell-1}\tau(\phi))d\phi(e_j)\Big). \end{aligned} \tag{1.7}$$

(2) The critical points of (1.6) are determined by

$$\begin{aligned} 0 = \tau_{2s+1}(\phi) :=& \bar{\Delta}^{2s}\tau(\phi) - R^N(\bar{\Delta}^{2s-1}\tau(\phi), d\phi(e_j))d\phi(e_j) \\ & - \sum_{\ell=1}^{s-1}\Big(R^N(\bar{\nabla}_{e_j}\bar{\Delta}^{s+\ell-1}\tau(\phi), \bar{\Delta}^{s-\ell-1}\tau(\phi))d\phi(e_j) \\ & \qquad - R^N(\bar{\Delta}^{s+\ell-1}\tau(\phi), \bar{\nabla}_{e_j}\bar{\Delta}^{s-\ell-1}\tau(\phi))d\phi(e_j)\Big) \\ & - R^N(\bar{\nabla}_{e_j}\bar{\Delta}^{s-1}\tau(\phi), \bar{\Delta}^{s-1}\tau(\phi))d\phi(e_j). \end{aligned} \tag{1.8}$$

Here, $\{e_j\}_{j=1,\ldots,\dim M}$ is a local orthornormal frame field tangent to M.

For the current status of research on polyharmonic maps we refer to [**BMOR1**]; polyharmonic hypersurfaces in Riemannian space forms have been investigated in [**MOR**]; this analysis has recently been extended to the pseudo-Riemannian case in

[**BMOR3**]. Unique continuation properties for polyharmonic maps were obtained in [**BMOR2**], and a structure theorem for polyharmonic maps from complete non-compact Riemannian manifolds was established in [**B4**].

Throughout this article we employ the following sign conventions. The Riemannian curvature tensor is defined by

$$R(X,Y)Z := [\nabla_X, \nabla_Y]Z - \nabla_{[X,Y]}Z,$$

where X, Y, Z are vector fields.

For the rough Laplacian on ϕ^*TN, we shall use the convention

$$\bar{\Delta} := -\operatorname{Tr}_g(\bar{\nabla}\bar{\nabla} - \bar{\nabla}_\nabla).$$

Hence, the sign of the Laplace operator acting on functions is $\Delta f = -f''$ on $\mathbb{R}$.

Whenever employing local coordinates we will use Latin letters to represent coordinates on M and Greek letters for coordinates on the target N.

For vector fields on N we will use the letters X, Y, Z, W and most often X will refer to a Killing vector field, while for vector fields on M we reserve the letters U, V.

Throughout this note we will employ the summation convention, i.e., we tacitly sum over repeated indices.

This article is organized as follows: In Section 2 we recall a number of results on conservation laws for harmonic and biharmonic maps which highlight their importance in differential geometry and geometric analysis. In Section 3 we first recall the stress-energy tensor for polyharmonic maps and, in a second step, we show how the existence of a Killing vector field on N leads to a conservation law for polyharmonic maps.

2. Conservation laws for harmonic and biharmonic maps

In this section we collect several known mathematical results which we will need throughout this article. These results will mostly apply to harmonic and biharmonic maps.

We choose local coordinates x^i on $\tilde{U} \subset M$ and y^α on $\tilde{V} \subset N$ such that $\phi(\tilde{U}) \subset \tilde{V}$. Recall that in our sign convention the tension field of a map can be expressed in terms of such local coordinates as follows

$$\tau^\alpha(\phi) = -\Delta\phi^\alpha + \Gamma^\alpha_{\beta\gamma}\langle d\phi^\beta, d\phi^\gamma\rangle, \tag{2.1}$$

where

$$\langle d\phi^\beta, d\phi^\gamma\rangle := g^{ij}\frac{\partial\phi^\beta}{\partial x^i}\frac{\partial\phi^\gamma}{\partial x^j}$$

and $\Gamma^\alpha_{\beta\gamma}$ represent the Christoffel symbols on the manifold N.

In the following we will frequently make use of the Lie derivative of the metric h on N with respect to a vector field X. This is defined as follows

$$(\mathcal{L}_X h)(Y,Z) = \langle\nabla_Y X, Z\rangle + \langle\nabla_Z X, Y\rangle, \tag{2.2}$$

where X, Y, Z are vector fields on N.

A vector field X is called *Killing vector field* if its Lie derivative with respect to the metric vanishes, that is,

$$\mathcal{L}_X h = 0. \tag{2.3}$$

As a nontrivial Killing vector field encodes information on the symmetries of a Riemannian manifold and as symmetries lead to conservation laws via Noether's theorem, it is clear that Killing vector fields will play an important role within this manuscript.

It directly follows that for a Killing vector field X we have

$$\langle \nabla_Y X, Z\rangle + \langle \nabla_Z X, Y\rangle = 0. \tag{2.4}$$

The next statement relates Killing vector fields to the Riemannian curvature tensor.

LEMMA 2.1. *Let X be a Killing vector field on a Riemann manifold (N, h). Then, the Riemann curvature tensor can be calculated as*

$$\nabla^2_{Y,Z} X = -R^N(X, Y)Z, \tag{2.5}$$

where Y, Z are vector fields on N.

PROOF. A proof can be found in [**P**, p. 242, Lemma 33]. □

In the following lemma we assume that the Lie derivative acts on the metric h on N and study the behavior of the Christoffel symbols under the action of the Lie derivative.

LEMMA 2.2. *The Lie derivative of the Christoffel symbols with respect to a vector field X is given by*

$$\mathcal{L}_X \Gamma^\alpha_{\beta\gamma} = \nabla_\beta \nabla_\gamma X^\alpha - R^\alpha_{\beta\gamma\delta} X^\delta, \tag{2.6}$$

where $R^\alpha_{\beta\gamma\delta}$ represent the components of the Riemann curvature tensor on N.

PROOF. For a proof we refer to [**Y**, p. 8]. □

Throughout this manuscript we use the notation $X \circ \phi$ to highlight that X is a vector field along the map ϕ.

2.1. The case of harmonic maps. To obtain a conservation law for harmonic maps we will calculate the Lie derivative of $|d\phi|^2$ with respect to a vector field X on N, where the Lie derivative is supposed to act on the metric h of N. More precisely, denoting by $\{e_i\}, i = 1, \ldots, m$ a local orthornormal frame tangent to M, we find

$$\begin{aligned}
\mathcal{L}_X |d\phi|^2 =& (\mathcal{L}_X h)(d\phi(e_i), d\phi(e_i)) \\
=& 2\langle d\phi(e_i), \nabla^N_{d\phi(e_i)} X\rangle \\
=& 2\langle d\phi(e_i), \bar{\nabla}_{e_i}(X \circ \phi)\rangle \\
=& 2\nabla_{e_i} \langle d\phi(e_i), X \circ \phi\rangle - 2\langle \tau(\phi), X \circ \phi\rangle \\
=& 2 \operatorname{div} J^1(\phi) - 2\langle \tau(\phi), X \circ \phi\rangle,
\end{aligned}$$

where the vector field $J^1(\phi)$ is defined as follows

$$J^1(\phi) = \langle d\phi(\cdot), X \circ \phi\rangle^\sharp. \tag{2.7}$$

Here, $\sharp$ represents the musical isomorphism.

Hence, we obtain the following statement which was first obtained in the seminal work of Hélein [**H2**].

PROPOSITION 2.3. *Let $\phi\colon M \to N$ be a harmonic map. Assume that N has a Killing vector field X. Then the vector field (2.7) is divergence-free.*

PROOF. We choose a local geodesic frame field $\{e_i\}, i = 1, \dots, \dim M$ such that at a fixed point $p \in M$ we have $\nabla_{e_j} e_i = 0, i, j = 1, \dots, \dim M$. We calculate

$$\operatorname{div}(J^1(\phi)) = \nabla_{e_i} \langle d\phi(e_i), X \circ \phi \rangle = \underbrace{\langle \tau(\phi)}_{=0}, X \circ \phi \rangle + \langle d\phi(e_i), \bar{\nabla}_{e_i}(X \circ \phi) \rangle = 0,$$

and the second term vanishes since X is a Killing vector field, that is a solution of equation (2.3). □

In the case of $u := \iota \circ \phi \colon M \to \mathbb{S}^n \subset \mathbb{R}^{n+1}$ the vector field (2.7) can be written as

$$J^1(u) = u \wedge \nabla u, \tag{2.8}$$

where $\wedge$ represents the exterior product on $\mathbb{R}^{n+1}$ and ι the canonical embedding of $\mathbb{S}^n$ into $\mathbb{R}^{n+1}$.

Note that for $\mathbb{S}^n \subset \mathbb{R}^{n+1}$ the Killing vector fields can be represented by antisymmetric matrices, if we restrict these to the sphere orthogonally, then these will represent Killing vector fields on $\mathbb{S}^n$.

Moreover, for a spherical target and in our sign convention the tension field acquires the form

$$d\iota(\tau(\phi)) = -\Delta u + |\nabla u|^2 u, \tag{2.9}$$

where $\iota \colon \mathbb{S}^n \to \mathbb{R}^{n+1}$.

The next proposition is well known but underlines the importance of conservation laws in the analysis of harmonic maps.

PROPOSITION 2.4. *Let $u \colon M \to \mathbb{S}^n \subset \mathbb{R}^{n+1}$ be a smooth map. Then the following two statements are equivalent:*

(1) *The map u is harmonic.*
(2) *The vector field $J^1(u)$ is divergence-free.*

PROOF. Suppose that $u \colon M \to \mathbb{S}^n \subset \mathbb{R}^{n+1}$ is a harmonic map, then the vector field $J^1(u)$ is divergence-free due to Proposition 2.3. For the converse direction suppose that $J^1(u)$ is divergence-free such that

$$u \wedge \Delta u = 0.$$

This implies that

$$\Delta u = \lambda u$$

for some nonvanishing $\lambda \in \mathbb{R}$. Taking the scalar product with u and using $\langle u, \Delta u \rangle = |\nabla u|^2$ we find $\lambda = |\nabla u|^2$. Consequently, u solves the equation for harmonic maps in the case of a spherical target (2.9). □

REMARK 2.5. The possibility of rewriting the harmonic map equation as a conservation law has led to strong mathematical results. For example, it is the key ingredient in the existence result of Shatah for wave maps from Minkowski space to spheres [**S**] and the regularity result of Hélein for harmonic maps from surfaces to spheres [**H1**].

Let us also make a remark on the relation between conservation laws and integrable systems.

REMARK 2.6. Assume that M is two-dimensional. Choosing local coordinates $x_i, i = 1,2$ on an open subset $\Omega \subset M$ and again considering the case of a spherical target, we may locally express $J^1(u)$ as follows

$$(J^1)_i^{\alpha\beta}(u) = u_i^\alpha u^\beta - u_i^\beta u^\alpha, \tag{2.10}$$

where we use the notation $u_i^\alpha := \frac{\partial u^\alpha}{\partial x^i}$.

By a direct calculation, exploiting the fact that we are considering a spherical target, we find that (2.10) satisfies

$$\partial_{x_1} J^1_{x_2} - \partial_{x_2} J^1_{x_1} = 2[J^1_{x_1}, J^1_{x_2}]. \tag{2.11}$$

The condition (2.11) can be identified as a zero curvature equation. Similar zero curvature conditions can also be found if the target is a symmetric space instead of a sphere. This construction points out a beautiful connection between harmonic maps to symmetric spaces and the theory of integrable systems; see, for example, the book [**H3**] for more details.

For additional results concerning conservation laws for harmonic maps we refer to [**B, BR**].

2.2. The biharmonic case. In this section we extend the previous analysis to the case of biharmonic maps between Riemannian manifolds. Some of these results have already been derived in [**B1, W**] where a different notation and sign convention was used.

LEMMA 2.7. *Let $\phi\colon M \to N$ be a smooth map. The Lie derivative of $|\tau(\phi)|^2$ with respect to a vector field X on N is given by*

$$\begin{aligned} \mathcal{L}_X|\tau(\phi)|^2 =& (\mathcal{L}_X h)(\tau(\phi),\tau(\phi)) - 2\langle X\circ\phi, \tau_2(\phi)\rangle \\ &+ 2\nabla_{e_j}\big(\langle\bar\nabla_{e_j}(X\circ\phi),\tau(\phi)\rangle - \langle X\circ\phi, \bar\nabla_{e_j}\tau(\phi)\rangle\big), \end{aligned} \tag{2.12}$$

where $\{e_i\}, i = 1,\ldots,\dim M$ is an orthornormal basis of TM.

PROOF. We calculate

$$\mathcal{L}_X|\tau(\phi)|^2 = (\mathcal{L}_X h)(\tau(\phi),\tau(\phi)) + 2h(\mathcal{L}_X\tau(\phi),\tau(\phi)).$$

Since the tension field depends on the metric h via the connection of N we will get a non-vanishing contribution from $\mathcal{L}_X\tau(\phi)$. Using the local expression of the tension field (2.1) we find

$$\begin{aligned} \mathcal{L}_X\tau^\alpha(\phi) =& \langle d\phi^\beta, d\phi^\gamma\rangle \mathcal{L}_X\Gamma^\alpha_{\beta\gamma} \\ =& \langle d\phi^\beta, d\phi^\gamma\rangle(\nabla_\beta\nabla_\gamma X^\alpha - R^\alpha_{\beta\gamma\delta}X^\delta), \end{aligned}$$

where we used (2.6) in the second step.

Hence, we obtain

$$\begin{aligned} h(\mathcal{L}_X\tau(\phi),\tau(\phi)) =& \langle\bar\nabla_{e_j}\bar\nabla_{e_j}(X\circ\phi),\tau(\phi)\rangle - \langle R^N(d\phi(e_j),X\circ\phi)d\phi(e_j),\tau(\phi)\rangle \\ =& -\langle\bar\Delta(X\circ\phi),\tau(\phi)\rangle - \langle R^N(d\phi(e_j),\tau(\phi))d\phi(e_j),X\circ\phi\rangle \\ =& \nabla_{e_j}\big(\langle\bar\nabla_{e_j}(X\circ\phi),\tau(\phi)\rangle - \langle X\circ\phi,\bar\nabla_{e_j}\tau(\phi)\rangle\big) \\ & - \langle X\circ\phi,\bar\Delta\tau(\phi)\rangle - \langle R^N(d\phi(e_j),\tau(\phi))d\phi(e_j),X\circ\phi\rangle \end{aligned}$$

completing the proof. □

Note that if X is a Killing vector field then the first term on the right-hand side in equation (2.12) vanishes. Moreover, if we also assume that ϕ is a biharmonic map then (2.12) immediately gives us a conservation law as follows.

PROPOSITION 2.8. *Let $\phi\colon M \to N$ be a smooth biharmonic map. Assume that N has a Killing vector field X. Then the following vector field is divergence-free*

$$J^2(\phi) = \langle \bar\nabla (X\circ\phi), \tau(\phi)\rangle^\sharp - \langle X\circ\phi, \bar\nabla\tau(\phi)\rangle^\sharp, \tag{2.13}$$

where $\sharp$ denotes the musical isomorphism.

PROOF. We choose a local geodesic frame field $\{e_i\}, i = 1,\ldots,\dim M$ such that at a fixed point $p\in M$ we have $\nabla_{e_j} e_i = 0, i,j = 1,\ldots,\dim M$. We calculate

$$\begin{aligned}
\operatorname{div} J^2(\phi) =&\nabla_{e_j}\big(\langle\bar\nabla_{e_j}(X\circ\phi),\tau(\phi)\rangle - \langle X\circ\phi,\bar\nabla_{e_j}\tau(\phi)\rangle\big)\\
=&\langle\bar\nabla_{e_j}\bar\nabla_{e_j}(X\circ\phi),\tau(\phi)\rangle - \langle X\circ\phi,\bar\nabla_{e_j}\bar\nabla_{e_j}\tau(\phi)\rangle\\
=&-\langle R^N(X\circ\phi, d\phi(e_j))d\phi(e_j),\tau(\phi)\rangle + \langle X\circ\phi,\bar\Delta\tau(\phi)\rangle\\
=&-\langle X\circ\phi,\tau_2(\phi)\rangle,
\end{aligned}$$

where we have used (2.5) after the third equals sign, proving the assertion. □

Now, let us briefly recall the structure of the biharmonic map equation for a spherical target $\mathbb{S}^n\subset\mathbb{R}^{n+1}$. Again, recall that in this case the tension field simplifies as

$$d\iota(\tau(\phi)) = -\Delta u + |\nabla u|^2 u.$$

Hence, for $u := \iota\circ\phi\colon M\to\mathbb{S}^n\subset\mathbb{R}^{n+1}$, the bi-energy acquires the simpler form

$$E_2(u) = \frac{1}{2}\int_M \left(|\Delta u|^2 - |\nabla u|^4\right)\,dv_g$$

and we now present an extrinsic characterization of its critical points.

PROPOSITION 2.9. *Let $u\colon M\to\mathbb{S}^n\subset\mathbb{R}^{n+1}$ be a smooth map. Then, the critical points of $E_2(u)$ are characterized by*

$$\Delta^2 u = (-|\Delta u|^2 + \Delta|\nabla u|^2 + 2\langle\nabla u,\nabla\Delta u\rangle - 2|\nabla u|^4)u - 2\nabla(|\nabla u|^2\nabla u). \tag{2.14}$$

PROOF. For a proof one can consult [**W**, Lemma 2.1]. □

In the case of a spherical target the next lemma shows that the equation for biharmonic maps (2.14) is equivalent to a conservation law; this result was first presented in [**W**, Lemma 2.2]. In this setup the vector field (2.13) acquires the form

$$J^2(u) = -\nabla\Delta u\wedge u + \Delta u\wedge\nabla u + 2|\nabla u|^2\nabla u\wedge u, \tag{2.15}$$

where $\wedge$ represents the exterior product in $\mathbb{R}^{n+1}$.

LEMMA 2.10. *Let $u\colon M\to\mathbb{S}^n\subset\mathbb{R}^{n+1}$ be a smooth map. Then the following two statements are equivalent*

1. *The map u is a biharmonic map, that is, a solution of* (2.14).
2. *The vector field $J^2(u)$ is divergence-free.*

PROOF. First of all, we note that by taking the wedge product of (2.14) with u we obtain the equation

$$\Delta^2 u \wedge u = -2\nabla(|\nabla u|^2 \nabla u) \wedge u = -2\nabla(|\nabla u|^2 \nabla u \wedge u). \tag{2.16}$$

Then, a direct calculation reveals that

$$\operatorname{div} J^2(u) = \Delta^2 u \wedge u + 2\nabla(|\nabla u|^2 \nabla u) \wedge u$$

showing that the vector field $J^2(u)$ defined in (2.15) is divergence-free whenever we have a solution of (2.14).

To show the converse direction we assume that the vector field $J^2(u)$ is divergence-free which means that (2.16) holds and also implies that

$$\Delta^2 u + 2\nabla(|\nabla u|^2 \nabla u) = \lambda u$$

for some nonvanishing $\lambda \in \mathbb{R}$. To determine λ we multiply the above equation with u and use the identities

$$\begin{aligned}\langle \Delta^2 u, u\rangle &= \Delta|\nabla u|^2 - |\Delta u|^2 + 2\langle \nabla u, \nabla \Delta u\rangle,\\ \langle \nabla(|\nabla u|^2 \nabla u), u\rangle &= -|\nabla u|^4\end{aligned}$$

which follow from differentiating $|u|^2 = 1$. The proof is now complete. □

We finish our analysis of (2.15) by making some comments if one can expect to find a zero-curvature representation for biharmonic maps to spheres similar to the results for harmonic maps presented in the previous section.

To this end, let $\Omega \subset M$ be a connected domain on which we choose local coordinates $x^i, 1 \leq i \leq \dim M$ and consider a spherical target. In this setup we can express (2.15) as

$$(J^2)_i^{\alpha\beta}(u) = u^\alpha_{jji} u^\beta - u^\beta_{jji} u^\alpha - u^\alpha_{jj} u^\beta_i + u^\beta_{jj} u^\alpha_i + 2u^\gamma_j u^\gamma_j u^\alpha_i u^\beta - 2u^\gamma_j u^\gamma_j u^\beta_i u^\alpha.$$

However, a direct but lengthy calculation shows that, in general, the above current does not satisfy the equation of a connection with zero-curvature, i.e.,

$$\partial_{x_i} J^2_{x_j} - \partial_{x_j} J^2_{x_i} \neq 2[J^2_{x_i}, J^2_{x_j}], \qquad 1 \leq i, j \leq \dim M. \tag{2.17}$$

This fact can easily be seen even without performing a detailed calculation. The order of derivatives on the left-hand side of (2.17) is the order of derivatives being contained in the current J^2 plus one additional derivative, while the order on the right-hand side is twice the order of derivatives appearing in the current as it enters quadratically into the equation. It is not hard to realize that in the case of harmonic maps the order of derivatives in the current is one, such that in the equation for a zero-curvature representation the order on both sides is two. In the biharmonic case the orders do not match up, on the left-hand side it is four while it is six on the right-hand side.

Although the vector field (2.17) does not help to find a zero curvature representation for biharmonic maps it can be considered as an important ingredient in the regularity theory for fourth order equations developed by Lamm and Rivière in [**LR**].

2.3. Conservation laws for biharmonic hypersurfaces in spheres. An important class of biharmonic maps which has received a lot of attention from differential geometers is the one of biharmonic hypersurfaces in spheres. The current results on biharmonic hypersurfaces available all support the conjecture that biharmonic hypersurfaces in spheres must have constant mean curvature (CMC) but the conjecture in its general form is still open.

In order to extend the general conservation law for biharmonic maps—given by Proposition 2.8—to biharmonic hypersurfaces in spheres, let us recall a number of geometric facts on hypersurfaces M^m in a Riemannian manifold N^{m+1}. The connections on M^m and N^{m+1} are related by the following formula:

$$\nabla^N_Y Z = \nabla^M_Y Z + \mathrm{I\!I}(Y, Z), \tag{2.18}$$

where $\mathrm{I\!I}$ represents the second fundamental form of the hypersurface and Y, Z are vector fields on M.

Let ν be the global unit normal of the hypersurface M^m, then its shape operator A is given by

$$\nabla^N_Z \nu = -A(Z), \tag{2.19}$$

where Z is a vector field on M.

The shape operator A and the second fundamental form $\mathrm{I\!I}$ are related by

$$\mathrm{I\!I}(Y, Z) = \langle A(Y), Z\rangle \nu. \tag{2.20}$$

If $\phi\colon M^m \to N^{m+1}$ is an isometric immersion the tension field acquires the form $\tau(\phi) = mf\nu$, where f represents the mean curvature function of the hypersurface. Recall that we can compute the mean curvature function by $f = \frac{1}{m}\operatorname{Tr} A$.

Also, it is well known that $\operatorname{Tr}_g(\nabla A)(\cdot,\cdot) = m \operatorname{grad} f$; see, for example, [**MOR**, Lemma 2.1].

A hypersurface in the Euclidean sphere $\mathbb{S}^{m+1}$ is biharmonic if the following system of equations holds

$$\Delta f = (m - |A|^2) f, \qquad A(\operatorname{grad} f) = -\frac{m}{2} f \operatorname{grad} f, \tag{2.21}$$

which arises from splitting the equation for a biharmonic hypersurface into its normal and its tangential part; see [**O**].

Noting that in the case of a hypersurface

$$\bar\nabla_U \tau(\phi) = mU(f)\nu - mfA(U),$$

for all vector fields U of M, we find that for a biharmonic hypersurface we have to define

$$J^{2,hyp}(\phi) := mf\langle \bar\nabla (X\circ\phi), \nu\rangle^\sharp + mf\langle X\circ\phi, A(\cdot)\rangle^\sharp, \tag{2.22}$$

where $\sharp$ denotes the musical isomorphism and X is a Killing vector field on $\mathbb{S}^{m+1}$.

Concerning the vector field (2.22) we prove the following result.

THEOREM 2.11. *Let $\phi\colon M^m \to \mathbb{S}^{m+1}$ be a biharmonic hypersurface. Then, the vector field $J^{2,hyp}(\phi)$ defined in (2.22) is divergence-free.*

PROOF. We choose a local geodesic frame field $\{e_i\}, i = 1, \ldots, \dim M$ such that at a fixed point $p \in M$ we have $\nabla_{e_j} e_i = 0, i, j = 1, \ldots, \dim M$. A direct calculation shows that

$$\begin{aligned}\operatorname{div} J^{2,hyp}(\phi) =& mf\langle\bar{\nabla}_{e_i}\bar{\nabla}_{e_i}(X\circ\phi),\nu\rangle + 2m\langle X\circ\phi, A(\operatorname{grad} f)\rangle\\ &+ mf\langle X\circ\phi, \nabla_{e_i}(A(e_i))\rangle.\end{aligned}$$

Since we have

$$R^{\mathbb{S}^{m+1}}(Y,Z)W = \langle Z,W\rangle Y - \langle Y,W\rangle Z$$

for all vector fields $Y, Z, W \in \Gamma(T\mathbb{S}^{m+1})$ we get using (2.5) that

$$\begin{aligned}\bar{\nabla}_{e_i}\bar{\nabla}_{e_i}(X\circ\phi) =& - R^{\mathbb{S}^{m+1}}((X\circ\phi), d\phi(e_i))d\phi(e_i)\\ =& - |d\phi|^2 X\circ\phi + \langle (X\circ\phi), d\phi(e_i))\rangle d\phi(e_i)\end{aligned}$$

and we may conclude that

$$\langle\bar{\nabla}_{e_i}\bar{\nabla}_{e_i}(X\circ\phi),\nu\rangle = 0.$$

Moreover, we find

$$\begin{aligned}\nabla_{e_i}(A(e_i)) =& (\nabla A)(e_i, e_i) + \mathrm{II}(e_i, A(e_i))\\ =& m \operatorname{grad} f + |A|^2\nu,\end{aligned}$$

where we used (2.20). Using these identities we get

$$\operatorname{div} J^{2,hyp}(\phi) = 2m\langle X\circ\phi, A(\operatorname{grad} f) + \frac{m}{2} f \operatorname{grad} f\rangle$$

and thus it is clear that the vector field $J^{2,hyp}(\phi)$ has vanishing divergence whenever the second equation of the system (2.21) holds. □

We would like to point out that the vector field (2.22), which is divergence-free for a biharmonic hypersurface in $\mathbb{S}^{m+1}$, is just a special case of the general statements detailed in Proposition 2.8.

3. Conservation laws for polyharmonic maps

In this section we present two conservation laws that arise in the analysis of polyharmonic maps. First, we recall the stress-energy tensor for polyharmonic maps, then we will obtain a conservation law by assuming the the manifold N has a Killing vector field. Note that the first conservation law arises from a symmetry on the domain while the second one originates from a symmetry of the target.

In order to obtain the stress-energy tensor we think of the energies (1.5), (1.6) for polyharmonic maps as functionals of the metrics on the domain such that we can vary them with respect to the latter.

3.1. The stress-energy tensor. The results in this section have been derived in full detail in [**B2**]— here we present a short summary of them.

For polyharmonic maps of even order (1.7), the stress-energy tensor is given by

$$\begin{aligned} S_{2s}(U,V) :=& g(U,V)\Big(\frac{1}{2}|\bar\Delta^{s-1}\tau(\phi)|^2 - \langle\tau(\phi),\bar\Delta^{2s-2}\tau(\phi)\rangle - \langle d\phi,\bar\nabla\bar\Delta^{2s-2}\tau(\phi)\rangle \\ &+\sum_{l=1}^{s-1}(-\langle\bar\Delta^{s-l}\tau(\phi),\bar\Delta^{s+l-2}\tau(\phi)\rangle + \langle\bar\nabla\bar\Delta^{s-l-1}\tau(\phi),\bar\nabla\bar\Delta^{s+l-2}\tau(\phi)\rangle)\Big) \\ &-\sum_{l=1}^{s-1}\Big(\langle\bar\nabla_U\bar\Delta^{s-l-1}\tau(\phi),\bar\nabla_V\bar\Delta^{s+l-2}\tau(\phi)\rangle \\ &+\langle\bar\nabla_V\bar\Delta^{s-l-1}\tau(\phi),\bar\nabla_U\bar\Delta^{s+l-2}\tau(\phi)\rangle\Big) \\ &+\langle d\phi(U),\bar\nabla_V\bar\Delta^{2s-2}\tau(\phi)\rangle + \langle d\phi(V),\bar\nabla_U\bar\Delta^{2s-2}\tau(\phi)\rangle, \end{aligned} \tag{3.1}$$

while for polyharmonic maps of odd order (1.8) it is defined as follows:

$$\begin{aligned} &S_{2s+1}(U,V) \\ &:= g(U,V)\Big(\frac{1}{2}|\bar\nabla\bar\Delta^{s-1}\tau(\phi)|^2 - \langle\tau(\phi),\bar\Delta^{2s-1}\tau(\phi)\rangle - \langle d\phi,\bar\nabla\bar\Delta^{2s-1}\tau(\phi)\rangle \\ &+\sum_{l=1}^{s-1}(-\langle\bar\Delta^{s-l}\tau(\phi),\bar\Delta^{s+l-1}\tau(\phi)\rangle + \langle\bar\nabla\bar\Delta^{s-l-1}\tau(\phi),\bar\nabla\bar\Delta^{s+l-1}\tau(\phi)\rangle)\Big) \\ &-\sum_{l=1}^{s-1}(\langle\bar\nabla_U\bar\Delta^{s-l-1}\tau(\phi),\bar\nabla_V\bar\Delta^{s+l-1}\tau(\phi)\rangle + \langle\bar\nabla_V\bar\Delta^{s-l-1}\tau(\phi),\bar\nabla_U\bar\Delta^{s+l-1}\tau(\phi)\rangle) \\ &+\langle d\phi(U),\bar\nabla_V\bar\Delta^{2s-1}\tau(\phi)\rangle + \langle d\phi(V),\bar\nabla_U\bar\Delta^{2s-1}\tau(\phi)\rangle \\ &-\langle\bar\nabla_U\bar\Delta^{s-1}\tau(\phi),\bar\nabla_V\bar\Delta^{s-1}\tau(\phi)\rangle. \end{aligned} \tag{3.2}$$

Here, U and V are vector fields on M.

It can be directly seen that the stress-energy tensors $S_{2s}(Y,Z)$ and $S_{2s+1}(Y,Z)$ are symmetric. Moreover, they satisfy the following conservation laws.

THEOREM 3.1. *Let $\phi\colon M \to N$ be a smooth map. Then, the stress-energy tensors* (3.1) *and* (3.2) *satisfy the following conservation laws:*

$$\begin{aligned} \operatorname{div} S_{2s} &= -\langle\tau_{2s}(\phi),d\phi\rangle, \\ \operatorname{div} S_{2s+1} &= -\langle\tau_{2s+1}(\phi),d\phi\rangle. \end{aligned}$$

In particular, S_{2s} is divergence-free whenever ϕ is a solution of (1.7) *and S_{2s+1} is divergence-free whenever ϕ is a solution of* (1.8).

PROOF. For the precise details, see [**B2**, Proof of Propositions 2.6 and 2.12]. □

In [**B2**, Subsection 2.3] it is shown how the fact that the stress-energy tensors for polyharmonic maps are conserved arises from the invariance of the energy functionals (1.5), (1.6) under diffeomorphisms on the domain and is thus a consequence of Noether's theorem.

A possible application of the stress-energy tensors considered above was pointed out in [**B2**, Section 4], where a classification result for finite energy solutions of the triharmonic map equation ((1.8) with $s=1$) from Euclidean space was obtained.

3.2. A conservation law arising from Killing vector fields. In this section we show how the presence of a Killing vector field on the target manifold N gives rise to a conservation law for polyharmonic maps. Again, we need to distinguish between the case of a polyharmonic map of even order and a polyharmonic map of odd order.

Recall that in the case of biharmonic maps, we were able to obtain a conservation law by calculating the Lie derivative of $|\tau(\phi)|^2$ with respect to a vector field X on N, where the Lie derivative is supposed to act on the metric h on N. However, it turns out that this approach is no longer feasible for k-harmonic maps of order $k\geq 3$ from a computational point of view.

Luckily, there is also a second constructive approach to derive a conservation law for polyharmonic maps inspired by the seminal work of Hélein for harmonic maps [**H1**].

To this end, let η be a smooth function on M with compact support, that is $\eta\in C_0^\infty(M)$ and X a Killing vector field on N. First, we focus on the case of polyharmonic maps of even order.

We test (1.7) with $\eta(X\circ\phi)$ and use integration by parts to manipulate all contributing terms

$$\begin{aligned}\int_M\langle\bar\Delta^{2s-1}\tau(\phi),X\circ\phi\rangle\eta\,dv_g=&\int_M\langle\bar\nabla\bar\Delta^{2s-2}\tau(\phi),\bar\nabla(X\circ\phi)\rangle\eta\,dv_g\\&+\int_M\langle\bar\nabla\bar\Delta^{2s-2}\tau(\phi),X\circ\phi\rangle\nabla\eta\,dv_g.\end{aligned}$$

Moreover, we find

$$\begin{aligned}-&\int_M\langle R^N(\bar\Delta^{2s-2}\tau(\phi),d\phi(e_j))d\phi(e_j),X\circ\phi\rangle\eta\,dv_g\\&=\int_M\langle\bar\nabla_{e_j}\bar\nabla_{e_j}(X\circ\phi),\bar\Delta^{2s-2}\tau(\phi)\rangle\eta\,dv_g\\&=-\int_M\langle\bar\nabla(X\circ\phi),\bar\nabla\bar\Delta^{2s-2}\tau(\phi)\rangle\eta\,dv_g-\int_M\langle\bar\nabla(X\circ\phi),\bar\Delta^{2s-2}\tau(\phi)\rangle\nabla\eta\,dv_g,\end{aligned}$$

where we used (2.5) in the first step.

Now, we manipulate

$$\begin{aligned}\int_M&\langle R^N(\bar\nabla_{e_j}\bar\Delta^{s+l-2}\tau(\phi),\bar\Delta^{s-l-1}\tau(\phi))d\phi(e_j),X\circ\phi\rangle\eta\,dv_g\\&=\int_M\langle\bar\nabla_{e_j}(\bar\nabla_{\bar\nabla_{e_j}\bar\Delta^{s+l-2}\tau(\phi)}(X\circ\phi)),\bar\Delta^{s-l-1}\tau(\phi)\rangle\eta\,dv_g\\&=-\int_M\langle\bar\nabla_{\bar\nabla\bar\Delta^{s+l-2}\tau(\phi)}(X\circ\phi),\bar\nabla\bar\Delta^{s-l-1}\tau(\phi)\rangle\eta\,dv_g\\&\quad-\int_M\langle\bar\nabla_{\bar\nabla\bar\Delta^{s+l-2}\tau(\phi)}(X\circ\phi),\bar\Delta^{s-l-1}\tau(\phi)\rangle\nabla\eta\,dv_g.\end{aligned}$$

Similarly, we find

$$
\begin{aligned}
-\int_M &\langle R^N(\bar\Delta^{s+l-2}\tau(\phi),\bar\nabla_{e_j}\bar\Delta^{s-l-1}\tau(\phi))d\phi(e_j),X\circ\phi\rangle\eta\,dv_g\\
&=\int_M\langle\bar\nabla_{e_j}(\bar\nabla_{\bar\nabla_{e_j}\bar\Delta^{s-l-1}\tau(\phi)}(X\circ\phi)),\bar\Delta^{s+l-2}\tau(\phi)\rangle\eta\,dv_g\\
&=-\int_M\langle\bar\nabla_{\bar\nabla\bar\Delta^{s-l-1}\tau(\phi)}(X\circ\phi),\bar\nabla\bar\Delta^{s+l-2}\tau(\phi)\rangle\eta\,dv_g\\
&\quad-\int_M\langle\bar\nabla_{\bar\nabla\bar\Delta^{s-l-1}\tau(\phi)}(X\circ\phi),\bar\Delta^{s+l-2}\tau(\phi)\rangle\nabla\eta\,dv_g.
\end{aligned}
$$

Adding up the different contributions and using the skew-symmetry of the Killing vector field (2.4) we find

$$
\begin{aligned}
\int_M\Big(&\langle\bar\nabla_{e_j}\bar\Delta^{2s-2}\tau(\phi),X\circ\phi\rangle-\langle\bar\nabla_{e_j}(X\circ\phi),\bar\Delta^{2s-2}\tau(\phi)\rangle\\
&+\sum_{l=1}^{s-1}\big(\langle\bar\nabla_{\bar\nabla_{e_j}\bar\Delta^{s+l-2}\tau(\phi)}(X\circ\phi),\bar\Delta^{s-l-1}\tau(\phi)\rangle\\
&+\langle\bar\nabla_{\bar\nabla_{e_j}\bar\Delta^{s-l-1}\tau(\phi)}(X\circ\phi),\bar\Delta^{s+l-2}\tau(\phi)\rangle\big)\Big)\nabla_{e_j}\eta\,dv_g\\
&=\int_M\langle J^{2s}(\phi),\nabla\eta\rangle\,dv_g.
\end{aligned}
$$

Hence, we can deduce that the vector field, which should be divergence-free for a polyharmonic map of even order, is precisely the following one

$$
\begin{aligned}
J^{2s}(\phi):=&\langle\bar\nabla\bar\Delta^{2s-2}\tau(\phi),X\circ\phi\rangle-\langle\bar\nabla(X\circ\phi),\bar\Delta^{2s-2}\tau(\phi)\rangle\\
&+\sum_{l=1}^{s-1}\big(\langle\bar\nabla_{\bar\nabla\bar\Delta^{s+l-2}\tau(\phi)}(X\circ\phi),\bar\Delta^{s-l-1}\tau(\phi)\rangle\\
&+\langle\bar\nabla_{\bar\nabla\bar\Delta^{s-l-1}\tau(\phi)}(X\circ\phi),\bar\Delta^{s+l-2}\tau(\phi)\rangle\big).
\end{aligned}
\tag{3.3}
$$

This is demonstrated by the following.

THEOREM 3.2. *Let $\phi\colon M\to N$ be a smooth map and assume that the manifold N has a Killing vector field X. Then, the vector field $J^{2s}(\phi)$ defined in* (3.3) *satisfies*

$$\operatorname{div}J^{2s}(\phi)=-\langle\tau_{2s}(\phi),X\circ\phi\rangle.$$

In particular, $J^{2s}(\phi)$ is divergence-free whenever $\phi\colon M\to N$ is a polyharmonic map of even order.

PROOF. We choose a local geodesic frame field $\{e_i\}, i=1,\dots,\dim M$ such that at a fixed point $p\in M$ we have $\nabla_{e_j}e_i=0, i,j=1,\dots,\dim M$. A direct calculation shows that

$$
\begin{aligned}
\bar\nabla_{e_i}\big(&\langle\bar\nabla_{e_i}\bar\Delta^{2s-2}\tau(\phi),X\circ\phi\rangle-\langle\bar\nabla_{e_i}(X\circ\phi),\bar\Delta^{2s-2}\tau(\phi)\rangle\big)\\
&=-\langle\bar\Delta^{2s-1}\tau(\phi),X\circ\phi\rangle+\langle\bar\Delta(X\circ\phi),\bar\Delta^{2s-2}\tau(\phi)\rangle\\
&=-\langle\bar\Delta^{2s-1}\tau(\phi),X\circ\phi\rangle+\langle R^N(X\circ\phi,d\phi(e_i))d\phi(e_i),\bar\Delta^{2s-2}\tau(\phi)\rangle,
\end{aligned}
$$

where we used (2.5) in the second step. Moreover, we find

$$\begin{aligned}
&\bar\nabla_{e_i}\big(\langle\bar\nabla_{\bar\nabla_{e_i}\bar\Delta^{s+l-2}\tau(\phi)}(X\circ\phi),\bar\Delta^{s-l-1}\tau(\phi)\rangle+\langle\bar\nabla_{\bar\nabla_{e_i}\bar\Delta^{s-l-1}\tau(\phi)}(X\circ\phi),\bar\Delta^{s+l-2}\tau(\phi)\rangle\big)\\
&\quad=\langle\bar\nabla_{e_i}\big(\bar\nabla_{\bar\nabla_{e_i}\bar\Delta^{s+l-2}\tau(\phi)}(X\circ\phi)\big),\bar\Delta^{s-l-1}\tau(\phi)\rangle\\
&\qquad+\langle\bar\nabla_{\bar\nabla_{e_i}\bar\Delta^{s+l-2}\tau(\phi)}(X\circ\phi),\bar\nabla_{e_i}\bar\Delta^{s-l-1}\tau(\phi)\rangle\\
&\qquad+\langle\bar\nabla_{e_i}\big(\bar\nabla_{\bar\nabla_{e_i}\bar\Delta^{s-l-1}\tau(\phi)}(X\circ\phi)\big),\bar\Delta^{s+l-2}\tau(\phi)\rangle\\
&\qquad+\langle\bar\nabla_{\bar\nabla_{e_i}\bar\Delta^{s-l-1}\tau(\phi)}(X\circ\phi),\bar\nabla_{e_i}\bar\Delta^{s+l-2}\tau(\phi)\rangle\\
&\quad=-\langle R^N(X\circ\phi,d\phi(e_i))\bar\nabla_{e_i}\bar\Delta^{s+l-2}\tau(\phi),\bar\Delta^{s-l-1}\tau(\phi)\rangle\\
&\qquad-\langle R^N(X\circ\phi,d\phi(e_i))\bar\nabla_{e_i}\bar\Delta^{s-l-1}\tau(\phi),\bar\Delta^{s+l-2}\tau(\phi)\rangle,
\end{aligned}$$

where we used both (2.4) and (2.5). The claim now follows by using the definition of $J^{2s}(\phi)$ and the Euler–Lagrange equation for polyharmonic maps of even order. □

In the case of polyharmonic maps of odd order we also have a conserved vector field, given by

$$\begin{aligned}
J^{2s+1}(\phi):=&\langle\bar\nabla\bar\Delta^{2s-1}\tau(\phi),X\circ\phi\rangle-\langle\bar\nabla(X\circ\phi),\bar\Delta^{2s-1}\tau(\phi)\rangle\\
&+\sum_{l=1}^{s-1}\big(-\langle\bar\nabla_{\bar\nabla\bar\Delta^{s-1}\tau(\phi)}(X\circ\phi),\bar\Delta^{s-1}\tau(\phi)\rangle\\
&+\langle\bar\nabla_{\bar\nabla\bar\Delta^{s+l-1}\tau(\phi)}(X\circ\phi),\bar\Delta^{s-l-1}\tau(\phi)\rangle\\
&+\langle\bar\nabla_{\bar\nabla\bar\Delta^{s-l-1}\tau(\phi)}(X\circ\phi),\bar\Delta^{s+l-1}\tau(\phi)\rangle\big).
\end{aligned}\tag{3.4}$$

We can construct the vector field (3.4) using the same arguments as in the even case such that we do not give the precise derivation here. Nonetheless, the vector field (3.4) satisfies the following conservation law.

THEOREM 3.3. *Let $\phi\colon M\to N$ be a smooth map and assume that the manifold N has a Killing vector field X. Then, the vector field $J^{2s+1}(\phi)$ defined in (3.4) satisfies*

$$\operatorname{div}J^{2s+1}(\phi)=-\langle\tau_{2s+1}(\phi),X\circ\phi\rangle.$$

In particular, $J^{2s+1}(\phi)$ is divergence-free whenever $\phi\colon M\to N$ is a polyharmonic map of odd order, that is a solution of (1.7).

PROOF. We choose a local geodesic frame field $\{e_i\}, i=1,\ldots,\dim M$ such that at a fixed point $p\in M$ we have $\nabla_{e_j}e_i=0, i,j=1,\ldots,\dim M$. We calculate

$$\begin{aligned}
&\bar\nabla_{e_i}\big(\langle\bar\nabla_{e_i}\bar\Delta^{2s-1}\tau(\phi),X\circ\phi\rangle-\langle\bar\nabla_{e_i}(X\circ\phi),\bar\Delta^{2s-1}\tau(\phi)\rangle\big)\\
&\quad=-\langle\bar\Delta^{2s}\tau(\phi),X\circ\phi\rangle+\langle\bar\Delta(X\circ\phi),\bar\Delta^{2s-1}\tau(\phi)\rangle\\
&\quad=-\langle\bar\Delta^{2s}\tau(\phi),X\circ\phi\rangle+\langle R^N(X\circ\phi,d\phi(e_i))d\phi(e_i),\bar\Delta^{2s-1}\tau(\phi)\rangle.
\end{aligned}$$

In addition, we find

$$\begin{aligned}
&\bar\nabla_{e_i}\big(-\langle\bar\nabla_{\bar\nabla_{e_i}\bar\Delta^{s-1}\tau(\phi)}(X\circ\phi),\bar\Delta^{s-1}\tau(\phi)\rangle+\langle\bar\nabla_{\bar\nabla_{e_i}\bar\Delta^{s+l-1}\tau(\phi)}(X\circ\phi),\bar\Delta^{s-l-1}\tau(\phi)\rangle\\
&\quad+\langle\bar\nabla_{\bar\nabla_{e_i}\bar\Delta^{s-l-1}\tau(\phi)}(X\circ\phi),\bar\Delta^{s+l-1}\tau(\phi)\rangle\big)\\
&=\underbrace{-\langle\bar\nabla_{e_i}\big(\bar\nabla_{\bar\nabla_{e_i}\bar\Delta^{s-1}\tau(\phi)}(X\circ\phi)\big),\bar\Delta^{s-1}\tau(\phi)\rangle
-\langle\bar\nabla_{\bar\nabla_{e_i}\bar\Delta^{s-1}\tau(\phi)}(X\circ\phi),\bar\nabla_{e_i}\bar\Delta^{s-1}\tau(\phi)\rangle}_{=0}\\
&\quad+\langle\bar\nabla_{e_i}\big(\bar\nabla_{\bar\nabla_{e_i}\bar\Delta^{s+l-1}\tau(\phi)}(X\circ\phi)\big),\bar\Delta^{s-l-1}\tau(\phi)\rangle\\
&\quad+\langle\bar\nabla_{\bar\nabla_{e_i}\bar\Delta^{s+l-1}\tau(\phi)}(X\circ\phi),\bar\nabla_{e_i}\bar\Delta^{s-l-1}\tau(\phi)\rangle\\
&\quad+\langle\bar\nabla_{e_i}\big(\bar\nabla_{\bar\nabla_{e_i}\bar\Delta^{s-l-1}\tau(\phi)}(X\circ\phi)\big),\bar\Delta^{s+l-1}\tau(\phi)\rangle\\
&\quad+\langle\bar\nabla_{\bar\nabla_{e_i}\bar\Delta^{s-l-1}\tau(\phi)}(X\circ\phi),\bar\nabla_{e_i}\bar\Delta^{s+l-1}\tau(\phi)\rangle\\
&=\langle R^N(X\circ\phi,d\phi(e_i))\bar\nabla_{e_i}\bar\Delta^{s-1}\tau(\phi),\bar\Delta^{s-1}\tau(\phi)\rangle\\
&\quad-\langle R^N(X\circ\phi,d\phi(e_i))\bar\nabla_{e_i}\bar\Delta^{s+l-1}\tau(\phi),\bar\Delta^{s-l-1}\tau(\phi)\rangle\\
&\quad-\langle R^N(X\circ\phi,d\phi(e_i))\bar\nabla_{e_i}\bar\Delta^{s-l-1}\tau(\phi),\bar\Delta^{s+l-1}\tau(\phi)\rangle.
\end{aligned}$$

The claim now follows from using the symmetries of the Riemannian curvature tensor and the concrete form of the equation for polyharmonic maps of odd order (1.8). □

REMARK 3.4. We want to point out that conserved quantities such as (3.3) and (3.4) turn out to be very useful in order to develop a general regularity theory for polyharmonic maps; see [**HL**] and [**LG**]. In order to establish such analytic results one only needs a local quantity that is conserved, hence one does not need to demand the existence of a Killing vector field on N but can use the Coulomb gauge construction to locally express the equation for polyharmonic maps as a conservation law.

References

[B] P. Baird, *Harmonic maps with symmetry, harmonic morphisms and deformations of metrics*, Research Notes in Mathematics, vol. 87, Pitman (Advanced Publishing Program), Boston, MA, 1983. MR716320

[B1] V. Branding, *On interpolating sesqui-harmonic maps between Riemannian manifolds*, J. Geom. Anal. **30** (2020), no. 1, 248–273, DOI 10.1007/s12220-018-00130-x. MR4058514

[B2] V. Branding, *The stress-energy tensor for polyharmonic maps*, Nonlinear Anal. **190** (2020), 111616, 17, DOI 10.1016/j.na.2019.111616. MR4007262

[B3] V. Branding, *Some analytic results on interpolating sesqui-harmonic maps*, Ann. Mat. Pura Appl. (4) **199** (2020), no. 5, 2039–2059, DOI 10.1007/s10231-020-00955-w. MR4142862

[B4] V. Branding, *A structure theorem for polyharmonic maps between Riemannian manifolds*, J. Differential Equations **273** (2021), 14–39, DOI 10.1016/j.jde.2020.11.046. MR4184658

[BR] P. Baird and A. Ratto, *Conservation laws, equivariant harmonic maps and harmonic morphisms*, Proc. London Math. Soc. (3) **64** (1992), no. 1, 197–224, DOI 10.1112/plms/s3-64.1.197. MR1132860

[BW] P. Baird and J. C. Wood, *Harmonic morphisms between Riemannian manifolds*, London Mathematical Society Monographs. New Series, vol. 29, The Clarendon Press, Oxford University Press, Oxford, 2003, DOI 10.1093/acprof:oso/9780198503620.001.0001. MR2044031

[BMOR1] V. Branding, S. Montaldo, C. Oniciuc, and A. Ratto, *Higher order energy functionals*, Adv. Math. **370** (2020), 107236, 60, DOI 10.1016/j.aim.2020.107236. MR4106647

[BMOR2] V. Branding, S. Montaldo, C. Oniciuc, and A. Ratto, *Unique continuation properties for polyharmonic maps between Riemannian manifolds*, Canad. J. Math. **75** (2023), no. 1, 1–28, DOI 10.4153/S0008414X21000420. MR4550874

[BMOR3] V. Branding, S. Montaldo, C. Oniciuc, and A. Ratto, *Polyharmonic hypersurfaces into pseudo-Riemannian space forms*, Ann. Mat. Pura Appl. (4) **202** (2023), no. 2, 877–899, DOI 10.1007/s10231-022-01263-1. MR4552081

[FO] D. Fetcu and C. Oniciuc, *Biharmonic and biconservative hypersurfaces in space forms*, Differential geometry and global analysis—in honor of Tadashi Nagano, Contemp. Math., vol. 777, Amer. Math. Soc., [Providence], RI, [2022] ©2022, pp. 65–90, DOI 10.1090/conm/777/15628. MR4410183

[LG] F. L. de Longueville and A. Gastel, *Conservation laws for even order systems of polyharmonic map type*, Calc. Var. Partial Differential Equations **60** (2021), no. 4, Paper No. 138, 18, DOI 10.1007/s00526-021-02009-2. MR4279397

[H1] F. Hélein, *Régularité des applications faiblement harmoniques entre une surface et une sphère* (French, with English summary), C. R. Acad. Sci. Paris Sér. I Math. **311** (1990), no. 9, 519–524. MR1078114

[H2] F. Hélein, *Regularity of weakly harmonic maps from a surface into a manifold with symmetries*, Manuscripta Math. **70** (1991), no. 2, 203–218, DOI 10.1007/BF02568371. MR1085633

[H3] F. Hélein, *Harmonic maps, conservation laws and moving frames*, 2nd ed., Cambridge Tracts in Mathematics, vol. 150, Cambridge University Press, Cambridge, 2002. Translated from the 1996 French original; With a foreword by James Eells, DOI 10.1017/CBO9780511543036. MR1913803

[HL] J. Hörter and T. Lamm, *Conservation laws for even order elliptic systems in the critical dimension—a new approach*, Calc. Var. Partial Differential Equations **60** (2021), no. 4, Paper No. 125, 23, DOI 10.1007/s00526-021-01995-7. MR4277329

[HW] F. Hélein and J. C. Wood, *Harmonic maps*, Handbook of global analysis, Elsevier Sci. B. V., Amsterdam, 2008, pp. 417–491, 1213, DOI 10.1016/B978-044452833-9.50009-7. MR2389639

[J] G. Y. Jiang, *2-harmonic maps and their first and second variational formulas* (Chinese), Chinese Ann. Math. Ser. A **7** (1986), no. 4, 389–402. An English summary appears in Chinese Ann. Math. Ser. B **7** (1986), no. 4, 523. MR886529

[LR] T. Lamm and T. Rivière, *Conservation laws for fourth order systems in four dimensions*, Comm. Partial Differential Equations **33** (2008), no. 1-3, 245–262, DOI 10.1080/03605300701382381. MR2398228

[MOR] S. Montaldo, C. Oniciuc, and A. Ratto, *Polyharmonic hypersurfaces into space forms*, Israel J. Math. **249** (2022), no. 1, 343–374, DOI 10.1007/s11856-022-2315-5. MR4462636

[O] C. Oniciuc, *Biharmonic maps between Riemannian manifolds*, An. Ştiinţ. Univ. Al. I. Cuza Iaşi. Mat. (N.S.) **48** (2002), no. 2, 237–248 (2003). MR2004799

[OC] Y.-L. Ou and B.-Y. Chen, *Biharmonic submanifolds and biharmonic maps in Riemannian geometry*, World Scientific Publishing Co. Pte. Ltd., Hackensack, NJ, [2020] ©2020. MR4265170

[P] P. Petersen, *Riemannian geometry*, 2nd ed., Graduate Texts in Mathematics, vol. 171, Springer, New York, 2006. MR2243772

[S] J. Shatah, *Weak solutions and development of singularities of the* SU(2) *σ-model*, Comm. Pure Appl. Math. **41** (1988), no. 4, 459–469, DOI 10.1002/cpa.3160410405. MR933231

[W] C. Wang, *Remarks on biharmonic maps into spheres*, Calc. Var. Partial Differential Equations **21** (2004), no. 3, 221–242, DOI 10.1007/s00526-003-0252-7. MR2094320

[Y] K. Yano, *The theory of Lie derivatives and its applications*, North-Holland Publishing Co., Amsterdam; P. Noordhoff Ltd., Groningen Interscience Publishers Inc., New York, 1957. MR88769

Faculty of Mathematics, University of Vienna, Oskar-Morgenstern-Platz 1, 1090 Vienna, Austria

Email address: volker.branding@univie.ac.at

Contemporary Mathematics
Volume **821**, 2025
https://doi.org/10.1090/conm/821/16435

Chen inequality for statistical submanifolds in Kenmotsu statistical manifolds of constant ϕ-sectional curvature

Simona Decu and Gabriel-Eduard Vîlcu

Dedicated to Professor Bang-Yen Chen on the occasion of his 80th birthday

ABSTRACT. In this paper, we establish the Chen first inequality in statistical submanifolds of Kenmotsu statistical manifolds of constant ϕ-sectional curvature. Moreover, we investigate the equality case of this inequality. We also reveal an illustrative example.

1. Introduction

The theory of Chen invariants, initiated by B.-Y. Chen in the 1990s, has proved to be of great interest for deepening research in submanifold geometry because these curvature invariants throw new light on some fundamental problems by original solutions. Moreover, this theory is still an inspiration for new studies and the development of new theories concerning the notions of intrinsic and extrinsic invariants of submanifolds.

The geometry of Riemannian manifolds has established their connection to the deeper understanding of the world in various scientific fields. In submanifold theory, J. F. Nash demonstrated in 1956 his celebrated embedding theorem [**26**] revealing that every Riemannian manifold can be isometrically embedded in some Euclidean space with sufficiently high codimension. Consequently, this result has generated the following fundamental problem in submanifold geometry [**6**].

PROBLEM 1.1. Establish simple relationships between main extrinsic invariants and main intrinsic invariants of a submanifold.

Ingenious solutions of this fundamental problem are given by B.-Y. Chen who introduced in the 1990s new types of intrinsic invariants called *δ-invariants* or *Chen invariants*, satisfying optimal inequalities for submanifolds in real space forms [**5**]. Later, many geometric inequalities involving the δ-invariants have been investigated by many geometers on various classes of submanifolds in various ambient spaces (see, e.g., [**14, 21, 25, 29, 30**]). Some comprehensive surveys on this research topic are provided by [**7, 10, 11, 13, 23, 24**].

On the other hand, new solutions of the above problem are focused on new types of geometric inequalities, like Casorati inequalities and Wintgen inequalities (see, e.g., [**1, 3, 4, 7–9, 11, 15, 16**]).

2020 *Mathematics Subject Classification.* Primary 53C05, 53C40.

Initiated in 1972 by K. Kenmotsu [**22**] as a branch of contact geometry, *Kenmotsu geometry* has leaded to a wide range of applications in physics (thermodynamics, classical mechanics, geometrical optics, geometric quantization, classical mechanics) and control theory [**28**]. In 1985, S. Amari introduced the concept of *statistical manifold* in the basic study on information geometry [**2**]. The *Kenmotsu statistical manifold*, defined by H. Furuhata in [**19**], is obtained locally as a warped product between a holomorphic statistical manifold and a real line. In [**16**], the authors established some Casorati inequalities for submanifolds in Kenmotsu statistical manifolds of constant ϕ-sectional curvature. Moreover, if the latter ambient space is endowed with semi-symmetric metric connection, some new Casorati inequalities for statistical submanifolds were obtained in [**17**]. The statistical manifolds endowed with almost product structures are investigated in [**31**].

In this article, we establish a basic inequality between the Chen first invariant (intrinsic invariant) and the mean curvature (extrinsic invariant) of statistical submanifolds in Kenmotsu statistical manifolds having a constant ϕ-sectional curvature. Moreover, we study the equality case of such inequality. An example is also provided.

2. Preliminaries

Let $(\bar{M}, \bar{g})$ be a Riemannian manifold, with $\bar{g}$ a Riemannian metric on $\bar{M}$ and let $\bar{\nabla}$ be an affine connection on $\bar{M}$. Then the triplet $(\bar{M}, \bar{g}, \bar{\nabla})$ is called a *statistical manifold* if the torsion tensor field of $\bar{\nabla}$ vanishes and $\bar{\nabla}g$ is totally symmetric [**18**]. The pair $(\bar{\nabla}, \bar{g})$ is called a statistical structure on $\bar{M}$. Let $\bar{\nabla}^*$ be an affine connection of $\bar{M}$ defined by

$$X\,\bar{g}(Y,Z) = \bar{g}(\bar{\nabla}_X Y, Z) + \bar{g}(Y, \bar{\nabla}^*_X Z),$$

for any X, Y, $Z \in \Gamma(T\bar{M})$, where $\Gamma(T\bar{M})$ is the set of smooth tangent vector fields on $\bar{M}$.

Then $\bar{\nabla}^*$ is named the *dual connection* of $\bar{\nabla}$ with respect to $\bar{g}$. Clearly, $(\bar{\nabla}^*)^* = \bar{\nabla}$. Moreover, the Levi-Civita connection on $\bar{M}$ is given by $\bar{\nabla}^0 = \frac{\bar{\nabla}+\bar{\nabla}^*}{2}$ [**32**]. If $(\bar{M}, \bar{g}, \bar{\nabla})$ is a statistical manifold, then $(\bar{M}, \bar{g}, \bar{\nabla}^*)$ is too.

Let M be a submanifold of a statistical manifold $(\bar{M}, \bar{g}, \bar{\nabla})$ with g the induced metric on M, and ∇ the induced connection on M. Then it is known that (M, g, ∇) is a statistical manifold as well.

Denote by h and h^* the *imbedding curvature tensor* of M in $\bar{M}$ with respect to $\bar{\nabla}$ and $\bar{\nabla}^*$, respectively. Then Gauss's formulas [**18**] are expressed by

$$\bar{\nabla}_X Y = \nabla_X Y + h(X,Y),$$
$$\bar{\nabla}^*_X Y = \nabla^*_X Y + h^*(X,Y),$$

for any $X, Y \in \Gamma(TM)$.

Furthermore, denote by R, $\bar{R}$, R^* and $\bar{R}^*$ the $(0,4)$-*curvature tensors* for the connections ∇, $\bar{\nabla}$, ∇^* and $\bar{\nabla}^*$, respectively. We can now define the *statistical curvature tensor field* [**18**] on M and $\bar{M}$, denoted by S and $\bar{S}$, respectively,

$$S(X,Y)Z = \frac{1}{2}\{R(X,Y)Z + R^*(X,Y)Z\}, \tag{2.1}$$

for any $X, Y, Z \in \Gamma(TM)$, and

$$\bar{S}(X,Y)Z = \frac{1}{2}\{\bar{R}(X,Y)Z + \bar{R}^*(X,Y)Z\}, \tag{2.2}$$

for any $X, Y, Z \in \Gamma(T\bar{M})$.

Set a tensor field $\bar{K} \in \Gamma(T\bar{M}^{(1,2)})$ by

$$\bar{K}_X Y = \bar{K}(X, Y) = \bar{\nabla}_X Y - \bar{\nabla}^0_X Y. \tag{2.3}$$

Furthermore, we have

$$\bar{K}(X, Y) = \bar{\nabla}^0_X Y - \bar{\nabla}^*_X Y = \frac{1}{2}(\bar{\nabla}_X Y - \bar{\nabla}^*_X Y).$$

Then $\bar{K}$ has the properties

$$\bar{K}(X, Y) = \bar{K}(Y, X),$$
$$g(\bar{K}(X, Y), Z) = g(Y, \bar{K}(X, Z)).$$

Next, we consider $(\bar{M}, \bar{g}, \phi, \xi)$ a $(2n+1)$-dimensional Kenmotsu manifold defined as an almost contact metric manifold $\bar{M}$ which satisfies for any $X, Y \in \Gamma(T\bar{M})$ the relations

$$(\bar{\nabla}^0_X \phi)(Y) = \bar{g}(\phi X, Y)\xi - \bar{\eta}(Y)\phi X,$$
$$\bar{\nabla}^0_X \xi = X - \bar{\eta}(X)\xi,$$

where $\phi \in \Gamma(T\bar{M}^{(1,1)})$, $\xi \in \Gamma(T\bar{M})$, $\bar{\eta}$ is a 1-form on $\bar{M}$ with $\bar{\eta}(X) = \bar{g}(X, \xi)$.

A Kenmotsu manifold $\bar{M}$ with a statistical structure $(\bar{\nabla}, \bar{g})$ is called a Kenmotsu statistical manifold [**19**] if the following formula holds for any $X, Y \in \Gamma(T\bar{M})$:

$$\bar{K}(X, \phi Y) = -\phi \bar{K}(X, Y),$$

where $\bar{K}$ is the tensor field defined in (2.3).

A Kenmotsu statistical manifold $(\bar{M}, \bar{\nabla}, \bar{g}, \phi, \xi)$ is said to be of constant ϕ-sectional curvature c if the statistical curvature tensor field $\bar{S}$ is given by [**19**]:

$$\begin{aligned}\bar{S}(X, Y)Z = {} & \frac{c-3}{4}\{\bar{g}(Y, Z)X - \bar{g}(X, Z)Y)\} \\ & + \frac{c+1}{4}\{\bar{g}(\phi Y, Z)\phi X - \bar{g}(\phi X, Z)\phi Y - 2\bar{g}(\phi X, Y)\phi Z \\ & - \bar{g}(Y, \xi)\bar{g}(Z, \xi)X + \bar{g}(X, \xi)\bar{g}(Z, \xi)Y \\ & + \bar{g}(Y, \xi)\bar{g}(Z, X)\xi - \bar{g}(X, \xi)\bar{g}(Z, Y)\xi\},\end{aligned} \tag{2.4}$$

for any $X, Y, Z \in \Gamma(T\bar{M})$.

Let M be an $(m+1)$-dimensional submanifold of a statistical manifold $\bar{M}$.

The *Gauss equations* are the following [**20**]:

$$\begin{aligned}\bar{g}(\bar{R}(X, Y)Z, W) = {} & g(R(X, Y)Z, W) - \bar{g}(h(Y, Z), h^*(X, W)) \\ & + \bar{g}(h(X, Z), h^*(Y, W)),\end{aligned} \tag{2.5}$$

$$\begin{aligned}2\bar{g}(\bar{S}(X, Y)Z, W) = {} & 2g(S(X, Y)Z, W) - \bar{g}(h(Y, Z), h^*(X, W)) \\ & + \bar{g}(h(X, Z), h^*(Y, W)) \\ & - \bar{g}(h^*(Y, Z), h(X, W)) + \bar{g}(h^*(X, Z), h(Y, W)),\end{aligned} \tag{2.6}$$

$$\begin{aligned}4\bar{g}(\bar{R}^0(X, Y)Z, W) = {} & 4g(R^0(X, Y)Z, W) \\ & - \bar{g}(h(Y, Z) + h^*(Y, Z), h(X, W) + h^*(X, W)) \\ & + \bar{g}(h(X, Z) + h^*(X, Z), h(Y, W) + h^*(Y, W)).\end{aligned} \tag{2.7}$$

If $p \in M$ and $\pi \subset T_pM$ is a nondegenerate 2-plane, then the *sectional curvature* is defined as [**18**]:

$$\mathcal{K}(X,Y) = \frac{g(S(X,Y)Y,X)}{g(X,X)g(Y,Y) - g^2(X,Y)}, \tag{2.8}$$

where $\{X,Y\}$ is a basis of π.

The *scalar curvature* τ of (M,∇,g) at a point $p \in M$ is defined by:

$$\tau(p) = \sum_{1\le i<j\le m+1} g(S(e_i,e_j)e_j,e_i), \tag{2.9}$$

where $\{e_1,\dots,e_{m+1}\}$ is an orthonormal basis at p.

The *mean curvature* vector fields of M are defined by, respectively:

$$H = \frac{1}{m+1}\sum_{i=1}^{m+1} h(e_i,e_i), \quad H^* = \frac{1}{m+1}\sum_{i=1}^{m+1} h^*(e_i,e_i).$$

It follows that we have $2h^0 = h + h^*$ and $2H^0 = H + H^*$, where h^0 and H^0 are the second fundamental form and the mean curvature field of M, respectively, with respect to the Levi-Civita connection ∇^0 on M.

Then, the *squared mean curvatures* of the submanifold M in $\bar{M}$ are given by

$$\|H\|^2 = \frac{1}{(m+1)^2}\sum_{\alpha=m+2}^{2n+1}\left(\sum_{i=1}^{m+1} h_{ii}^{\alpha}\right)^2, \quad \|H^*\|^2 = \frac{1}{(m+1)^2}\sum_{\alpha=m+2}^{2n+1}\left(\sum_{i=1}^{m+1} h_{ii}^{*\alpha}\right)^2,$$

where $h_{ij}^{\alpha} = g(h(e_i,e_j),e_\alpha)$ and $h_{ij}^{*\alpha} = g(h^*(e_i,e_j),e_\alpha)$, for $i,j \in \{1,\dots,m+1\}$, $\alpha \in \{m+2,\dots,2n+1\}$.

Next, we consider the following result which we will use later.

LEMMA 2.1 ([**12**]). *Let $n \geq 3$ be an integer and let $a_1, a_2, \dots, a_n$ be real numbers. Then we have:*

$$\sum_{1\le i<j\le n} a_i a_j - a_1 a_2 \le \frac{n-2}{2(n-1)}\left(\sum_{i=1}^{n} a_i\right)^2. \tag{2.10}$$

Moreover, the equality holds if and only if

$$a_1 + a_2 = a_3 = \cdots = a_n.$$

3. Main Inequality

Let $(\bar{M},\bar{\nabla},\bar{g},\phi,\xi)$ be a $(2n+1)$-dimensional Kenmotsu statistical manifold of constant ϕ-sectional curvature c, and let M be an $(m+1)$-dimensional statistical submanifold of $(\bar{M},\bar{\nabla},\bar{g},\phi,\xi)$.

Let π be a plane section at $p \in M$ and an orthonormal basis $\{e_1,e_2\}$ of π. Moreover, consider $\{e_1,\dots,e_{m+1}\}$ and $\{e_{m+2},\dots,e_{2n+1}\}$ orthonormal bases of T_pM and $T_p^{\perp}M$, respectively. We denote by

$$\mathcal{K}_0 = \mathcal{K}_0(e_1,e_2) = g(R^0(e_1,e_2)e_2,e_1)$$

the sectional curvature of the Levi-Civita connection ∇^0 on M and by h^0 the second fundamental form of M. We consider the tensor field Q of type $(1,3)$ on $\bar{M}$ defined by [**27**]

$$Q(X,Y) = [\bar{K}_X,\bar{K}_Y], \; X,Y \in \Gamma(T\bar{M}).$$

Then the sectional Q-curvature $\mathcal{K}^Q(\pi)$ of the plane section π is given by [**27**]

$$\begin{aligned}\mathcal{K}^Q(\pi) &= g(Q(e_1,e_2)e_2,e_1)\\ &= \frac{1}{2}\{g(R(e_1,e_2)e_2,e_1)+g(R^*(e_1,e_2)e_2,e_1)-2g(R^0(e_1,e_2)e_2,e_1)\}.\end{aligned} \tag{3.1}$$

Then $\mathcal{K}^Q(\pi)$ can be written also:

$$\mathcal{K}^Q(\pi) = g(S(e_1,e_2)e_2,e_1) - g(R^0(e_1,e_2)e_2,e_1). \tag{3.2}$$

From the equations (2.4) and (2.6), we get

$$\begin{aligned}\mathcal{K}^Q(\pi) = \frac{c-3}{4} &+ \frac{c+1}{4}\{\bar{g}(e_1,\phi e_1)\bar{g}(e_2,\phi e_2)\\ &- 3\bar{g}(e_1,\phi e_2)\bar{g}(e_2,\phi e_1) - \bar{g}^2(e_1,\xi) - \bar{g}^2(e_2,\xi)\}\\ &+ \frac{1}{2}\sum_{\alpha=m+2}^{2n+1}(h^\alpha_{22}h^{*\alpha}_{11} + h^\alpha_{11}h^{*\alpha}_{22} - 2h^\alpha_{12}h^{*\alpha}_{12}) - \mathcal{K}_0(\pi).\end{aligned} \tag{3.3}$$

By virtue of the relations developed in [**12**], the equation (3.3) becomes

$$\begin{aligned}\mathcal{K}^Q(\pi) = \frac{c-3}{4} &+ \frac{c+1}{4}\{\bar{g}(e_1,\phi e_1)\bar{g}(e_2,\phi e_2)\\ &- 3\bar{g}(e_1,\phi e_2)\bar{g}(e_2,\phi e_1) - \bar{g}^2(e_1,\xi) - \bar{g}^2(e_2,\xi)\}\\ &+ \frac{1}{2}\sum_{\alpha=m+2}^{2n+1}[(h^\alpha_{11}+h^{*\alpha}_{11})(h^\alpha_{22}+h^{*\alpha}_{22}) - h^\alpha_{11}h^\alpha_{22} - h^{*\alpha}_{11}h^{*\alpha}_{22}\\ &- (h^\alpha_{12}+h^{*\alpha}_{12})^2 + (h^\alpha_{12})^2 + (h^{*\alpha}_{12})^2] - \mathcal{K}_0(\pi).\end{aligned} \tag{3.4}$$

Moreover, because $2h^0 = h + h^*$, the latter equation is given by

$$\begin{aligned}\mathcal{K}^Q(\pi) = \frac{c-3}{4} &+ \frac{c+1}{4}\{\bar{g}(e_1,\phi e_1)\bar{g}(e_2,\phi e_2)\\ &- 3\bar{g}(e_1,\phi e_2)\bar{g}(e_2,\phi e_1) - \bar{g}^2(e_1,\xi) - \bar{g}^2(e_2,\xi)\}\\ &+ \sum_{\alpha=m+2}^{2n+1}\{2[h^{0\alpha}_{11}h^{0\alpha}_{22} - (h^{0\alpha}_{12})^2] - \frac{1}{2}[h^\alpha_{11}h^\alpha_{22} - (h^\alpha_{12})^2]\\ &- \frac{1}{2}[h^{*\alpha}_{11}h^{*\alpha}_{22} - (h^{*\alpha}_{12})^2]\} - \mathcal{K}_0(\pi).\end{aligned} \tag{3.5}$$

From Gauss equation (2.7), we obtain

$$\sum_{\alpha=m+2}^{2n+1}[h^{0\alpha}_{11}h^{0\alpha}_{22} - (h^{0\alpha}_{12})^2] = \mathcal{K}_0(\pi) - \bar{\mathcal{K}}_0(\pi), \tag{3.6}$$

where $\bar{\mathcal{K}}_0$ is the sectional curvature of the Levi-Civita connection $\bar{\nabla}^0$ on $\bar{M}$. Using the expression (3.6) in the relation (3.5), we get

$$\begin{aligned}\mathcal{K}^Q(\pi) = \frac{c-3}{4} &+ \frac{c+1}{4}\{\bar{g}(e_1,\phi e_1)\bar{g}(e_2,\phi e_2) - 3\bar{g}(e_1,\phi e_2)\bar{g}(e_2,\phi e_1)\\ &- \bar{g}^2(e_1,\xi) - \bar{g}^2(e_2,\xi)\} - \frac{1}{2}\sum_{\alpha=m+2}^{2n+1}[h^\alpha_{11}h^\alpha_{22} - (h^\alpha_{12})^2]\\ &- \frac{1}{2}\sum_{\alpha=m+2}^{2n+1}[h^{*\alpha}_{11}h^{*\alpha}_{22} - (h^{*\alpha}_{12})^2] + \mathcal{K}_0(\pi) - 2\bar{\mathcal{K}}_0(\pi).\end{aligned} \tag{3.7}$$

On the other hand, we study the Q-scalar curvature of M, denoted by τ^Q, corresponding to the sectional Q-curvature. We have

$$\begin{aligned}\tau^Q &= \frac{1}{2}\sum_{1\leq i<j\leq m+1}[g(R(e_i,e_j)e_j,e_i)+g(R^*(e_i,e_j)e_j,e_i)-2g(R^0(e_i,e_j)e_j,e_i)]\\ &= \sum_{1\leq i<j\leq m+1} g(S(e_i,e_j)e_j,e_i)-\sum_{1\leq i<j\leq m+1} g(R^0(e_i,e_j)e_j,e_i).\end{aligned}$$

Thus, we derive

$$\tau^Q = \tau - \tau_0, \tag{3.8}$$

where τ is the statistical scalar curvature of ∇ on M and τ_0 is the scalar curvature of the Levi-Civita connection ∇^0 on M.

Next, we denote $\phi X = PX + QX$, where PX is the tangent component of ϕX and QX is the normal component of ϕX. Moreover, denote *trace* P and *trace* P^2 as follows:

$$trace\ P = \sum_{i=1}^{m+1} \bar{g}(Pe_i, e_i),$$

$$trace\ P^2 = \sum_{i=1}^{m+1} \bar{g}(P^2e_i, e_i). \tag{3.9}$$

Then the equations (2.4) and (2.6) imply

$$\begin{aligned}\tau = {} & \frac{m(m+1)(c-3)}{8}+\frac{c+1}{4}[\frac{1}{2}(trace\ P)^2\\ &+\sum_{1\leq i\leq m+1}\bar{g}^2(e_i,Pe_i)-\frac{3}{2}trace\ P^2-m]\\ &+\frac{1}{2}\sum_{\alpha=m+2}^{2n+1}\sum_{1\leq i<j\leq m+1}(h_{ii}^{*\alpha}h_{jj}^{\alpha}+h_{ii}^{\alpha}h_{jj}^{*\alpha}-2h_{ij}^{\alpha}h_{ij}^{*\alpha}).\end{aligned} \tag{3.10}$$

Furthermore, summing over $1 \leq i < j \leq m+1$ for $X = W = e_i$ and $Y = Z = e_j$ in the Gauss equation (2.7), we get

$$\begin{aligned}2\tau_0 = {} & 2\bar{\tau}_0+\frac{1}{2}\sum_{\alpha=m+2}^{2n+1}\sum_{1\leq i<j\leq m+1}[h_{ii}^{\alpha}h_{jj}^{\alpha}-(h_{ij}^{\alpha})^2]\\ &+\frac{1}{2}\sum_{\alpha=m+2}^{2n+1}\sum_{1\leq i<j\leq m+1}[h_{ii}^{*\alpha}h_{jj}^{*\alpha}-(h_{ij}^{*\alpha})^2]\\ &+\frac{1}{2}\sum_{\alpha=m+2}^{2n+1}\sum_{1\leq i<j\leq m+1}(h_{jj}^{\alpha}h_{ii}^{*\alpha}+h_{ii}^{\alpha}h_{jj}^{*\alpha}-2h_{ij}^{\alpha}h_{ij}^{*\alpha}).\end{aligned} \tag{3.11}$$

Then, using equations (3.10) and (3.11) in equation (3.8), we obtain

$$\begin{aligned}\tau^Q &= (\tau - 2\tau_0) + \tau_0 \\ &= \frac{m(m+1)(c-3)}{8} + \frac{c+1}{4}[\frac{1}{2}(trace\ P)^2 \\ &+ \sum_{1\le i<j\le m+1} \bar{g}^2(e_i, Pe_i) - \frac{3}{2}trace\ P^2 - m] + \tau_0 - 2\bar{\tau}_0 \\ &- \frac{1}{2}\sum_{\alpha=m+2}^{2n+1}\sum_{1\le i<j\le m+1}[h^\alpha_{ii}h^\alpha_{jj} - (h^\alpha_{ij})^2] \\ &- \frac{1}{2}\sum_{\alpha=m+2}^{2n+1}\sum_{1\le i<j\le m+1}[h^{*\alpha}_{ii}h^{*\alpha}_{jj} - (h^{*\alpha}_{ij})^2].\end{aligned} \tag{3.12}$$

By subtracting the equation (3.7) from the equation (3.12), we get

$$\begin{aligned}&(\tau^Q - \mathcal{K}^Q(\pi)) - (\tau_0 - \mathcal{K}_0(\pi)) \\ &= \frac{(m-1)(m+2)(c-3)}{8} + \frac{c+1}{4}[\frac{1}{2}(trace\ P)^2 \\ &+ \sum_{1\le i<j\le m+1} \bar{g}^2(e_i, Pe_i) - \frac{3}{2}trace\ P^2 - m - \bar{g}(e_1, Pe_1)\bar{g}(e_2, Pe_2) \\ &- 3\bar{g}(e_1, Pe_2)\bar{g}(e_2, Pe_1) - \bar{g}^2(e_1, \xi) - \bar{g}^2(e_2, \xi)] \\ &- \frac{1}{2}\sum_{\alpha=m+2}^{2n+1}\sum_{1\le i<j\le m+1}\{[h^\alpha_{ii}h^\alpha_{jj} - (h^\alpha_{ij})^2] + [h^{*\alpha}_{ii}h^{*\alpha}_{jj} - (h^{*\alpha}_{ij})^2]\} \\ &+ \frac{1}{2}\sum_{\alpha=m+2}^{2n+1}[h^\alpha_{11}h^\alpha_{22} - (h^\alpha_{12})^2] + [h^{*\alpha}_{11}h^{*\alpha}_{22} - (h^{*\alpha}_{12})^2] - 2\bar{\tau}_0 + 2\bar{\mathcal{K}}_0(\pi).\end{aligned} \tag{3.13}$$

Next, by virtue of the Lemma 2.1, we obtain

$$\sum_{1\le i<j\le m+1} h^\alpha_{ii}h^\alpha_{jj} - h^\alpha_{11}h^\alpha_{22} \le \frac{m-1}{2m}(\sum_{i=1}^{m+1} h^\alpha_{ii})^2 = \frac{(m-1)(m+1)^2}{2m}(H^\alpha)^2, \tag{3.14}$$

$$\sum_{1\le i<j\le m+1} h^{*\alpha}_{ii}h^{*\alpha}_{jj} - h^{*\alpha}_{11}h^{*\alpha}_{22} \le \frac{m-1}{2m}(\sum_{i=1}^{m+1} h^{*\alpha}_{ii})^2 = \frac{(m-1)(m+1)^2}{2m}(H^{*\alpha})^2. \tag{3.15}$$

Applying the expressions of (3.14) and (3.15) in the equation (3.13), we reveal the Chen first inequality for arbitrary statistical submanifolds of a Kenmotsu statistical manifold of constant ϕ-sectional curvature c, as follows.

THEOREM 3.1. *Let $(\bar{M}, \bar{\nabla}, \bar{g}, \phi, \xi)$ be a $(2n+1)$-dimensional Kenmotsu statistical manifold of constant ϕ-sectional curvature c, and M is an $(m+1)$-dimensional* *statistical submanifold of* $(\bar{M}, \bar{\nabla}, \bar{g}, \phi, \xi)$*. Then the following inequality holds*

$$
\begin{aligned}
(3.16)\quad & (\tau^Q - \tau_0) - (\mathcal{K}^Q(\pi) - \mathcal{K}_0(\pi)) \\
& \geq \frac{(m-1)(m+2)(c-3)}{8} \\
& + \frac{c+1}{4}\Big[\frac{1}{2}(trace\ P)^2 + \sum_{i=1}^{m+1} \bar{g}^2(e_i, Pe_i) - \frac{3}{2} trace\ P^2 - m \\
& - \bar{g}(e_1, Pe_1)\bar{g}(e_2, Pe_e) - 3\bar{g}(e_1, Pe_2)\bar{g}(e_2, Pe_1) - \bar{g}^2(e_1, \xi) - \bar{g}^2(e_2, \xi)\Big] \\
& - \frac{(m-1)(m+1)^2(\|H\|^2 + \|H^*\|^2)}{4m} - 2(\bar{\tau}_0 - \bar{\mathcal{K}}_0(\pi)),
\end{aligned}
$$

for any plane section π *at* p *in* M*, where* $\{e_1, e_2\}$ *is an orthonormal basis of* π*.*

Furthermore, the equality holds in (3.16) *if and only if*

$$
\begin{aligned}
h_{11}^{\alpha} + h_{22}^{\alpha} &= h_{33}^{\alpha} = \cdots = h_{m+1\ m+1}^{\alpha}, \\
h_{11}^{*\alpha} + h_{22}^{*\alpha} &= h_{33}^{*\alpha} = \cdots = h_{m+1\ m+1}^{*\alpha}, \\
h_{ij}^{\alpha} &= h_{ij}^{*\alpha} = 0,
\end{aligned}
$$

for any $1 \leq i \neq j \leq m+1$, $m+2 \leq \alpha \leq 2n+1$.

4. An example

Let $(H^{2n+1}, \bar{\nabla} = \bar{\nabla}^0 + \bar{K}, \bar{g}, \phi, \xi)$ be a $(2n+1)$-dimensional Kenmotsu statistical manifold (see [**19**, Examples 3.3 and 3.10]). We recall that

$$H^{2n+1} = \{(x_1, \ldots, x_n, y_1, \ldots, y_n, z) \in \mathbb{R}^{2n+1} | z > 0\}$$

and the structure tensors (g, ϕ, ξ) are defined by

$$
\begin{aligned}
&\bar{g} = \frac{1}{z^2}\{(dx_1)^2 + \cdots + (dx_n)^2 + (dy_1)^2 + \cdots + (dy_n)^2 + (dz)^2\}, \\
&\phi\frac{\partial}{\partial x_\alpha} = \frac{\partial}{\partial y_\alpha},\ \phi\frac{\partial}{\partial y_\alpha} = -\frac{\partial}{\partial x_\alpha},\ \phi\frac{\partial}{\partial z} = 0,\ for\ \alpha \in \{1, \ldots, n\}
\end{aligned}
$$

and

$$\xi = -z\frac{\partial}{\partial z}.$$

For any $X, Y \in \Gamma(TH^{2n+1})$, we set the $(1,2)$-tensor field $\bar{K}$ as

$$\bar{K}(X, Y) = \nu\, \bar{\eta}(X)\, \bar{\eta}(Y)\, \xi,$$

where $\nu \in C^\infty(H^{2n+1})$ and $\bar{\eta}$ is the 1-form on H^{2n+1} dual to ξ, that is $\bar{\eta}(X) = g(X, \xi)$.

Then $(H^{2n+1}, \bar{\nabla} = \bar{\nabla}^0 + \bar{K}, g, \phi, \xi)$ is a Kenmotsu statistical manifold with constant ϕ-sectional curvature $c = -1$ (see ([**19**]). Moreover, $M = H^{2k+1}$ (where $0 < k < n$) is statistical submanifold of the Kenmotsu statistical manifold H^{2n+1} for which the inequality (3.16) is satisfied.

5. Conclusions

This paper is focused on a Chen inequality for statistical submanifolds in Kenmotsu statistical manifolds of constant ϕ-sectional curvature. In addition, we studied the equality case of this inequality and we provided a nontrivial example to illustrate the result.

References

[1] H. Alodan, B.-Y. Chen, S. Deshmukh, and G.-E. Vîlcu, *A generalized Wintgen inequality for quaternionic CR-submanifolds*, Rev. R. Acad. Cienc. Exactas Fís. Nat. Ser. A Mat. RACSAM **114** (2020), no. 3, Paper No. 129, 14, DOI 10.1007/s13398-020-00866-8. MR4098167

[2] S.-i. Amari, *Differential-geometrical methods in statistics*, Lecture Notes in Statistics, vol. 28, Springer-Verlag, New York, 1985, DOI 10.1007/978-1-4612-5056-2. MR0788689

[3] M. Aquib, M. S. Lone, C. Neacşu, and G.-E. Vîlcu, *On δ-Casorati curvature invariants of Lagrangian submanifolds in quaternionic Kähler manifolds of constant q-sectional curvature*, Rev. R. Acad. Cienc. Exactas Fís. Nat. Ser. A Mat. RACSAM **117** (2023), no. 3, Paper No. 107, 21, DOI 10.1007/s13398-023-01438-2. MR4581817

[4] N. D. Brubaker and B. D. Suceavă, *A geometric interpretation of Cauchy-Schwarz inequality in terms of Casorati curvature*, Int. Electron. J. Geom. **11** (2018), no. 1, 48–51. MR3798868

[5] B.-Y. Chen, *Some pinching and classification theorems for minimal submanifolds*, Arch. Math. (Basel) **60** (1993), no. 6, 568–578, DOI 10.1007/BF01236084. MR1216703

[6] B.-Y. Chen, *Relations between Ricci curvature and shape operator for submanifolds with arbitrary codimensions*, Glasg. Math. J. **41** (1999), no. 1, 33–41, DOI 10.1017/S0017089599970271. MR1689730

[7] B.-Y. Chen, *Pseudo-Riemannian geometry, δ-invariants and applications*, World Scientific Publishing Co. Pte. Ltd., Hackensack, NJ, 2011. With a foreword by Leopold Verstraelen, DOI 10.1142/9789814329644. MR2799371

[8] B.-Y. Chen, *Recent developments in δ-Casorati curvature invariants*, Turkish J. Math. **45** (2021), no. 1, 1–46, DOI 10.3906/mat-2009-8. MR4245492

[9] B.-Y. Chen, *Recent developments in Wintgen inequality and Wintgen ideal submanifolds*, Int. Electron. J. Geom. **14** (2021), no. 1, 6–45, DOI 10.36890/iejg.838446. MR4250303

[10] B.-Y. Chen and A. M. Blaga, *Recent developments on Chen-Ricci inequalities in differential geometry*, Geometry of submanifolds and applications, Infosys Sci. Found. Ser., Springer, Singapore, 2024, pp. 1–61, DOI 10.1007/978-981-99-9750-3_1. MR4759264

[11] B.-Y. Chen, A. M. Blaga, and G.-E. Vîlcu, *Differential Geometry of Submanifolds in Complex Space Forms Involving δ-Invariants*, Mathematics, **10** (2022), 591.

[12] B.-Y. Chen, A. Mihai, and I. Mihai, *A Chen first inequality for statistical submanifolds in Hessian manifolds of constant Hessian curvature*, Results Math. **74** (2019), no. 4, Paper No. 165, 11, DOI 10.1007/s00025-019-1091-y. MR3998760

[13] B.-Y. Chen and G.-E. Vîlcu, *Recent Developments on the First Chen Inequality in Differential Geometry*, Mathematics, **11** (2023), 4186.

[14] S. Decu, *Optimal inequalities for submanifolds in quaternion-space-forms with semi-symmetric metric connection*, Bull. Transilv. Univ. Braşov Ser. III **2(51)** (2009), 175–184. MR2642506

[15] S. Decu and S. Haesen, *Chen Inequalities for Spacelike Submanifolds in Statistical Manifolds of Type Para-Kähler Space Forms*, Mathematics, **10** (2022), 330.

[16] S. Decu, S. Haesen, L. Verstraelen, and G.-E. Vîlcu, *Curvature invariants of statistical submanifolds in Kenmotsu statistical manifolds of constant φ-sectional curvature*, Entropy **20** (2018), no. 7, Paper No. 529, 15, DOI 10.3390/e20070529. MR3891159

[17] S. Decu and G.-E. Vîlcu, *Casorati inequalities for statistical submanifolds in Kenmotsu statistical manifolds of constant φ-sectional curvature with semi-symmetric metric connection*, Entropy **24** (2022), no. 6, Paper No. 800, 16, DOI 10.3390/e24060800. MR4450476

[18] H. Furuhata and I. Hasegawa, *Submanifold theory in holomorphic statistical manifolds*, Geometry of Cauchy-Riemann submanifolds, Springer, Singapore, 2016, pp. 179–215. MR3525856

[19] H. Furuhata, I. Hasegawa, Y. Okuyama, and K. Sato, *Kenmotsu statistical manifolds and warped product*, J. Geom. **108** (2017), no. 3, 1175–1191, DOI 10.1007/s00022-017-0403-1. MR3721595

[20] H. Furuhata, I. Hasegawa, and N. Satoh, *Chen invariants and statistical submanifolds*, Commun. Korean Math. Soc. **37** (2022), no. 3, 851–864, DOI 10.4134/CKMS.c210185. MR4468733

[21] S. Haesen and L. Verstraelen, *Ideally embedded space-times*, J. Math. Phys. **45** (2004), no. 4, 1497–1510, DOI 10.1063/1.1668333. MR2043839

[22] K. Kenmotsu, *A class of almost contact Riemannian manifolds*, Tohoku Math. J. (2) **24** (1972), 93–103, DOI 10.2748/tmj/1178241594. MR319102
[23] A. Mihai, *Modern topics in submanifold theory*, Editura Universităţii din Bucureşti, Bucharest, 2006. MR2263812
[24] I. Mihai, *Statistical manifolds and their submanifolds. Results on Chen-like invariants*, Geometry of submanifolds, Contemp. Math., vol. 756, Amer. Math. Soc., [Providence], RI, 2020, pp. 163–172, DOI 10.1090/conm/756/15206. MR4186947
[25] A. Mihai and I. Mihai, *The $\delta(2,2)$-invariant on statistical submanifolds in Hessian manifolds of constant Hessian curvature*, Entropy **22** (2020), no. 2, Paper No. 164, 8, DOI 10.3390/e22020164. MR4075763
[26] J. Nash, *The imbedding problem for Riemannian manifolds*, Ann. of Math. (2) **63** (1956), 20–63, DOI 10.2307/1969989. MR75639
[27] B. Opozda, *A sectional curvature for statistical structures*, Linear Algebra Appl. **497** (2016), 134–161, DOI 10.1016/j.laa.2016.02.021. MR3466639
[28] G. Pitiş, *Contact forms in geometry and topology*, Topics in modern differential geometry, Atlantis Trans. Geom., vol. 1, Atlantis Press, [Paris], 2017, pp. 79–94. MR3588308
[29] B. Suceavă, *Some remarks on B.-Y. Chen's inequality involving classical invariants*, An. Ştiinţ. Univ. Al. I. Cuza Iaşi. Mat. (N.S.) **45** (1999), no. 2, 405–412 (2000). MR1811740
[30] B. D. Suceavă and M. B. Vajiac, *Estimates of B.-Y. Chen's $\hat{\delta}$-invariant in terms of Casorati curvature and mean curvature for strictly convex Euclidean hypersurfaces*, Int. Electron. J. Geom. **12** (2019), no. 1, 26–31. MR3937521
[31] G.-E. Vîlcu, *Almost product structures on statistical manifolds and para-Kähler-like statistical submersions*, Bull. Sci. Math. **171** (2021), Paper No. 103018, 21, DOI 10.1016/j.bulsci.2021.103018. MR4285860
[32] P. W. Vos, *Fundamental equations for statistical submanifolds with applications to the Bartlett correction*, Ann. Inst. Statist. Math. **41** (1989), no. 3, 429–450, DOI 10.1007/BF00050660. MR1032591

Department of Applied Mathematics, Bucharest University of Economic Studies, 6 Piaţa Romană, 010374 Bucharest, Romania; and Romanian Academy, Center of Mountain Economy (CE-MONT), "Costin C. Kiriţescu" National Institute of Economic Research, 13 Calea 13 Septembrie, 030508 Bucharest, Romania

Email address: simona.decu@gmail.com, simona.marinescu@csie.ase.ro

Department of Mathematics and Informatics, Faculty of Applied Sciences, University Politehnica of Bucharest, Splaiul Independenţei 313, 060042 Bucharest, Romania; and Romanian Academy, "Gheorghe Mihoc-Caius Iacob" Institute of Mathematical Statistics and Applied Mathematics, 050711 Bucharest, Romania

Email address: gabriel.vilcu@upb.ro

Contemporary Mathematics
Volume **821**, 2025
https://doi.org/10.1090/conm/821/16436

Thurston geometries in dimension four from a Riemannian perspective

Marie D'haene

Abstract. We focus on a special class of homogeneous manifolds called Thurston geometries. We give special attention to the four-dimensional Thurston geometries with 4– or 5-dimensional isometry group which are not a product, except for F^4. These are the manifolds Sol_0^4, Sol_1^4, $\mathrm{Sol}_{m,n}^4$, Nil^4. We give a description of each of the spaces and we exhibit all Riemannian metrics which are invariant under the action of their associated group.

1. Introduction

Riemannian homogeneous spaces are a fundamental class of manifolds whose study requires techniques from geometry, algebra and group theory. Thurston geometries are a subset of Riemannian homogeneous spaces, famous due to Thurston's geometrization for 3-manifolds where these appear as fundamental building blocks. This geometrization theorem concerns compact orientable 3-manifolds and it is very similar to the uniformization theorem for compact orientable surfaces. Namely, it says that any 3-manifold can be cut up into pieces along 2-spheres and 2-tori, and the interior of each of these pieces admits a Riemannian metric which is locally isometric to one of the eight 3-dimensional Thurston geometries. These are $\mathbb{R}^3$, $\mathbb{S}^3$, $\mathbb{H}^3$, $\mathbb{S}^2 \times \mathbb{R}$, $\mathbb{H}^2 \times \mathbb{R}$, $\widetilde{\mathrm{SL}}(2,\mathbb{R})$, Nil^3 and Sol^3 (see Table 1). Hence, understanding these model spaces is an essential part of the study of 3-manifolds and as such they have already received a lot of attention in the literature. In the context of isometrically immersed submanifolds, we refer to the following selection of works: Daniel [**Dan07**] studied the fundamental equations for surfaces in 3-dimensional homogeneous spaces, which include $\mathbb{S}^2 \times \mathbb{R}$, $\mathbb{H}^2 \times \mathbb{R}$, $\widetilde{\mathrm{SL}}(2,\mathbb{R})$ and Nil^3. Parallel and totally geodesic surfaces in these spaces were classified in [**BDI02**], while the classification in all remaining three-dimensional homogeneous spaces, including Sol^3, can be found in [**IVdV07**] and [**IVdV08**]. Moreover, totally geodesic and parallel submanifolds of the space forms were covered in [**Law69**], [**Tak81**], and [**BR83**].

In this article, however, the main focus is on the 4-dimensional Thurston geometries, which were classified in the PhD thesis of Filipkiewicz [**Fil83**] (see Table 2). Although no geometrization is known in dimension 4, the study of 4-dimensional

2020 *Mathematics Subject Classification.* Primary 53C42.
The author was supported by Methusalem grant METH/21/03-long term structural funding of the Flemish Government.

TABLE 1. Based on [**Mai98**]. Three-dimensional Thurston geometries ordered by their stabilizer of the *connected component of the identity* of the group G that acts on them.

Thurston geometry	Stabilizer	dimension of G
$\mathbb{R}^3$, $\mathbb{S}^3$, $\mathbb{H}^3$	SO(3)	6
$\mathbb{S}^2 \times \mathbb{R}$, $\mathbb{H}^2 \times \mathbb{R}$, $\widetilde{\mathrm{SL}}(2,\mathbb{R})$, Nil^3	SO(2)	4
Sol^3	$\{1\}$	3

TABLE 2. Based on [**Wal86**]. Four-dimensional Thurston geometries ordered by their stabilizer of the *connected component of the identity* of the group G that acts on them. In $\mathrm{Sol}^4_{m,n}$, m and n are positive integers and when $m = n$, $\mathrm{Sol}^4_{m,n} \cong \mathrm{Sol}^3 \times \mathbb{R}$.

Thurston geometry	Stabilizer	dimension of G
$\mathbb{R}^4$, $\mathbb{S}^4$, $\mathbb{H}^4$	SO(4)	10
$\mathbb{C}\mathrm{P}^2$, $\mathbb{C}\mathrm{H}^2$	U(2)	8
$\mathbb{S}^3 \times \mathbb{R}$, $\mathbb{H}^3 \times \mathbb{R}$	SO(3)	7
$\mathbb{S}^2 \times \mathbb{R}^2$, $\mathbb{S}^2 \times \mathbb{S}^2$ $\mathbb{H}^2 \times \mathbb{S}^2$, $\mathbb{H}^2 \times \mathbb{R}^2$, $\mathbb{H}^2 \times \mathbb{H}^2$	SO(2) $\times$ SO(2)	6
Sol^4_0, F^4, $\widetilde{\mathrm{SL}}(2,\mathbb{R}) \times \mathbb{R}$, $\mathrm{Nil}^3 \times \mathbb{R}$	SO(2)	5
Nil^4, Sol^4_1, $\mathrm{Sol}^4_{m,n}$	$\{1\}$	4

geometries and their submanifolds is interesting on its own from a Riemannian geometric point of view. It provides essential insight in the structure of these spaces. In many of the 4-dimensional geometries—more specifically, the real and complex space forms and products of lower-dimensional geometries, ample research has been done regarding their isometrically immersed submanifolds. Hence in this article we focus on other 4-dimensional geometries, Sol^4_0, Sol^4_1, $\mathrm{Sol}^4_{m,n}$, and Nil^4. Note that there is a remaining Thurston geometry, called F^4, which does not appear in our present study, as this space requires different techniques than the ones mentioned earlier and hence does not fit in the scope of this article.

An important remark is that the references mentioned above study Thurston geometries but, more often than not, they are not approached as such. By this we mean that there is no systematic study preceding the study of submanifolds in these spaces regarding the family of Riemannian metrics which is invariant under the action of the corresponding group. The aim of the present article is to provide such a framework for the aforementioned subset of 4-dimensional Thurston spaces, mentioned in Table 3 with the associated groups. The results we present are not deep, but they are essential for the study of submanifolds.

TABLE 3. Based on **[Hil02]** and **[AHW85]**. A selection of 4-dimensional Thurston geometries with 4- or 5-dimensional isometry group. In the second column we provide the group G that acts on the manifolds in the first column. The manifolds together with the group form Thurston geometries. The group $\mathrm{SL}^{\pm}(2,\mathbb{R})$ consists of matrices of determinant $+1$ or -1.

Thurston geometry	G
Sol_0^4	$\mathrm{Sol}_0^4 \rtimes (\mathrm{O}(2) \times \mathbb{Z}/2\mathbb{Z})$
Sol_1^4	$\mathrm{Sol}_1^4 \rtimes D_4$
$\mathrm{Sol}_{m,n}^4 (m \neq n)$	$\mathrm{Sol}_{m,n}^4 \rtimes (\mathbb{Z}/2\mathbb{Z})^3$
Nil^4	$\mathrm{Nil}^4 \rtimes (\mathbb{Z}/2\mathbb{Z})^2$
F^4	$\mathbb{R}^2 \rtimes \mathrm{SL}^{\pm}(2,\mathbb{R})$

2. Thurston geometries

The origin of Thurston geometries can be traced back to a very fundamental question: "How does the topology of a surface relate to its geometry and vice versa?" In this context, geometry refers to a Riemannian structure.

An example of the interplay between geometry and topology is the famous theorem of Gauss and Bonnet. Namely, for a surface S we have $\int_S K\, dA = 2\pi\chi(S)$, relating the total Gaussian curvature on the left-hand side to a multiple of the Euler characteristic $\chi(S)$, a purely topological quantity, on the right-hand side. This result shows that the topology of a surface prescribes its curvature and vice versa. Another theorem that links the topology of a manifold to its geometry is the theorem of Bonnet and Myers, which states that a Riemannian manifold with sectional curvature (a generalization of the Gaussian curvature of a surface) bounded from below by a positive constant must be compact and have finite first fundamental group, which again highlights the connection between curvature and topology.

The theorems mentioned above are very strong results, however, the most beautiful interplay between geometry and topology is illustrated by the theorem of Klein and Poincaré: any compact orientable surface admits a Riemannian metric with constant Gaussian curvature k. Moreover, if we denote by g the genus of S, then $k > 0$ if and only if $g = 0$, $k = 0$ if and only if $g = 1$ and $k < 0$ if and only if $g > 2$. This is often referred to as the uniformization theorem of compact orientable surfaces. A nice corollary of this statement is that any such surface admits a Riemannian universal covering by $\mathbb{R}^2$ if $g = 1$, by $\mathbb{S}^2$ if $g = 0$ or by $\mathbb{H}^2$ if $g > 2$. In this sense we say that the Riemannian manifolds $\mathbb{R}^2$, $\mathbb{S}^2$ and $\mathbb{H}^2$ are models for surfaces. Upon encountering this theorem, one is immediately inclined to wonder if a similar result holds for 3-manifolds. However, the geometry of 3-manifolds is much richer and more complicated than that of surfaces, since we easily see that not every compact orientable 3-manifold can be endowed with a metric of constant sectional curvature, for example, one can easily show that $\mathbb{S}^2 \times \mathbb{S}^1$ cannot be endowed with a constant curvature metric. Therefore, the list of 3-dimensional model geometries includes more spaces than $\mathbb{R}^3$, $\mathbb{S}^3$, and $\mathbb{H}^3$. In an attempt to state an analogue of the 2-dimensional geometrization for 3-manifolds, Thurston pinpointed the special properties of the spaces $\mathbb{R}^2$, $\mathbb{S}^2$, and $\mathbb{H}^2$ and the definition of a Thurston geometry was born. Following [**TL97**] and [**Bon02**], a Thurston geometry is a pair (X, G) of a smooth manifold X and a Lie group $G \subset \mathrm{Diff}(X)$ satisfying the following properties.

(a) X is connected and simply connected.
(b) G acts transitively on X with compact stabilizers. Note that this implies the existence of a G-invariant Riemannian metric on X.
(c) G is maximal, in the sense that there does not exist a larger Lie group containing G that also acts transitively on X with compact stabilizers.
(d) There exists a finite volume manifold which is modelled by the pair (X, G), meaning that there exists a finite volume Riemannian manifold (M, g) that admits $(X, \tilde{g})$ as a Riemannian universal cover (here $\tilde{g}$ is any G-invariant Riemannian metric on X).

REMARK 2.1. The definition of Thurston in [**TL97**] requires the existence of a *compact* manifold modelled by the pair (X, G), while Filipkiewicz [**Fil83**] relaxes this to *finite volume*. For 3-dimensional geometries, changing the *finite volume* condition to compactness does not alter the spaces satisfying the definition of a Thurston geometry, as all 3-dimensional Thurston geometries are models for compact manifolds. However, in the 4-dimensional case, making this change does alter the list, in the sense that the space F^4 only models finite volume manifolds. This is the only 4-dimensional geometry that does not model any compact manifold.

There are exactly eight pairs (X, G) with $\dim X = 3$ that satisfy the definition above, as Thurston shows in [**TL97**]. These comprise the eight 3-dimensional Thurston geometries, listed in Table 1. They are an essential part of the 3-dimensional analogue of the uniformization of surfaces, which is called the Thurston geometrization conjecture. The proof is in large part due to Thurston and was finished in 2003 by Perelman, making it the first Millennium Prize Problem to be solved. Indeed, the Poincaré conjecture is a consequence of the Thurston geometrization conjecture and given that the Poincaré conjecture had been an open problem for a hundred years, the proof of Perelman was an enormous breakthrough in geometry. Furthermore, the Thurston geometrization conjecture gives a deep understanding of the interplay between the topology and geometry of 3-manifolds.

Four-manifolds are not as well understood as 3-manifolds, which is reflected in the lack of a geometrization-type theorem for 4-manifolds. However, in an attempt to understand these, we still study 4-dimensional Thurston geometries. As mentioned earlier, they are pinpointed by Filipkiewicz [**Fil83**], and can be found in Table 2.

3. A Riemannian description of some Thurston geometries

Since Thurston geometries are Riemannian homogeneous spaces enjoying many symmetries, they form an interesting class of manifolds in which to study isometrically immersed submanifolds. The first step in this process is to determine all Riemannian metrics which are compatible with the group structure, i.e., for a pair (X, G) we have to determine the family of G-invariant metrics. Only after we have accomplished this, we can study submanifolds of a Thurston geometry. Up until today, no systematic study of G-invariant metrics has been performed in the non-product Thurston 4-spaces with 4– or 5-dimensional isometry group (see the last two rows of Table 2). These are the spaces Sol_0^4, F^4, $\mathrm{Sol}_{m,n}^4$, Sol_1^4, and Nil^4, which can be found in Table 3 along with the corresponding group. In this section we treat each geometry separately (except F^4). For the definition of the underlying manifolds, as well as the associated groups, we refer to [**Hil02**], [**AHW85**] and [**Fil83**] for all spaces except Sol_0^4, for which we refer to previous work of the author [**DIVdV24**].

Sol_0^4. The manifold associated to this Thurston geometry is a solvable Lie group containing the following matrices:

$$\begin{pmatrix} e^t & 0 & 0 & x \\ 0 & e^t & 0 & y \\ 0 & 0 & e^{-2t} & z \\ 0 & 0 & 0 & 1 \end{pmatrix}$$

where $t, x, y, z \in \mathbb{R}$. The Lie algebra is spanned by the basis $\{e_1, e_2, e_3, e_4\}$ satisfying the following commutation relations:

$$[e_1, e_2] = e_2, \quad [e_1, e_3] = e_3, \quad [e_1, e_4] = -2e_4,$$
$$[e_2, e_3] = [e_2, e_4] = [e_3, e_4] = 0.$$

The left invariant vector fields determined by e_1, e_2, e_3, and e_4 are

$$E_1 = \frac{\partial}{\partial t}, \quad E_2 = e^t \frac{\partial}{\partial x}, \quad E_3 = e^t \frac{\partial}{\partial y}, \quad E_4 = e^{-2t} \frac{\partial}{\partial z}.$$

The group associated with this Thurston geometry is $G = \mathrm{Sol}_0^4 \rtimes (\mathrm{O}(2) \times \mathbb{Z}/2\mathbb{Z})$ where Sol_0^4 acts by left translation, $\mathrm{O}(2)$ acts in the xy-plane, and $\mathbb{Z}/2\mathbb{Z}$ acts by reflecting the z-coordinate.

To determine all G-invariant Riemannian metrics on the manifold, we note that $\mathrm{Sol}_0^4 = G/K$ is a *reductive* homogeneous space, where K denotes the stabilizer. This means that there exists a subspace $\mathfrak{p}$ of the Lie algebra $\mathfrak{g}$ of G such that $\mathfrak{g} = \mathfrak{k} \oplus \mathfrak{p}$ with the property that $\mathfrak{p}$ is invariant under the adjoint action of K. For a reductive homogeneous space there is a bijection between Ad_K-invariant inner products on $\mathfrak{p}$ and G-invariant metrics on G/K. Using this correspondence we find that, up to homothety, there is a unique G-invariant Riemannian metric on Sol_0^4 given by

$$dt^2 + e^{-2t}(dx^2 + dy^2) + e^{4t} dz^2.$$

Note that the frame $\{E_1, E_2, E_3, E_4\}$ is orthonormal with respect to this metric.

As mentioned by Wall [**AHW85**], Sol_0^4 admits a complex structure compatible with the group action of the identity component. In fact, it is also compatible with the metric above, and it is unique up to isomorphism of the group G. The structure is moreover integrable, making Sol_0^4 a complex manifold (see [**DIVdV24**]). Note however that it is not Kähler, but one can show that the metric given above is globally conformal to a Kähler metric (the conformal factor is e^{2t}).

$\mathbf{Sol}^4_{m,n}$. The manifold of this Thurston geometry is also a solvable Lie group consisting of the matrices

$$\begin{pmatrix} e^{at} & 0 & 0 & x \\ 0 & e^{bt} & 0 & y \\ 0 & 0 & e^{ct} & z \\ 0 & 0 & 0 & 1 \end{pmatrix}$$

where $t, x, y, z \in \mathbb{R}$ and $a < b < c$ such that e^{at}, e^{bt} and e^{ct} are the roots of the polynomial $x^3 - mx^2 + nx - 1$. Here, m and n are positive integers such that the roots are all *real* and *distinct*. The associated Lie algebra is spanned by $\{e_1, e_2, e_3, e_4\}$ satisfying the commutation relations

$$\begin{aligned} &[e_1, e_2] = ae_2, \quad [e_1, e_3] = be_3, \quad [e_1, e_4] = ce_4, \\ &[e_2, e_3] = [e_2, e_4] = [e_3, e_4] = 0. \end{aligned}$$

The left invariant vector fields associated to e_1, e_2, e_3 and e_4 are

$$E_1 = \frac{\partial}{\partial t}, \quad E_2 = e^{at}\frac{\partial}{\partial x}, \quad E_3 = e^{bt}\frac{\partial}{\partial y}, \quad E_4 = e^{ct}\frac{\partial}{\partial z}.$$

Note that if $m = n$ one of the roots equals 1 and $\mathrm{Sol}^4_{m,n}$ can be associated with $\mathrm{Sol}^3 \times \mathbb{R}$. We will not consider this case here, since we are only concerned with non-product spaces. Hence from now on we will assume that m and n are different. In that case, the associated group is $\mathrm{Sol}^4_{m,n} \rtimes (\mathbb{Z}/2\mathbb{Z})^3$. Each $\mathbb{Z}/2\mathbb{Z}$ factor is generated by reflecting either the x-, y- or z-coordinate. Another important remark to make is that if we allow m and n to be such that two of the roots are equal, we can identify $\mathrm{Sol}^4_{m,n}$ with Sol^4_0. In this case there is an additional symmetry in the x- and y-coordinates, corresponding to the $\mathrm{O}(2)$ factor in the group associated to Sol^4_0.

Also $\mathrm{Sol}^4_{m,n}$ is a *reductive* homogeneous space and using the same technique as for Sol^4_0, we find that with $m \neq n$ there is a unique metric up to homothety which is invariant under the action of the associated group, given by

$$dt^2 + e^{-2at}dx^2 + e^{-2bt}dy^2 + e^{-2ct}dz^2.$$

Note that the frame $\{E_1, E_2, E_3, E_4\}$ is orthonormal with respect to this metric.

In [**AHW85**] it is shown that $\mathrm{Sol}^4_{m,n}$ does not admit any compatible complex structures.

$\mathbf{Sol}^4_1$. The underlying manifold of this Thurston geometry is again a solvable Lie group and it consists of matrices of the form

$$M(t, x, y, z) = \begin{pmatrix} 1 & x & z \\ 0 & t & y \\ 0 & 0 & 1 \end{pmatrix}$$

where $t, x, y, z \in \mathbb{R}$ and $t > 0$. The Lie algebra is spanned by the basis $\{e_1, e_2, e_3, e_4\}$ satisfying the following commutation relations:

$$\begin{aligned} &[e_1, e_2] = -e_2, \quad [e_1, e_3] = e_3, \quad [e_2, e_3] = e_4, \\ &[e_1, e_4] = [e_2, e_4] = [e_3, e_4] = 0. \end{aligned}$$

The left invariant vector fields determined by e_1, e_2, e_3 and e_4 are

$$E_1 = t\frac{\partial}{\partial t} + x\frac{\partial}{\partial x}, \quad E_2 = \frac{\partial}{\partial x}, \quad E_3 = t\frac{\partial}{\partial y} + x\frac{\partial}{\partial z}, \quad E_4 = \frac{\partial}{\partial z}.$$

The group that acts on Sol_1^4 making it a Thurston geometry is $\mathrm{Sol}_1^4 \rtimes D_4$ (where D_4 denotes the dihedral group of 8 elements). Sol_1^4 acts by left translations while the action of D_4 is given by the action of its generators:

$$s : M(t,x,y,z) \mapsto M(t,x,-y,-z), \quad r : M(t,x,y,z) \mapsto M(1/t, -y/t, x/t, z - xy/t).$$

Again using that Sol_1^4 is a reductive homogeneous space, we find, up to homothety, a two-parameter family of G-invariant metrics given by

$$\frac{1}{t^2}\Big((x^2+\tau_1)dt^2 + t^2dx^2 + (1+\tau_2x^2)dy^2 + \tau_2t^2dz^2 - 2tx(dt\,dx + \tau_2dy\,dz)\Big),$$

where $\tau_1, \tau_2 \in \mathbb{R}^+$. Note that the frame

$$\left\{\frac{E_1}{\sqrt{\tau_1}}, E_2, E_3, \frac{E_4}{\sqrt{\tau_2}}\right\}$$

is orthonormal with respect to this metric.

In [**AHW85**] it is shown that Sol_1^4 admits a compatible complex structure, and there are two isomorphism classes. It is not Kähler.

Nil^4. The manifold associated to this Thurston geometry is a nilpotent Lie group and is most easily described as a semi-direct product $\mathbb{R} \ltimes_\theta \mathbb{R}^3$ where for each $t \in \mathbb{R}$ we have the following automorphism of $\mathbb{R}^3$:

$$\theta(t) = \begin{pmatrix} 1 & t & \frac{t^2}{2} \\ 0 & 1 & t \\ 0 & 0 & 1 \end{pmatrix} \in \mathrm{Nil}^3.$$

The Lie algebra is spanned by the basis $\{e_1, e_2, e_3, e_4\}$ with the following brackets:

$$[e_1,e_2] = 0, \quad [e_1,e_3] = e_2, \quad [e_2,e_3] = e_3,$$
$$[e_2,e_3] = [e_2,e_4] = [e_3,e_4] = 0.$$

The left invariant vector fields determined by e_1, e_2, e_3 and e_4 are

$$E_1 = \frac{\partial}{\partial t}, \quad E_2 = \frac{\partial}{\partial x}, \quad E_3 = t\frac{\partial}{\partial x} + \frac{\partial}{\partial y}, \quad E_4 = \frac{t^2}{2}\frac{\partial}{\partial x} + t\frac{\partial}{\partial y} + \frac{\partial}{\partial z},$$

where x, y, z are the coordinates on $\mathbb{R}^3$ and t is the coordinate on $\mathbb{R}$ in $\mathbb{R} \ltimes \mathbb{R}^3$.

The group associated to Nil^4 is $\mathrm{Nil}^4 \rtimes (\mathbb{Z}/2\mathbb{Z})^2$ where Nil^4 acts on itself by left translations and the copies of $\mathbb{Z}/2\mathbb{Z}$ are generated by the maps

$$s_1 : (t,x,y,z) \mapsto (t,-x,-y,-z), \quad s_2 : (t,x,y,z) \mapsto (-t,x,-y,z).$$

The homogeneous space Nil^4 is also reductive, and thus we find a 4-parameter family of G-invariant Riemannian metrics,

$$\tau_1dt^2 + dx^2 + (t^2+\tau_2)dy^2 + \left(\frac{t^4}{4} + t^2(\alpha+\tau_2) + \tau_3\right)dz^2$$
$$- 2t\,dx\,dy + (t^2+2\alpha)dx\,dz - (t^3 + 2t(\alpha+\tau_2))dy\,dz,$$

where $\tau_1, \tau_2, \tau_3 \in \mathbb{R}^+$ and $\alpha \in \mathbb{R} \setminus \{\pm\sqrt{\tau_3}\}$. Note that the frame

$$\left\{\frac{E_1}{\sqrt{\tau_1}}, E_2, \frac{E_3}{\sqrt{\tau_2}}, \frac{E_4 - \alpha E_2}{\sqrt{\tau_3 - \alpha^2}}\right\}$$

is orthonormal with respect to these metrics.

In [**AHW85**] it is shown that Nil^4 does not admit any compatible complex structures.

4. Future research on 4-dimensional Thurston geometries

In this section we collect some ideas for future research in 4-dimensional Thurston geometries, mainly focussed on isometrically immersed submanifolds of these spaces.

(a) Determine if the families of G-invariant metrics for Nil^4 and Sol_1^4 can be reduced, i.e., if it is possible to eliminate any of the parameters in these families.
(b) In this article the Thurston geometry F^4 was not described and no family of G-invariant metrics was exhibited. This is an obvious course for future research.
(c) Study parallel and totally geodesic submanifolds of the Thurston geometries Sol_1^4, $\mathrm{Sol}_{m,n}^4$ and F^4. Note that parallel and totally geodesic hypersurfaces of Sol_0^4 are classified in [**DIVdV24**]. In Nil^4 these types of hypersurfaces are classified in [**DHMCB23**] for just one metric in the family of G-invariant metrics we exhibited in this article.
(d) Formulate the fundamental equations for hypersurfaces in the 4-dimensional Thurston geometries with 4- or 5-dimensional isometry group. It might be an essential step to first develop a unified description of these spaces, in the same spirit as the $\mathbb{E}(\kappa,\tau)$ spaces that Daniel used in [**Dan07**].

References

[AHW85] C. T. C. Wall, *Geometries and geometric structures in real dimension* 4 *and complex dimension* 2, Geometry and topology (College Park, Md., 1983/84), Lecture Notes in Math., vol. 1167, Springer, Berlin, 1985, pp. 268–292, DOI 10.1007/BFb0075230. MR827276

[BDI02] M. Belkhelfa, F. Dillen, and J.-I. Inoguchi, *Surfaces with parallel second fundamental form in Bianchi-Cartan-Vranceanu spaces*, PDEs, submanifolds and affine differential geometry (Warsaw, 2000), Banach Center Publ., vol. 57, Polish Acad. Sci. Inst. Math., Warsaw, 2002, pp. 67–87, DOI 10.4064/bc57-0-5. MR1972463

[Bon02] F. Bonahon, *Geometric structures on 3-manifolds*, Handbook of geometric topology, North-Holland, Amsterdam, 2002, pp. 93–164. MR1886669

[BR83] E. Backes and H. Reckziegel, *On symmetric submanifolds of spaces of constant curvature*, Math. Ann. **263** (1983), no. 4, 419–433, DOI 10.1007/BF01457052. MR707240

[Dan07] B. Daniel, *Isometric immersions into 3-dimensional homogeneous manifolds*, Comment. Math. Helv. **82** (2007), no. 1, 87–131, DOI 10.4171/CMH/86. MR2296059

[DHMCB23] N. Djellali, A. Hasni, A. M. Cherif, and M. Belkhelfa, *Classification of Codazzi and note on minimal hypersurfaces in* Nil^4, Int. Electron. J. Geom. **16** (2023), no. 2, 707–714, DOI 10.36890/iejg.1256112. MR4664761

[DIVdV24] M. D'haene, J.-i. Inoguchi, and J. Van der Veken, *Parallel and totally umbilical hypersurfaces of the four-dimensional Thurston geometry* Sol_0^4, Math. Nachr. **297** (2024), no. 5, 1879–1891, DOI 10.1002/mana.202300372. MR4755741

[Fil83] R. Filipkiewicz, *Four-dimensional geometries*, Ph.D. Thesis, University of Warwick, 1983.

[Hil02] J. A. Hillman, *Four-manifolds, geometries and knots*, Geometry & Topology Monographs, vol. 5, Geometry & Topology Publications, Coventry, 2002. MR1943724

[IVdV07] J.-i. Inoguchi and J. Van der Veken, *Parallel surfaces in the motion groups* $E(1,1)$ *and* $E(2)$, Bull. Belg. Math. Soc. Simon Stevin **14** (2007), no. 2, 321–332. MR2341567

[IVdV08] J.-i. Inoguchi and J. Van der Veken, *A complete classification of parallel surfaces in three-dimensional homogeneous spaces*, Geom. Dedicata **131** (2008), 159–172, DOI 10.1007/s10711-007-9222-0. MR2369197

[Law69] H. B. Lawson Jr., *Local rigidity theorems for minimal hypersurfaces*, Ann. of Math. (2) **89** (1969), 187–197, DOI 10.2307/1970816. MR238229

[Mai98] S. Maier, *Conformal flatness and self-duality of Thurston-geometries*, Proc. Amer. Math. Soc. **126** (1998), no. 4, 1165–1172, DOI 10.1090/S0002-9939-98-04686-3. MR1476377

[Tak81] M. Takeuchi, *Parallel submanifolds of space forms*, Manifolds and Lie groups (Notre Dame, Ind., 1980), Progr. Math., vol. 14, Birkhäuser, Boston, MA, 1981, pp. 429–447. MR642871

[TL97] W. P. Thurston, *Three-dimensional geometry and topology. Vol. 1*, Princeton Mathematical Series, vol. 35, Princeton University Press, Princeton, NJ, 1997. Edited by Silvio Levy. MR1435975

[Wal86] C. T. C. Wall, *Geometric structures on compact complex analytic surfaces*, Topology **25** (1986), no. 2, 119–153, DOI 10.1016/0040-9383(86)90035-2. MR837617

Department of Mathematics, KU Leuven, Celestijnenlaan 200B–Box 2400, 3001 Leuven, Belgium

Email address: marie.dhaene@kuleuven.be

Contemporary Mathematics
Volume **821**, 2025
https://doi.org/10.1090/conm/821/16437

Recent results in Ricci and Yamabe solitons in contact geometry

Amalendu Ghosh, Ramesh Sharma, and Rahul Poddar

Dedicated to the 80th birthday of Professor Bang-Yen Chen

ABSTRACT. Generalizing the corresponding result for the Sasakian case, we prove that, if a Ricci solton is a K-contact manifold whose Ricci operator commutes with the fundamental collineation φ, then it is either Einstein, or $\mathcal{D}$-homothetically fixed η-Einstein, expanding, its Ricci tensor is Killing and furthermore, the potential vector field leaves the fundamental collineation φ invariant and is an infinitesimal contact transformation. Finally, we provide a survey of the results known so far, on Yamabe solitons as Sasakian and K-contact manifolds.

1. Introduction

Ricci soliton is a self-similar solution of the Hamilton's Ricci flow [**16**]:

$$\frac{\partial}{\partial t}g(t) = -2Ric(t)$$

where $Ric(t)$ is the Ricci tensor of the time-dependent Riemannian metric $g(t)$.

DEFINITION 1.1. A Riemannian metric g on a smooth manifold M is said to be a Ricci soliton if there exists a smooth vector field V called the potential vector field, satisfying

$$\frac{1}{2}\pounds_V g + Ric = \lambda g, \tag{1.1}$$

where $\pounds_V$ denotes the Lie derivative along the vector field V and λ is a real constant.

Actually, a Ricci soliton is a generalized fixed point of the Hamilton's Ricci flow, viewed as a dynamical system on the space of Riemannian metrics modulo diffeomorphisms and scalings. If the potential vector field V of the Ricci soliton (M, g, V, λ) of dimension > 2 is Killing, then g is Einstein. Thus, a Ricci soliton can be considered as a generalization of an Einstein metric, and is said to be *expanding* if $\lambda < 0$, *steady* if $\lambda = 0$, and *shrinking* if $\lambda > 0$. For details, we refer to

---

2020 *Mathematics Subject Classification.* Primary 53C25, Secondary 53E50, 53C21.

Key words and phrases. Sasakian manifold, K-contact manifold, Ricci soliton, $\mathcal{D}$-homothetically fixed η-Einstein, Killing tensor, Yamabe soliton.

R. Sharma was partially supported by a Sabbatical Leave granted by the University of New Haven. R. Poddar was funded by the University Grants Commission (UGC), India, in the form of a Senior Research Fellowship.

Chow et al. [**7**]. A Ricci soliton is called a gradient Ricci soliton if V is the gradient of a scalar function f on M. Gradient Ricci solitons are important in understanding the Hamilton's Ricci flow, and they arise often as singularity models of the Ricci flow.

An odd dimensional analogue of the Kaehler geometry is the Sasakian geometry. Sasakian geometry has been extensively studied since its recently perceived relevance in string theory. Sasakian Einstein metrics have received a lot of attention in physics, for example, p-brane solutions in superstring theory, Maldacena conjecture (AdS/CFS duality) [**21**]. For details, we refer to Boyer, Galicki, and Matzeu [**4**].

Ricci soliton was studied in contact geometry by Sharma in [**28**] who showed that if a K-contact (in particular, Sasakian) metric is a gradient Ricci soliton, then it is Einstein. For non-gradient Ricci soliton, Sharma and Ghosh [**32**] proved that if a 3-dimensional Sasakian metric represents a non-trivial (i.e., non-Einstein) Ricci soliton, then it is homothetic to the standard Sasakian metric on nil^3. Subsequently, Ghosh and Sharma [**13**] showed the existence of an expanding non-trivial Ricci soliton on a Sasakian manifold, and proved the following result "If a Sasakian manifold $M(\eta, \xi, g, \varphi)$ is a non-trivial Ricci soliton, then (i) M is null η-Einstein (i.e., $\mathcal{D}$-homothetically fixed and transverse Calabi-Yau), (ii) the Ricci soliton is expanding, and (iii) the potential vector field V leaves the structure tensor φ invariant, and is an infinitesimal contact transformation".

We note that a Sasakian manifold is K-contact. Also, the Ricci operator Q and the fundamental collineation φ commute (i.e., $Q\varphi = \varphi Q$) on a Sasakian manifold, but this need not hold for a K-contact manifold. Intrigued by this fact, we obtain the following generalization of the above mentioned theorem on a K-contact manifold satisfying $Q\varphi = \varphi Q$.

THEOREM 1.2. *If a Ricci soliton (M, g) is a K-contact manifold whose Ricci operator commutes with its fundamental collineation φ, then it is either Einstein, or $\mathcal{D}$-homothetically fixed η-Einstein. In the second case, (i) it is expanding, (ii) its Ricci tensor is Killing, (iii) the potential vector field V leaves the structure tensor φ invariant, and (iv) V is an infinitesimal contact transformation.*

Combining the following result from [**30**]: "If a complete connected η-Einstein K-contact manifold M of dimension > 3 admits a non-isometric conformal vector field, then M is isometric to a unit sphere", with Theorem 1.2, and taking into account the result [**27**]: "A homothetic vector field on a K-contact manifold is Killing", we obtain the following characterization of an odd-dimensional sphere.

COROLLARY 1.3. *If a Ricci soliton (M, g) is a complete connected K-contact manifold of dimension > 3, whose Ricci operator commutes with its fundamental collineation φ, and admits a non-isometric conformal vector field, then M is isometric to a unit sphere.*

Next, we recall that the second order Killing tensor occurs in general theory of relativity (Wald [**36**]). A Killing tensor is defined as a symmetric tensor T on a semi-Riemannian manifold (M, g) satisfying

$$(\nabla_X T)(Y, Z) + (\nabla_Y T)(Z, X) + (\nabla_Z T)(X, Y) = 0 \tag{1.2}$$

for any arbitrary vector fields X, Y, Z on M. The quadratic first integrals of the geodesic equation are associated with second rank Killing tensors. Indeed, if T is a second order Killing tensor, then $T(V,V)$ is constant along an affine parametrized geodesic with tangent vector V (Eisenhart [**12**]). Trivial examples of Killing tensors are the metric tensor and symmetrized tensor products of two Killing vectors. Second order Killing tensors have also appeared in the stability analysis of generalized black holes in $D = 11$ supergravity (Gibbons, Hartnoll, and Pope [**14**]). For details, we refer to Heil, Moroianu, and Semmelmann [**18**], and Jelonek [**19**]. Manifolds with Ricci tensor as a Killing tensor were studied by Gray [**15**] who provided nontrivial examples such as $O(4)/O(2)$ with a bi-invariant metric and the 3-sphere with many homogeneous metrics.

Replacing the commutativity $Q\varphi = \varphi Q$ condition, with Ricci tensor being Killing tensor, we obtain the following corollary to the Proof of Theorem 1.2.

COROLLARY 1.4. *If a Ricci soliton is a K-contact manifold such that its Ricci tensor is Killing, then all the conclusions of the Proof of Theorem* 1.2 *hold.*

Its proof is based on the following lemma which will be proved after the proof of Lemma 3.1 in Section 3.

LEMMA 1.5. *If the Ricci tensor of a K-contact manifold is Killing, then its Ricci operator commutes with its fundamental collineation φ.*

2. A brief review of contact geometry

A $(2n+1)$-dimensional smooth manifold is said to be a contact manifold if it admits a global 1-form η such that $\eta\wedge(d\eta)^n$ is nonvanishing everywhere on M. This 1-form is called a contact 1-form. For such a 1-form η one can find a unit vector field ξ, called the Reeb vector field, such that $\eta(\xi) = 1$ and $d\eta(\xi, .) = 0$. A contact structure is regular, if each point has a neighborhood such that any integral curve of ξ passing through the neighborhood, passes through it only once. Polarizing $d\eta$ on the contact subbundle $\eta = 0$ [**1**], we obtain a Riemannian metric g and a $(1,1)$-tensor field φ (called the fundamental collineation) such that

$$d\eta(X,Y) = g(X,\varphi Y), \quad \eta(X) = g(X,\xi), \quad \varphi^2 = -I + \eta\otimes\xi. \tag{2.1}$$

The quadruple (φ, η, ξ, g) defines a contact metric structure on M. The equations in (2.1) imply

$$\varphi\xi = 0, \quad \eta\circ\varphi = 0, \quad g(\varphi X, \varphi Y) = g(X,Y) - \eta(X)\eta(Y). \tag{2.2}$$

If ξ is Killing, then the contact metric manifold M is said to be K-contact, for which the following formulas hold:

$$\nabla_X\xi = -\varphi X, \tag{2.3}$$

$$Q\xi = 2n\xi, \tag{2.4}$$

$$R(X,\xi)\xi = X - \eta(X)\xi, \tag{2.5}$$

$$R(\xi, X)Y = (\nabla_X\varphi)Y, \tag{2.6}$$

where R denotes the Riemann curvature tensor of g. A smooth contact submanifold of a K-contact manifold (M,η) invariant under the Reeb flow of ξ is called a K-contact submanifold of (M,η), and have been studied by Nazawa [**23**]. K-contact manifolds as hypersurfaces of a real space form were classified and characterized by Takahashi and Tanno [**33**], and as submanifolds of a real space form (in particular, a sphere) were classified and characterized by Chen and Mihai [**6**]. A contact metric structure on M is said to be Sasakian if the almost Kaehler structure induced on the cone manifold $(M \times R^{+}, r^2 g + dr^2)$ over M, is Kaehler. For a Sasakian manifold,

$$(\nabla_X \varphi)Y = g(X,Y)\xi - \eta(Y)X. \tag{2.7}$$

Sasakian manifolds are K-contact and 3-dimensional K-contact manifolds are Sasakian. If a contact metric manifold M is Sasakian, then $Q\varphi = \varphi Q$, however the converse need not be true. Let us recall the following result of Davidov and Muskarov [**10**]: "There exist compact almost Kaehler manifolds whose Ricci operator commutes with the corresponding almost complex structure tensor J", and also the following result of Deng [**11**]: "A compact regular contact manifold with base B a symplectic manifold and usual structure tensors, satisfies $Q\varphi = \varphi Q$ if and only if the Ricci operator of B commutes with J." In view of these results and a well-known theorem of Boothby and Wang [**2**], Koufogiorgos [**20**] showed that there exist non-Sasakian K-contact manifolds with $Q\varphi = \varphi Q$.

A vector field V on a contact metric manifold M is said to be an infinitesimal contact transformation if $\pounds_V \eta = \sigma\eta$ for some smooth function σ on M (Tanno [**34**]).

A contact metric manifold M is said to be η-Einstein (a nice generalization of an Einstein metric) if

$$Ric(X,Y) = \alpha g(X,Y) + \beta\eta(X)\eta(Y) \tag{2.8}$$

for some smooth functions α and β on M. For a K-contact manifold of dimension > 3, α and β are constants. Under a $\mathcal{D}$-homothetic deformation (see Tanno [**35**]):

$$\bar{\eta} = a\eta, \quad \bar{\xi} = \frac{1}{a}\xi, \quad \bar{\varphi} = \varphi, \quad \bar{g} = ag + a(a-1)\eta \otimes \eta$$

for some positive constant a, a K-contact η-Einstein manifold transforms to another K-contact η-Einstein manifold in which $\bar{\alpha} = \frac{\alpha+2-2a}{a}$ and $\bar{\beta} = 2n - \bar{\alpha}$. An η-Einstein Sasakian manifold with $\alpha = -2$ and $\beta = 2(n+1)$ is known as a null η-Einstein (Boyer, Galicki and Matzeu [**4**]). *We highlight the fact that the particular value: $\alpha = -2$ remains fixed under a $\mathcal{D}$-homothetic deformation. So, an η-Einstein K-contact manifold with $\alpha = -2$ is said to be $\mathcal{D}$-homothetically fixed.* For details on contact metric manifolds we refer to the standard monograph [**1**] of Blair.

3. Proofs of the results

First we need the following lemma.

LEMMA 3.1. *On a K-contact manifold, we have $\nabla_\xi Q = Q\varphi - \varphi Q$.*

PROOF. As ξ is Killing on a K-contact manifold, it preserves the Ricci operator, i.e., $\pounds_\xi Q = 0$ which can be expressed as $[\xi, QX] = Q[\xi, X]$, i.e., $\nabla_\xi QX - \nabla_{QX}\xi = Q(\nabla_\xi X - \nabla_X \xi)$. Writing this as $(\nabla_\xi Q)X = \nabla_{QX}\xi - Q\nabla_X\xi$ and using formula (2.3) yields $(\nabla_\xi Q)X = Q\varphi X - \varphi QX$, completing the proof. □

PROOF THEOREM 1.2. Covariant differentiation of equation (1.1) gives

$$(\nabla_X \pounds_V g)(Y,Z) + 2(\nabla_X Ric)(Y,Z) = 0. \tag{3.1}$$

Use of this in the following commutation formula (see Yano [**37**, p. 23])

$$\begin{aligned}(\pounds_V \nabla_X g - \nabla_X \pounds_V g - \nabla_{[V,X]} g)(Y,Z) \\ = -g((\pounds_V \nabla)(X,Y), Z) - g((\pounds_V \nabla)(X,Z), Y),\end{aligned} \tag{3.2}$$

followed by cyclically permuting X, Y, Z twice and a straightforward combinatorial computation shows

$$g((\pounds_V\nabla)(X,Y),Z) = (\nabla_Z Ric)(X,Y) - (\nabla_X Ric)(Y,Z) - (\nabla_Y Ric)(X,Z). \tag{3.3}$$

Next, differentiating (2.4) and using (2.3) provides

$$(\nabla_X Q)\xi = Q\varphi X - 2n\varphi X. \tag{3.4}$$

Invoking Lemma 3.1 and the hypothesis $Q\varphi = \varphi Q$ shows that

$$\nabla_\xi Ric = 0. \tag{3.5}$$

Substituting ξ for Y in (3.3) and using equations (3.4) and (3.5) we find

$$(\pounds_V\nabla)(X,\xi) = -2Q\varphi X + 4n\varphi X. \tag{3.6}$$

Differentiating (3.6) and using formula (2.3) provides

$$\begin{aligned}(\nabla_Y\pounds_V\nabla)(X,\xi) - (\pounds_V\nabla)(X,\varphi Y)&\\ = -2(\nabla_Y Q)\varphi X - 2Q(\nabla_Y\varphi)X + 4n(\nabla_Y\varphi)X.&\end{aligned} \tag{3.7}$$

The use of (3.7) in the commutation formula [**37**]:

$$(\pounds_V R)(X,Y)Z = (\nabla_X\pounds_V\nabla)(Y,Z) - (\nabla_Y\pounds_V\nabla)(X,Z) \tag{3.8}$$

leads us to

$$\begin{aligned}(\pounds_V R)(X,Y)\xi = (\pounds_V\nabla)(Y,\varphi X) - 2(\nabla_X Q)\varphi Y - 2Q(\nabla_X\varphi)Y + 4n(\nabla_X\varphi)Y&\\ - (\pounds_V\nabla)(X,\varphi Y) + 2(\nabla_Y Q)\varphi X + 2Q(\nabla_Y\varphi)X - 4n(\nabla_Y\varphi)X.&\end{aligned} \tag{3.9}$$

Further, substituting ξ for Y and taking into account the equations (2.3), (3.5), (3.6) and $\nabla_\xi\varphi = 0$ (follows from (2.6)) we obtain the equation

$$(\pounds_V R)(X,\xi)\xi = 4QX - 8nX. \tag{3.10}$$

On the other hand, Lie differentiating formula (2.5) along V and using it in (3.10), we have

$$4QX - 8nX + R(X,\pounds_V\xi)\xi + R(X,\xi)\pounds_V\xi = -(\pounds_V\eta)(X)\xi - \eta(X)\pounds_V\xi. \tag{3.11}$$

Contracting (3.11) over X shows that

$$4r - 8n(2n+1) + 2Ric(\pounds_V\xi,\xi) = -(\pounds_V\eta)\xi - \eta(\pounds_V\xi) = 0. \tag{3.12}$$

Now, Lie differentiating $\eta(\xi) = g(\xi,\xi) = 1$ along V and using (1.1) and (2.4) gives $\eta(\pounds_V\xi) = 2n - \lambda$. Using these consequences in (3.12) we obtain

$$r = 2n^2 + 2n + n\lambda. \tag{3.13}$$

Hence r is a constant. The inner product of (3.11) with Y, the use of the Bianchi first identity on the third term of the left-hand side, along with (2.6), provides

$$\begin{aligned}4g(QX,Y) - 8ng(X,Y) + g((\nabla_{\pounds_V\xi}\varphi)X,Y) + 2g((\nabla_X\varphi)Y,\pounds_V\xi)&\\ = -\eta(Y)(\pounds_V\eta)(X) - \eta(X)g(\pounds_V\xi,Y).&\end{aligned} \tag{3.14}$$

Symmetrizing (3.14) with respect to X and Y, and noting the anti-self-adjointness of φ, we have

$$\begin{aligned}8g(QX,Y) - 16ng(X,Y) + 2g((\nabla_X\varphi)Y + (\nabla_Y\varphi)X,\pounds_V\xi)&\\ = -\eta(Y)(\pounds_V\eta)(X) - \eta(X)g(\pounds_V\xi,Y)&\\ - \eta(X)(\pounds_V\eta)(Y) - \eta(Y)g(\pounds_V\xi,X).&\end{aligned} \tag{3.15}$$

Our next strategy is to recall the following formula ([**1**, p. 116]) for a contact metric manifold with $h = 0$ (as M is K-contact):

$$(\nabla_X \varphi)Y + (\nabla_{\varphi X}\varphi)\varphi Y = 2g(Y,X)\xi - \eta(Y)(X + \eta(X)\xi). \tag{3.16}$$

Replacing X with φX and Y with φY in (3.15), adding the resulting equation to (3.15), and then using (3.16), the hypothesis $Q\varphi = \varphi Q$ and the equation $\eta(\pounds_V \xi) = 2n - \lambda$ obtained earlier, we obtain

$$\begin{aligned} 16g(QX,Y) - 32ng(X,Y) + 8(2n-\lambda)g(X,Y) - 4(2n-\lambda)\eta(X)\eta(Y) \\ = -\eta(Y)(\pounds_V\eta)(X) + \eta(X)g(\pounds_V\xi, Y) \\ - \eta(X)(\pounds_V\eta)(Y) + \eta(Y)g(\pounds_V\xi, X). \end{aligned} \tag{3.17}$$

Lie differentiating $\eta(X) = g(X,\xi)$ along V and using (1.1) in conjunction with (2.4) we get $(\pounds_V\eta)X - g(\pounds_V\xi, X) = 2(\lambda - 2n)\eta(X)$. This equation enables us to reduce (3.17) to

$$Ric(X,Y) = \left(n + \frac{\lambda}{2}\right)g(X,Y) + \left(n - \frac{\lambda}{2}\right)\eta(X)\eta(Y). \tag{3.18}$$

At this point, we appeal to the following integrability formula for a Ricci soliton [**28**]

$$\pounds_V r = \Delta r + 2(|Q|^2 - \lambda r),$$

where $\Delta = div(grad)$. Since r is constant, we get $|Ric|^2 = \lambda r$. On the other hand, writing (3.18) as $QX = (n + \frac{\lambda}{2})X + (n - \frac{\lambda}{2})\eta(X)\xi$, operating it by Q, using the K-contact formula (2.4), and then contracting the resulting equation, we obtain $|Ric|^2 = |Q|^2 = (n + \frac{\lambda}{2})r + 2n(n - \frac{\lambda}{2})$. Comparing the two values of $|Ric|^2$ that have been obtained, we conclude that

$$(r + 2n)(2n - \lambda) = 0.$$

Hence, either $\lambda = 2n$, or $r = -2n$. If $\lambda = 2n$, then it follows from (3.18) that $Ric = 2ng$, i.e., M is Einstein, proving the first case of the theorem. If $r = -2n$, we have the second case and (3.13) shows that $\lambda = -2(n+2)$ and hence by virtue of (3.18) we obtain

$$Ric(X,Y) = -2g(X,Y) + 2(n+1)\eta(X)\eta(Y). \tag{3.19}$$

Hence M is a $\mathcal{D}$-homothetically fixed K-contact η-Einstein manifold. Substituting $r = -2n$ back in (3.13) shows that $\lambda = -2(n+2)$, and hence the Ricci soliton is expanding, proving part (i). By taking the covariant derivative of (3.19) along an arbitrary vector field Z and using (2.3), provides

$$(\nabla_Z Ric)(X,Y) = -2(n+1)\{\eta(X)g(Y,\varphi Z) + \eta(Y)g(X,\varphi Z)\}. \tag{3.20}$$

Its cyclic sum over X, Y, Z shows that the Ricci tensor is Killing, proving part (ii).

Next, we use equation (3.19) in (3.3) in order to get

$$(\pounds_V\nabla)(X,Y) = 4(n+1)\{\eta(X)\varphi Y + \eta(Y)\varphi X\}. \tag{3.21}$$

Using this, equation (2.3), and formula (3.8), we get

$$\begin{aligned} (\pounds_V R)(Z,X)Y = 4(n-1)\{2g(Z,\varphi X)\varphi Y + g(Y,\varphi X)\varphi Z - g(Y,\varphi Z)\varphi X \\ + \eta(X)(\nabla_Z\varphi)Y - \eta(Z)(\nabla_X\varphi)Y + \eta(Y)((\nabla_Z\varphi)X - (\nabla_X\varphi)Z)\}. \end{aligned}$$

Contracting this equation over Z and using the following formula (Olszak [**25**]) for a contact metric manifold

$$\sum_{i=1}^{2n+1} g((\nabla_{e_i}\varphi)Y, e_i) = -2n\eta(Y),$$

we have that

$$(\pounds_V Ric)(X,Y) = 8(n+1)\{g(X,Y) - (2n+1)\eta(X)\eta(Y)\}. \tag{3.22}$$

Further, (3.19) reduces the soliton equation (1.1) to the form

$$(\pounds_V g)(X,Y) = -4(n+1)\{g(X,Y) + \eta(X)\eta(Y)\}. \tag{3.23}$$

On the other hand, Lie-differentiating (3.19) along V and using equations (3.22) and (3.23) gives

$$\begin{aligned}(\pounds_V Ric)(X,Y) &= 8(n+1)\{g(X,Y) + \eta(X)\eta(Y)\}\\ &\quad + 2(n+1)\{\eta(X)(\pounds_V\eta)(Y) + \eta(Y)(\pounds_V\eta)X\}.\end{aligned} \tag{3.24}$$

Comparing the equation (3.22) with (3.24) and substituting ξ for Y gives

$$\pounds_V\eta = -4(n+1)\eta. \tag{3.25}$$

Thus, V is an infinitesimal contact transformation, proving part (iv). Moreover, Lie differentiating $\eta(X) = g(X,\xi)$ along V, using (3.23) and (3.25) immediately yields $\pounds_V\xi = 4(n+1)\xi$. Applying d (exterior differentiation) on (3.25) and noting that d commutes with $\pounds_V$, provides

$$(\pounds_V d\eta)(X,Y) = -4(n+1)d\eta(X,Y).$$

Finally, taking the Lie derivative of $d\eta(X,Y) = g(X,\varphi Y)$ along V and using the above equation, we obtain $\pounds_V\varphi = 0$, proving part (iii). This completes the proof. □

PROOF OF LEMMA 1.5. By hypothesis, the Ricci tensor Ric is Killing, i.e.,

$$(\nabla_X Ric)(Y,Z) + (\nabla_Y Ric)(Z,X) + (\nabla_Z Ric)(X,Y) = 0.$$

Substituting ξ for Z and taking into account the equation (3.4) and Lemma 3.1, we find that

$$g(Q\varphi X,Y) - g(\varphi QX,Y) = 0.$$

Hence the Ricci operator Q and φ commute, completing the proof. □

REMARK 3.2.

(1) Let us recall from [**7**] that a Kaehler manifold (M,g,J), i.e., an even dimensional smooth manifold M with a Riemannian metric g and a $(1,1)$-tensor J such that $J^2 = -I, g(JX,JY) = g(X,Y), \nabla J = 0$, is said to be a Kaehler Ricci soliton if it is a Ricci soliton on (M,g) such that the potential vector field V is an infinitesimal automorphism of the complex structure tensor J, i.e., $\pounds_V J = 0$. The last condition requires that a solution of the Ricci flow that starts with a Kaehler metric remains Kaehler with respect to the same complex structure. In the non-Einstein case of Theorem 1.2, it automatically turns out that $\pounds_V\varphi = 0$ and so, we need not assume this as an extra condition.

(2) For the Einstein case of Theorem 1.2, if we had assumed M to be complete, then by the following result (Boyer and Galicki [**3**], and Morimoto [**22**]) "A compact K-contact Einstein manifold is Sasakian", and noting that Einstein constant is $2n$, thanks to Myers' theorem, M would be compact and hence Sasakian. For the nontrivial case assumed in Theorem 1.2, M turned out to be η-Einstein (equation (2.8)) with Ricci tensor given by the equation (3.18). So $\alpha = -2$. Now, by the following result ([**3**], [**22**]): "A compact K-contact η-Einstein manifold with $\alpha = -2$ is Sasakian", we would be tempted to expect that M would be Sasakian if it is complete. However, the following result of Hasegawa and Seino [**17**] "A complete Sasakian manifold for which $Ric \geq \delta > -2$ is compact, this is valid for K-contact case as well", poses a marginal obstruction to our expectation because an eigenvalue of Ric in our case is -2, and therefore M (if assumed complete) need not be compact and hence need not be Sasakian.

4. Yamabe solitons in contact geometry

The evolution of a Riemannian metric g on a smooth manifold M to a metric $g(t)$ in time t through the equation

$$\frac{\partial}{\partial t} g(t) = -r(t)g(t), \quad g(0) = g,$$

where $r(t)$ denotes the scalar curvature of $g(t)$, is called the Yamabe flow, and it was introduced by Hamilton [**16**]. The Yamabe flow is a natural geometric deformation to metrics of constant scalar curvature. Just as a Ricci soliton is a special solution of the Ricci flow, a Yamabe soliton is a special solution of the Yamabe flow, and it is defined by the equation

$$\pounds_V g = 2(c - r)g, \tag{4.1}$$

where g is a Riemannian metric on a smooth manifold M, V is a vector field on M, r the scalar curvature of g, and c is a real constant. The Yamabe soliton is said to be shrinking, steady, or expanding when $c > 0, = 0, < 0$ respectively. In particular, if $V = Df$ for a smooth function f, where D denotes the gradient operator of g, then a Yamabe soliton is called a gradient Yamabe soliton in which case (4.1) assumes the form

$$\nabla\nabla f = (c - r)g. \tag{4.2}$$

The gradient Yamabe soliton is trivial when f is a constant and r is a constant. We recall the following result for the compact case: "A Yamabe soliton on a compact Riemannian n-dimensional manifold with $n \geq 3$, has constant scalar curvature", which was proved by Chow, Lu, and Ni in [**8**], by Cerbo and Disconzi in [**5**], and by Daskalopoulos and Sesum in [**9**] for the gradient case.

In [**29**], Sharma studied Yamabe solitons in contact Riemannian geometry, and proved that a 3-dimensional Yamabe soliton whose metric is Sasakian has constant scalar curvature and the soliton vector field V is Killing. This result was extended and generalized in any dimension by Poddar, Balasubramanian, and Sharma in [**26**] in the form of the following result.

THEOREM 4.1 (P-B-S). *Let (M, g) be a Yamabe soliton with soliton vector field V. If g is a Sasakian metric, then it has constant scalar curvature and V is Killing.*

It was also shown in [**26**] that the same conclusion (as in the above theorem) holds on a $(2n+1)$-dimensional K-contact manifold M (a generalization of Sasakian manifold) under certain conditions specified in the following result.

THEOREM 4.2 (P-B-S). *Let (M, g) be a $(2n + 1)$-dimensional Yamabe soliton with soliton vector field V, and g be a K-contact metric. Then its scalar curvature is constant and V is Killing, if any one of the following conditions hold:*

(i) *the scalar curvature is constant along V;*
(ii) *V is an eigenvector of the Ricci operator with eigenvalue $2n$;*
(iii) *V is gradient.*

In case (iii), *the Yamabe soliton becomes trivial (i.e., f is a constant).*

The proofs of the above-mentioned two results use the fact that the defining equation (4.1) of a Yamabe soliton means that the associated vector field V is a conformal vector field with conformal scale function $2(c-r)$. A vector field V on an m-dimensional Riemannian manifold (M, g) is said to be a conformal vector field if

$$\pounds_V g = 2\rho g \tag{4.3}$$

for a smooth function ρ on M. Denoting the gradient vector field of ρ by $D\rho$, and the Laplacian $-divD$ by Δ, we have the following integrability conditions for the conformal vector field V (Yano [**37**]):

$$\begin{aligned}(\pounds_V R)(X, Y, Z) \\ = -(\nabla\nabla\rho)(Y, Z)X + (\nabla\nabla\rho)(X, Z)Y - g(Y, Z)\nabla_X D\rho + g(X, Z)\nabla_Y D\rho,\end{aligned} \tag{4.4}$$

$$(\pounds_V Ric)(X, Y) = -(m-2)(\nabla\nabla\rho)(X, Y) + (\Delta\rho)g(X, Y), \tag{4.5}$$

$$\pounds_V r = 2(m-1)\Delta\rho - 2r\rho. \tag{4.6}$$

PROOF. We include a shorter proof of Theorem 4.1 than the one given in [**26**]. First, we note that Theorem 4.1 was already proved in dimension 3 by Sharma [**29**]. In higher dimensions, the crucial idea of the proof is based on the following results of Okumura [**24**, Theorems 3.1 and 3.3]: "A conformal vector field V (defined by (4.3) on a Sasakian manifold of dimension $2n+1 > 3$ differs from $-D\rho$ by a Killing vector field, i.e., $V = W - D\rho$ where W is a Killing vector field on M, and the conformal scale function ρ satisfies the concircular equation: $\nabla\nabla\rho = -\rho g$." For a Yamabe soliton equation (4.1), we note that $\rho = c - r$, and therefore the preceding equation can be expressed as $\nabla\nabla r = (c-r)g$. But we know from Sharma and Deshmukh [**31**] that if the scalar curvature r of a Riemannian manifold (M, g) satisfies the foregoing equation, then r is constant. Hence r is a constant, completing the proof. □

Concluding remarks.

(1) Condition (ii) in Theorem 4.2 is motivated by the fact that, if we take V as the Reeb vector field ξ, then the Yamabe soliton equation (4.1) implies $r = c$ because ξ is Killing for a K-contact metric.

(2) In the second case of Theorem 1.2, V turned out to be an infinitesimal contact transformation for which we know from [**1**] that $V = -\frac{1}{2}\varphi Df + f\xi$ where $f = \eta(V)$. We also found that V preserves φ (stronger conclusion than V being an infinitesimal contact transformation). Therefore, it would be interesting to find some more information on f through this conclusion.

(3) Just as we did in the case of a Ricci soliton, can we study and possibly classify/characterize a Yamabe soliton as a K-contact manifold whose Ricci operator commutes with the fundamental collineation, i.e., $Q\varphi = \varphi Q$?

References

[1] D. E. Blair, *Riemannian geometry of contact and symplectic manifolds*, 2nd ed., Progress in Mathematics, vol. 203, Birkhäuser Boston, Ltd., Boston, MA, 2010, DOI 10.1007/978-0-8176-4959-3. MR2682326

[2] W. M. Boothby and H. C. Wang, *On contact manifolds*, Ann. of Math. (2) **68** (1958), 721–734, DOI 10.2307/1970165. MR112160

[3] C. P. Boyer and K. Galicki, *Einstein manifolds and contact geometry*, Proc. Amer. Math. Soc. **129** (2001), no. 8, 2419–2430, DOI 10.1090/S0002-9939-01-05943-3. MR1823927

[4] C. P. Boyer, K. Galicki, and P. Matzeu, *On eta-Einstein Sasakian geometry*, Comm. Math. Phys. **262** (2006), no. 1, 177–208, DOI 10.1007/s00220-005-1459-6. MR2200887

[5] L. F. Di Cerbo and M. M. Disconzi, *Yamabe solitons, determinant of the Laplacian and the uniformization theorem for Riemann surfaces*, Lett. Math. Phys. **83** (2008), no. 1, 13–18, DOI 10.1007/s11005-007-0195-6. MR2377942

[6] B.-Y. Chen and I. Mihai, *Isometric immersions of contact Riemannian manifolds in real space forms*, Houston J. Math. **31** (2005), no. 3, 743–764. MR2148814

[7] B. Chow, S.-C. Chu, D. Glickenstein, C. Guenther, J. Isenberg, T. Ivey, D. Knopf, P. Lu, F. Luo, and L. Ni, *The Ricci flow: techniques and applications. Part I*, Mathematical Surveys and Monographs, vol. 135, American Mathematical Society, Providence, RI, 2007. Geometric aspects, DOI 10.1090/surv/135. MR2302600

[8] B. Chow, P. Lu, and L. Ni, *Hamilton's Ricci flow*, Graduate Studies in Mathematics, vol. 77, American Mathematical Society, Providence, RI; Science Press Beijing, New York, 2006, DOI 10.1090/gsm/077. MR2274812

[9] P. Daskalopoulos and N. Sesum, *The classification of locally conformally flat Yamabe solitons*, Adv. Math. **240** (2013), 346–369, DOI 10.1016/j.aim.2013.03.011. MR3046313

[10] J. Davidov and O. Muškarov, *Twistor spaces with Hermitian Ricci tensor*, Proc. Amer. Math. Soc. **109** (1990), no. 4, 1115–1120, DOI 10.2307/2048144. MR1017845

[11] S. R. Deng, *Variational problems on contact manifolds*, Thesis, Michigan State University, 1991.

[12] L. P. Eisenhart, *Riemannian Geometry*, Princeton University Press, Princeton, NJ, 1949. MR0035081

[13] A. Ghosh and R. Sharma, *Sasakian metric as a Ricci soliton and related results*, J. Geom. Phys. **75** (2014), 1–6, DOI 10.1016/j.geomphys.2013.08.016. MR3126930

[14] G. W. Gibbons, S. A. Hartnoll, and C. N. Pope, *Bohm and Einstein-Sasaki metrics, black holes, and cosmological event horizons*, Phys. Rev. D (3) **67** (2003), no. 8, 084024, 24, DOI 10.1103/PhysRevD.67.084024. MR1995313

[15] A. Gray, *Einstein-like manifolds which are not Einstein*, Geom. Dedicata **7** (1978), no. 3, 259–280, DOI 10.1007/BF00151525. MR505561

[16] R. S. Hamilton, *The Ricci flow on surfaces*, Mathematics and general relativity (Santa Cruz, CA, 1986), Contemp. Math., vol. 71, Amer. Math. Soc., Providence, RI, 1988, pp. 237–262, DOI 10.1090/conm/071/954419. MR954419

[17] I. Hasegawa and M. Seino, *Some remarks on Sasakian geometry—applications of Myers' theorem and the canonical affine connection*, J. Hokkaido Univ. Ed. Sect. II A **32** (1981/82), no. 1, 1–7. MR638798

[18] K. Heil, A. Moroianu, and U. Semmelmann, *Killing and conformal Killing tensors*, J. Geom. Phys. **106** (2016), 383–400, DOI 10.1016/j.geomphys.2016.04.014. MR3508929

[19] W. Jelonek, *Killing tensors and Einstein-Weyl geometry*, Colloq. Math. **81** (1999), no. 1, 5–19, DOI 10.4064/cm-81-1-5-19. MR1716184

[20] T. Koufogiorgos, *Contact metric manifolds*, Ann. Global Anal. Geom. **11** (1993), no. 1, 25–34, DOI 10.1007/BF00773361. MR1201408

[21] J. Maldacena, *The large-N limit of superconformal field theories and supergravity*, Internat. J. Theoret. Phys. **38** (1999), no. 4, 1113–1133, DOI 10.1023/A:1026654312961. Quantum gravity in the southern cone (Bariloche, 1998). MR1705508

[22] S. Morimoto, *Almost complex foliations and its application to contact geometry*, Natur. Sci. Rep. Ochanomizu Univ. **43** (1992), no. 1, 11–23. MR1186577

[23] H. Nazawa, *Five dimensional K-contact manifold of rank 2*, arXiv:0907.0208v1 [math.SG], 1 July 2009.

[24] M. Okumura, *On infinitesimal conformal and projective transformations of normal contact spaces*, Tohoku Math. J. (2) **14** (1962), 398–412, DOI 10.2748/tmj/1178244076. MR146774

[25] Z. Olszak, *On contact metric manifolds*, Tohoku Math. J. (2) **31** (1979), no. 2, 247–253, DOI 10.2748/tmj/1178229842. MR538923

[26] R. Poddar, S. Balasubramanian, and R. Sharma, *Yamabe solitons in contact geometry*, New Zealand J. Math. **54** (2023), 49–55, DOI 10.53733/286. MR4683603

[27] R. Sharma, *Addendum: "Conformal motion of contact manifolds with characteristic vector field in the k-nullity distribution" [Illinois J. Math.* **40** *(1996), no. 4, 553–563; MR1415018 (97i:53054)] by Sharma and D. E. Blair*, Illinois J. Math. **42** (1998), no. 4, 673–677. MR1649889

[28] R. Sharma, *Certain results on K-contact and (k,μ)-contact manifolds*, J. Geom. **89** (2008), no. 1-2, 138–147, DOI 10.1007/s00022-008-2004-5. MR2457028

[29] R. Sharma, *A 3-dimensional Sasakian metric as a Yamabe soliton*, Internat. J. Geom. Methods Mod. Phys. **9** (2012) 1220003 (5 pages).

[30] R. Sharma, *Conformal and projective characterizations of an odd dimensional unit sphere*, Kodai Math. J. **42** (2019), no. 1, 160–169, DOI 10.2996/kmj/1552982511. MR3934618

[31] R. Sharma and S. Deshmukh, *Remarks on scalar curvature and concircular field equation*, Int. Electron. J. Geom. **14** (2021), no. 1, 121–124, DOI 10.36890/iejg.906792. MR4250311

[32] R. Sharma and A. Ghosh, *Sasakian 3-manifold as a Ricci soliton represents the Heisenberg group*, Int. J. Geom. Methods Mod. Phys. **8** (2011), no. 1, 149–154, DOI 10.1142/S021988781100504X. MR2782881

[33] T. Takahashi and S. Tanno, *K-contact Riemannian manifolds isometrically immersed in a space of constant curvature*, Tohoku Math. J. (2) **23** (1971), 535–539, DOI 10.2748/tmj/1178242598. MR310807

[34] S. Tanno, *Note on infinitesimal transformations over contact manifolds*, Tohoku Math. J. (2) **14** (1962), 416–430, DOI 10.2748/tmj/1178244078. MR145441

[35] S. Tanno, *The topology of contact Riemannian manifolds*, Illinois J. Math. **12** (1968), 700–717. MR234486

[36] R. M. Wald, *General relativity*, University of Chicago Press, Chicago, IL, 1984, DOI 10.7208/chicago/9780226870373.001.0001. MR0757180

[37] K. Yano, *Integral formulas in Riemannian geometry*, Pure and Applied Mathematics, vol. No. 1, Marcel Dekker, Inc., New York, 1970. MR0284950

Chandernagore College, Hooghly, 712136, West Bengal, India
Current address: Bethune College, Kolkata, West Bengal 700006, India
Email address: aghosh_70@yahoo.com

University of New Haven, West Haven, Connecticut 06516
Email address: rsharma@newhaven.edu

Sri Sathya Sai Institute of Higher Learning, Prasanthi Nilayam 515134, Andhra Pradesh, India
Current address: Harish-Chandra Research Institute, Prayagraj 211019, Uttar Pradesh, India
Email address: rahulpoddar@sssihl.edu.in; rahulpoddar@hri.res.in

Contemporary Mathematics
Volume **821**, 2025
https://doi.org/10.1090/conm/821/16438

Understanding of constant p-mean curvature hypersurfaces in the Heisenberg groups

Hsiao-Fan Liu

ABSTRACT. This article introduces the concept of p-minimal or constant p-mean curvature (pCMC) surfaces in Heisenberg groups H_n. We review known results, discuss analogous CMC or pCMC questions in H_n, and propose interesting problems for further study.

1. Introduction

The concept of minimal surfaces or constant mean curvature surfaces plays an important role in differential geometry for studying fundamental geometry and topology of manifolds. Besides mathematical significance, it also has practical applications in many scientific disciplines, such as physical and biological applications, material science, computer graphics ([**SY**, **SSPC**, **HP**, **TD**, **BHA**]), etc. There is an analogous concept in pseudo-hermitian manifolds, called p-minimal or constant p-mean curvature (pCMC) surfaces. The study of constant mean curvature (CMC) hypersurfaces in Heisenberg groups became prominent in the 1990s. Tomter discovered a minimal surface invariant in H_1 under isometry group action ([**T**]) following the general method of Hsiang and Lawson ([**HL**]). As this project had been continued by many mathematicians, Figueroa, Mercuri, and Pedrosa [**FMP**] classified all surfaces with constant mean curvature (including zero curvature) of H_1 under a certain left-invariant metric. Much later, César Rosales generalized to describe complete stable surfaces with constant mean curvature without singularity in Sasakian sub-Riemannian 3-folds [**R**].

The purpose of the present article is to address some results in analogous CMC or pCMC questions in H_n and some interesting problems to study. The organization is as follows. We briefly review the Heisenberg groups we will study in Section 2, and then introduce related results in H_n, for either $n = 1$ or for general n, in Section 3. In Section 4, we introduce some ongoing works and open problems.

2020 *Mathematics Subject Classification.* Primary 53A10, 53C42; Secondary 53C22, 34A26.
Key words and phrases. Differential geometry, algebraic geometry.
The author was supported in part by NSTC 112-2628-M-032-001-MY4.

2. Basic settings of the Heisenberg groups

The Heisenberg group H_n is the space $\mathbb{R}^{2n+1}$ associated with the group multiplication

$$(x,y,t)\circ(\tilde{x},\tilde{y},\tilde{t}) = (x+\tilde{x}, y+\tilde{y}, t+\tilde{t}+y\tilde{x}-x\tilde{y}), \tag{2.1}$$

where $x=(x_1,\dots,x_n), y=(y_1,\dots,y_n), \tilde{x}=(\tilde{x}_1,\dots,\tilde{x}_n), \tilde{y}=(\tilde{y}_1,\dots,\tilde{y}_n)$. It can be checked that H_n is a Lie group of dimension $2n+1$ with the left invariant vector fields spanned by

$$\mathring{e}_j = \frac{\partial}{\partial x_j} + y_j\frac{\partial}{\partial t},\quad \mathring{e}_{n+j} = \frac{\partial}{\partial y_j} - x_j\frac{\partial}{\partial t},\quad T=\frac{\partial}{\partial t},\quad 1\le j\le n, \tag{2.2}$$

and H_n is equipped with the standard pseudohermitian structure (J,θ). Throughout this article, we consider the following contact form of H_n defined by

$$\theta = dt + \sum_{j=1}^{n} x_j dy_j - y_j dx_j,$$

and its contact bundle $\xi = \ker\theta \in TH_n$, or equivalently, is spanned by $\mathring{e}_n$ and $\mathring{e}_{n+j}$. The CR structure now is the endomorphism $J:\xi\to\xi$ defined by

$$J(\mathring{e}_j) = \mathring{e}_{n+j},\quad J(\mathring{e}_{n+j}) = -\mathring{e}_j.$$

Let $X\subset H_n$ be a hypersurface of H_n. The first fundamental form I of X is a Riemannian metric on X defined to be the restriction of the adapted metric.

DEFINITION 2.1. A singular point $p\in X$ is a point at which the tangent plane coincides with the contact plane.

The singular set is the set containing all singular points of X whereas the regular part of X is defined to be the complement of the singular set in X.

For any point of the regular part in X, there exists a locally orthonormal frame of $TX\cap\xi$,

$$\{e_1, e_2, \dots, e_{2n-1}\}$$

such that $J(e_j) = e_{n+j}$, where $1\le j\le n-1$.

We then have the invariant subbundle $\xi'\in TX\cap\xi$, where

$$\xi' = \text{span}\{e_1,\dots,e_{n-1},e_{n+1},\dots,e_{2n-1}\}.$$

The bundle $TX\cap\xi$ can be decomposed into $\xi'\oplus\mathbb{R}e_n$. Such e_n is uniquely determined up to a sign and it defines a one-dimensional foliation of X. Denote $e_{2n}=J(e_n)$, then TX has a natural frame $\{e_1,e_2,\dots,e_{2n},T\}$, however, in which not every vector is tangent to X. In [**CHMY1**], an associated orthonormal (w.r.t. I) tangent frame was first introduced to be $\{\hat{e}_1,\dots,\hat{e}_{2n}\}$ defined as

$$\hat{e}_j = e_j,\quad \hat{e}_{2n} = \frac{\alpha e_{2n}+T}{\sqrt{1+\alpha^2}},\quad 1\le j\le 2n-1,$$

for some function α. Such α is called the fundamental function on X, or simply, we call it α-function throughout this proposal.

The second fundamental form II is then defined by

$$\text{II} = -<\nabla e_{2n},\cdot>$$

with the matrix representation

$$h_{jk} = \text{II}(e_j,e_k) = -<\nabla_{e_k}e_{2n}, e_j>, 1\le j,k\le 2n-1.$$

The *p(horizontal)-mean curvature* H of X is defined to be the trace of (h_{jk}). From the geometric point of view, J.-H. Cheng, J.-F. Hwang, A. Malchiodi and P. Yang proposed this notion of mean curvature to generalize the one introduced first by S. Pauls in H_1 for graphs over the xy-plane [**P**].

DEFINITION 2.2. The partially normal p-mean curvature of X is defined to be h_{nn} and denoted by l. Moreover, X is called a constant p-mean curvature (pCMC) hypersurface if H is constant and a p-minimal hypersurface if $H = 0$.

The structure equations of H_n give the Codazzi-like equation ([**CHMY3, CL1**]) as follows:

(2.3)

$$\hat{e}_n\hat{e}_n\alpha + 6\alpha(\hat{e}_n\alpha) + 4\alpha^3 - \frac{\hat{e}_{2n}l}{\sqrt{1+\alpha^2}} + \alpha|J(\nabla_{\hat{e}}e_{2n} + l\hat{e}_n)|^2 + 3(J(\nabla_{\hat{e}}e_{2n} + l\hat{e}_n)\alpha) = 0.$$

3. Some results in H_n

Classic results about CMC surfaces in H_1 were presented in the works [**T**] and [**R, RR1**] without considering the CR structure. Tomter ([**T**]) proved the classification theorem on constant mean curvature surfaces in the Heisenberg group by applying a suitable symmetry group to reduce the constant mean curvature equation to an ordinary differential equation on an orbit space with singularities, that was generalized from Hsiang and Lawson's method [**HL, H1, H2, HH**].

Rosales said ([**R**]) that a complete surface in H_1 without singular points is a vertical plane, and further if one defines the energy to be

$$E = \frac{x^{2n-1}t'}{\sqrt{x^2x'^2 + t'^2}} + \frac{H}{2n}x^{2n}, \tag{3.1}$$

where $\gamma = (x(s), t(s)), x \geq 0$ is a generating curve of a C^2 rotationally invariant hypersurface in H_n and the arc-length parameter s, then such an energy E is constant, provided that H is constant. Moreover, Ritoré and Rosales ([**RR1**]) classified rotationally invariant hypersurfaces with constant mean curvature in H_n as the following five categories,

$$\{H = 0, E = 0\}, \{H = 0, E \neq 0\}, \{H \neq 0, E = 0\}, \{HE > 0\}, \text{ and } \{HE < 0\},$$

depending on the signs of H and E.

Taking the CR structure into account, a natural question arises as to how to distinguish all pCMC hypersurfaces. The answer to this in H_1 was based on the work [**CL2**] by investigating properties of the Codazzi-like equation (2.3), which was discovered for the p-area element along the characteristic curves and has been used to study the structure of the singular set ([**CHMY3, CL1**]), including the local configuration of the singular set and the characteristic curves of H_n.

THEOREM 3.1 ([**CL2**]). *The existence of a constant p-mean curvature surface in H_1 is equivalent to the existence of a solution to the equation* (2.3).

This theorem suggests that the four invariants, $\mathbf{I}, \mathbf{II}, \alpha, H$, of a surface, can be reduced to α, H by choosing *suitable* frame fields. (By being suitable, we mean a normal coordinate system. In fact, we showed that there exists *normal coordinates* for each pCMC surface.) It turns out that the information of α, H is sufficient to recover a pCMC surface in H_1.

THEOREM 3.2 ([**CL2**]). *Let $\alpha(x,y), H(x,y)$ be two arbitrary functions on a coordinate neighborhood $(U;x,y) \subset \mathbb{R}^2$. If they satisfy the following integrability condition*

(3.2)
$$\left(h(y) - \int H\alpha \left(e^{\int 2\alpha dx}\right) dx\right) H_x + e^{k(y)} H_y = \left(e^{\int 2\alpha dx}\right) (\alpha_{xx} + 6\alpha\alpha_x + 4\alpha^3 + \alpha H^2),$$

for some functions $k(y)$ and $h(y)$ in the variable y, then there exists an embedding $X : U \to H_1$ (provided that U is small enough) such that the surface $\Sigma = X(U)$ has H and α as its p-mean curvature and α-function, respectively, and $\hat{e}_1 = \frac{\partial}{\partial x}$, $\hat{e}_2 = a(x,y)\frac{\partial}{\partial x} + b(x,y)\frac{\partial}{\partial y}$, where

$$a = \frac{e^{-\int 2\alpha dx}}{(1+\alpha^2)^{\frac{1}{2}}} \left(h(y) - \int H\alpha \left(e^{\int 2\alpha dx}\right) dx\right),$$
$$|b| = e^{k(y)} \frac{e^{-\int 2\alpha dx}}{(1+\alpha^2)^{\frac{1}{2}}},$$

for some functions $k(y)$ and $h(y)$.

All p-minimal surfaces can be classified into several classes, which are vertical, special type I, special type II and general type (see [**CL2**]). Each type of these p-minimal surfaces is open without singular points. Generically, each p-minimal surface is a union of these types of surfaces.

In addition, such embeddings are unique, up to a Heisenberg rigid motion. Notably, if H is constant, the integrability condition (3.2) becomes

$$\alpha_{xx} + 6\alpha\alpha_x + 4\alpha^3 + \alpha H^2 = 0, \tag{3.3}$$

which is the Codazzi-like equation (2.3) under the chosen frame fields $\hat{e}_1$ and $\hat{e}_2$ in Theorem 3.2. Moreover, the structure of the singular set in a surface of class C^2 was studied by J.-H. Cheng and J.-F. Hwang ([**CH**]), P. Yang and Andrea Malchiodi later ([**CHMY2, M**]) to investigate the p-minimal surfaces in H_1. It turns out that the Bernstein-type problem, first shown in [**CHMY1**], has the following results. That is,

$$u(x,y) = Ax + By + C,$$

for some constants $A, B, C \in \mathbb{R}$ and

$$u(x,y) = -ABx^2 + (A^2 - B^2)xy + ABy^2 + g(-Bx + Ay),$$

for some constants $A, B \in \mathbb{R}$ and $A^2 + B^2 = 1$, and $g \in C^\infty$ are the only two classes of entire p-minimal graphs in H_1. The Bernstein-type problem in H_1 has then intensely been studied recently (cf. [**CH, P, ACV, RR2, DGNP**]). The Bernstein conjecture remains open in higher dimensions.

4. Open problems

The final section will highlight the current state of research by addressing the existing open problems in the field. Some related continued or generalizing problems are listed below. One problem we are concerned about is that

> *Must the global minimal surfaces without singular points in H_n be vertical planes?*

J.-H. Cheng et al ([**CH**]) using methods of analysis and later Chiu et al ([**CL2**]) showed that the global minimal surfaces without singular points must be vertical planes for $n = 1$. It follows immediately from these works to ask what happens in the case $n \geq 2$. In addition, we try to investigate the existence and classification of pCMC in H_n by understanding relations between Codazzi-like equations and pCMC in H_n. One difficulty one should overcome first is to choose natural frame fields for pCMC in H_n (in $\mathbb{R}^{2n+1}$) so that the Codazzi-like equation is completely integrable. The underlying geometry of pCMC in H_n should then be revealed via the solutions to the Codazzi-like equation.

On one hand, the Alexandrov theorem says that a closed (compact with no boundary) hypersurface of positive constant mean curvature in $\mathbb{R}^n$ must be a sphere. One can ask if an Alexandrov-type theorem holds for H_n $(n > 1)$ as Ritoré and Rosales showed that the theorem holds for $n = 1$ ([**RR2**]. On the other hand, Henry C. Wente ([**W**]) answered the H. Hopf's conjecture:

must an immersion of an oriented, closed, hypersurface with constant mean curvature $H \neq 0$ in $\mathbb{R}^n$ be the embedded sphere $\mathbb{S}^{n-1}$?

by providing a counterexample, the Wente torus. By analogy with the Euclidean situation, we would like to seek the Wente-type torus in the Heisenberg situation using the following information.

We see that the parametrization of pCMC surfaces given in [**CLL**] describes all pCMC surfaces in H_1. Moreover, the complete invariants obtained from the solution α to (3.3) suggest that either the surfaces won't extend to a surface with singular points, or the surface must be a closed constant p-mean curvature surface without singular points if the surface is closed, which means the surface is of type of torus. It indicates that it is possible to find a Wente-type torus in this class of surfaces.

References

[ACV] V. Barone Adesi, F. Serra Cassano, and D. Vittone, *The Bernstein problem for intrinsic graphs in Heisenberg groups and calibrations*, Calc. Var. Partial Differential Equations **30** (2007), no. 1, 17–49, DOI 10.1007/s00526-006-0076-3. MR2333095

[BHA] A. I. Bobenko, T. Hoffmann, and B. A. Springborn, *Minimal surfaces from circle patterns: geometry from combinatorics*, Ann. of Math. (2) **164** (2006), no. 1, 231–264, DOI 10.4007/annals.2006.164.231. MR2233848

[CH] J.-H. Cheng and J.-F. Hwang, *Properly embedded and immersed minimal surfaces in the Heisenberg group*, Bull. Austral. Math. Soc. **70** (2004), no. 3, 507–520, DOI 10.1017/S0004972700034766. MR2103983

[CHMY1] J.-H. Cheng, J.-F. Hwang, A. Malchiodi, and P. Yang, *Minimal surfaces in pseudohermitian geometry*, Ann. Sc. Norm. Super. Pisa Cl. Sci. (5) **4** (2005), no. 1, 129–177. MR2165405

[CHMY2] J.-H. Cheng, J.-F. Hwang, A. Malchiodi, and P. Yang, *Minimal surfaces in pseudohermitian geometry*, Ann. Sc. Norm. Super. Pisa Cl. Sci. (5) **4** (2005), no. 1, 129–177. MR2165405

[CHMY3] J.-H. Cheng, J.-F. Hwang, A. Malchiodi, and P. Yang, *A Codazzi-like equation and the singular set for C^1 smooth surfaces in the Heisenberg group*, J. Reine Angew. Math. **671** (2012), 131–198, DOI 10.1515/crelle.2011.159. MR2983199

[CL1] H.-L. Chiu and S.-H. Lai, *The fundamental theorem for hypersurfaces in Heisenberg groups*, Calc. Var. Partial Differential Equations **54** (2015), no. 1, 1091–1118, DOI 10.1007/s00526-015-0818-1. MR3385193

[CL2] H.-L. Chiu and H.-F. Liu, *A characterization of constant p-mean curvature surfaces in the Heisenberg group H_1*, Adv. Math. **405** (2022), Paper No. 108514, 50, DOI 10.1016/j.aim.2022.108514. MR4437615

[CLL] H.-L. Chiu, S.-H. Lai, and H.-F. Liu, *On invariants of constant p-mean curvature surfaces in the Heisenberg group* H_1, SIGMA Symmetry Integrability Geom. Methods Appl. **21** (2025), Paper No. 011, 25, DOI 10.3842/SIGMA.2025.011. MR4865601

[DGNP] D. Danielli, N. Garofalo, D.-M. Nhieu, and S. D. Pauls, *The Bernstein problem for embedded surfaces in the Heisenberg group* $\mathbb{H}^1$, Indiana Univ. Math. J. **59** (2010), no. 2, 563–594, DOI 10.1512/iumj.2010.59.4291. MR2648078

[FMP] C. B. Figueroa, F. Mercuri, and R. H. L. Pedrosa, *Invariant surfaces of the Heisenberg groups*, Ann. Mat. Pura Appl. (4) **177** (1999), 173–194, DOI 10.1007/BF02505908. MR1747630

[H1] W.-Y. Hsiang, *Minimal cones and the spherical Bernstein problem. I*, Ann. of Math. (2) **118** (1983), no. 1, 61–73, DOI 10.2307/2006954. MR707161

[H2] W.-Y. Hsiang, *Minimal cones and the spherical Bernstein problem. II*, Invent. Math. **74** (1983), no. 3, 351–369, DOI 10.1007/BF01394241. MR724010

[HH] W.-t. Hsiang and W.-Y. Hsiang, *On the construction of constant mean curvature imbeddings of exotic and/or knotted spheres into* $S^n(1)$, Invent. Math. **82** (1985), no. 3, 423–445, DOI 10.1007/BF01388863. MR811544

[HL] W.-y. Hsiang and H. B. Lawson Jr., *Minimal submanifolds of low cohomogeneity*, J. Differential Geometry **5** (1971), 1–38. MR298593

[HP] R. K. Hladky and S. D. Pauls, *Minimal surfaces in the roto-translation group with applications to a neuro-biological image completion model*, J. Math. Imaging Vision **36** (2010), no. 1, 1–27, DOI 10.1007/s10851-009-0167-9. MR2579306

[M] A. Malchiodi, *Minimal surfaces in three dimensional pseudo-Hermitian manifolds*, Subelliptic PDE's and applications to geometry and finance, Lect. Notes Semin. Interdiscip. Mat., vol. 6, Semin. Interdiscip. Mat. (S.I.M.), Potenza, 2007, pp. 195–207. MR2384646

[P] S. D. Pauls, *Minimal surfaces in the Heisenberg group*, Geom. Dedicata **104** (2004), 201–231, DOI 10.1023/B:GEOM.0000022861.52942.98. MR2043961

[R] C. Rosales, *Complete stable CMC surfaces with empty singular set in Sasakian sub-Riemannian 3-manifolds*, Calc. Var. Partial Differential Equations **43** (2012), no. 3-4, 311–345, DOI 10.1007/s00526-011-0412-0. MR2875642

[RR1] M. Ritoré and C. Rosales, *Rotationally invariant hypersurfaces with constant mean curvature in the Heisenberg group* $\mathbb{H}^n$, J. Geom. Anal. **16** (2006), no. 4, 703–720, DOI 10.1007/BF02922137. MR2271950

[RR2] M. Ritoré and C. Rosales, *Area-stationary surfaces in the Heisenberg group* $\mathbb{H}^1$, Adv. Math. **219** (2008), no. 2, 633–671, DOI 10.1016/j.aim.2008.05.011. MR2435652

[SSPC] Louis Salkin, Alexandre Schmit, Pascal Panizza, and Laurent Courbin, *Influence of boundary conditions on the existence and stability of minimal surfaces of revolution made of soap films*, American Journal of Physics, vol. 82, No. 9, 2014, pp. 839–847.

[SY] R. Schoen and S. T. Yau, *On the proof of the positive mass conjecture in general relativity*, Comm. Math. Phys. **65** (1979), no. 1, 45–76. MR526976

[T] P. Tomter, *Constant mean curvature surfaces in the Heisenberg group*, Differential geometry: partial differential equations on manifolds (Los Angeles, CA, 1990), Proc. Sympos. Pure Math., vol. 54, Amer. Math. Soc., Providence, RI, 1993, pp. 485–495, DOI 10.1090/pspum/054.1/1216601. MR1216601

[TD] S. Torquato and A. Donev, *Minimal surfaces and multifunctionality*, Proc. R. Soc. Lond. Ser. A Math. Phys. Eng. Sci. **460** (2004), no. 2047, 1849–1856, DOI 10.1098/rspa.2003.1269. MR2067862

[W] H. C. Wente, *Counterexample to a conjecture of H. Hopf*, Pacific J. Math. **121** (1986), no. 1, 193–243. MR815044

Department of Applied Mathematics and Data Science, Tamkang University, Taiwan; and Department of Applied Mathematics, National Chung Hsing University, Taiwan

Email address: `hfliu@mail.tku.edu.tw`

Contemporary Mathematics
Volume **821**, 2025
https://doi.org/10.1090/conm/821/16439

Open problems on compact constant mean curvature surfaces with boundary

Rafael López

ABSTRACT. We present a collection of easily stated open problems in the theory of compact constant mean curvature surfaces with boundary. We also survey the current status of answering them.

1. Introduction and preliminaries

In this paper, we collect a number of open problems that we have found along the years during my research, have struggled with, and which are still open. At the same time, we present the current state of knowledge of these problems, indicating what the difficulties in answering them are. We hope this collection of open questions may stimulate younger minds, especially because most of the problems are easily formulated and any mathematician with a basic background in classical differential geometry can understand them.

Surfaces with constant mean curvature (cmc surfaces, in short; or H-surfaces, where H is the constant value of the mean curvature) belongs to the field of classical differential geometry of surfaces. A cmc surface in Euclidean 3-dimensional space $\mathbb{R}^3$ has the property its area is stationary under all volume-preserving variations in compact domains. For this reason, cmc surfaces are models of soap films and soap bubbles ([**27**]), hence its use in physics, chemistry, biology, and engineering. From another perspective, the mean curvature equation is a quasilinear elliptic equation and it serves as a model in the theory of PDEs of elliptic type [**20**]. Hence, surfaces with constant mean curvature are of interest in other fields of mathematics. As a recent sample button, overdetermined problems in PDEs initiated by Serrin, Berestycki, Cafarelli, and Nirenberg [**10**,**19**,**52**] have recently received a great boost using cmc surfaces; we refer to [**16**] and references therein for the interested reader.

The main open problems proposed in this paper come from the three celebrated characterizations of the sphere in the family of the closed cmc surfaces.

2020 *Mathematics Subject Classification.* Primary 53A10.

The author is a member of the IMAG and of the Research Group "Problemas variacionales en geometría", Junta de Andalucía (FQM 325). This research has been partially supported by MINECO/MICINN/FEDER grant no. PID2023-150727NB-I00, and by the "Maria de Maeztu" Excellence Unit IMAG, reference CEX2020-001105- M, funded by MCINN/AEI/10.13039/501100011033/ CEX2020-001105-M. Finally, I thank the editors of the CONM volume for the possibility to contribute.

THEOREM 1.1.

(1) *Spheres are the only* **embedded** *closed cmc surfaces (Alexandrov* [**2**]*).*
(2) *Spheres are the only* **genus zero** *cmc surfaces (Hopf* [**25**]*).*
(3) *Spheres are the only* **stable** *closed cmc surfaces (Barbosa-do Carmo* [**8**]*).*

There exist closed cmc surfaces others than spheres. It was Wente [**55**] who first proved the existence of an immersed torus in $\mathbb{R}^3$ with constant mean curvature. After that work, new examples of closed cmc surfaces with higher genus were discovered [**1, 11, 29, 30, 46**].

It is natural to ask for the boundary versions of these characterizations. The simplest case of boundary curve is a circle $\mathbb{S}^1$. If we intersect a sphere by a plane, we obtain two surfaces, called spherical caps, whose common boundary curve is a circle $\mathbb{S}^1$ (Figure 1.1). Both surfaces have nonzero constant mean curvature. Furthermore, the planar disk determined by $\mathbb{S}^1$ is a minimal ($H = 0$) surface. Thus, planar disks and spherical caps are examples of cmc surfaces spanning a circle.

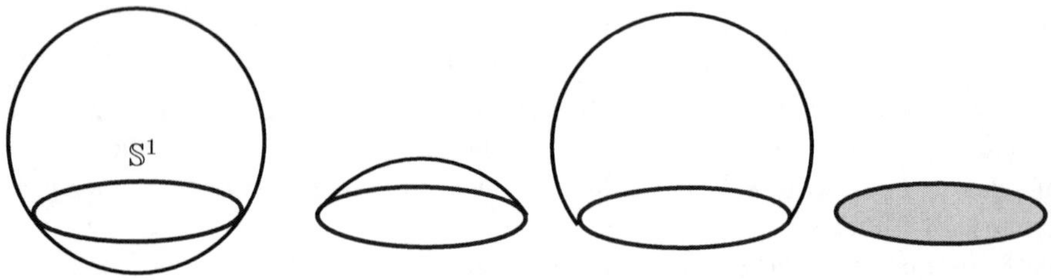

FIGURE 1.1. Surfaces with constant mean curvature spanning a circle $\mathbb{S}^1$ obtained by intersecting a sphere with a plane (left). The middle surfaces are (small and big) spherical caps; the right surface is a planar disk.

Following Theorem 1.1, the analogous hypothesis to consider in the boundary version is whether the surface is embedded, is a topological disk (genus zero), or is stable. There are the following three conjectures.

CONJECTURE 1. *Planar disks and spherical caps are the only compact embedded cmc surfaces in $\mathbb{R}^3$ with circular boundary.*

CONJECTURE 2. *Planar disks and spherical caps are the only cmc disks immersed in $\mathbb{R}^3$ with circular boundary.*

CONJECTURE 3. *Planar disks and spherical caps are the only compact stable cmc surfaces in $\mathbb{R}^3$ with circular boundary.*

As far as we know, the first time when similar questions appeared was in [**15**]; see Problem 1.4 proposed by Gulliver and Kusner in [**15**]. They asked about the existence of cmc surfaces spanning a circle. They proposed the hypothesis that the surface is embedded (Conjecture 1) or it is a topological disk (Conjecture 2). For the initial question of existence of examples, Kapouleas proved that there are compact cmc surfaces of genus $g > 2$ spanning a circle [**29**]. These examples appeared in the mentioned context of finding closed cmc surfaces.

Some of the problems proposed in this paper around these conjectures are well known by people working in cmc surfaces, particularly around the 1980s and '90s. In one way or another they appear in the works of Barbosa, Brito, Kapouleas, Koiso, Korevaar, Kusner, Meeks, Palmer, Sa Earp, and Rosenberg, among others. Perhaps because of their difficulty, these problems were gradually abandoned and forgotten but I think that they deserve a wider dissemination. It is also clear that the simplicity in the formulations will make famous—in this small world of differential geometry—the one who can prove or disprove the above conjectures.

We use the following notations. If $\Psi : \Sigma \to \mathbb{R}^3$ is an isometric immersion of a compact surface Σ with nonempty boundary $\partial\Sigma$ and $C \subset \mathbb{R}^3$ is a closed curve, we say that Σ has C as a boundary if $\Psi_{|\partial\Sigma}\colon \partial\Sigma \to C$ is a diffeomorphism. If C is a planar curve, we denote by P the plane containing C. We will assume that P is horizontal, that is, parallel to the xy-plane when we consider (x, y, z) the standard coordinates of $\mathbb{R}^3$. In the case where C is a round circle, we use $\mathbb{S}^1$ instead of C. If the boundary is planar, then any minimal surface spanning C is an open set of the plane containing C. Hence the situation $H = 0$ can be viewed as "trivial". In this paper we will consider surfaces in Euclidean 3-dimensional space $\mathbb{R}^3$, although many of the problems can be formulated in other space forms as well as for hypersurfaces in arbitrary dimensions.

Just like Google reviews allows one to rate the quality of a site, we have rated—from 1 to 5 stars—the importance, in the author's opinion, of each of the problems. The problems will be classified in three categories: the circular boundary case, the arbitrary boundary curve, and the Dirichlet problem.

2. Part I: Circular boundary

In this section we will investigate the case when the boundary of the surface is a circle $\mathbb{S}^1$. Spheres are rotational surfaces. Rotational surfaces of constant mean curvature are called Delaunay surfaces and they are well known: planes, catenoids, spheres, cylinders, unduloids, and nodoids [**17**]. The profile curves of the Delaunay surfaces are the trace of the focus of a conic which is rolled without slipping along a line and rotating this curve around that line. Planes and catenoids are the only ones with $H = 0$. Taking compact pieces of Delaunay surfaces, planar disks and spherical caps are the only compact rotational cmc surfaces spanning a circle.

The first problem is the boundary version of the Alexandrov theorem.

PROBLEM 2.1 (5*). Is a compact embedded cmc surface of $\mathbb{R}^3$ spanning a circle a planar disk or a spherical cap?

Let us show the difference from the closed case. The technique employed in the Alexandrov theorem is called the reflection method [**2**]. If $\Sigma \subset \mathbb{R}^3$ is a closed embedded surface, then Σ encloses a 3-domain $W \subset \mathbb{R}^3$. Let N be the orientation pointing towards W. Fix an arbitrary direction of $\mathbb{R}^3$, say, the x-direction. Let Π_t be the plane of equation $x = t$. Denote by $\Sigma_t^+ = \Sigma \cap \{x > t\}$, $\Sigma_t^- = \Sigma \cap \{x < t\}$, and Σ_t^* the reflection of Σ_t^+ with respect to Π_t. Coming from $t = +\infty$ towards Σ, let Π_{t_0} be the first plane touching Σ. By embeddedness of Σ, $\Sigma_t^* \subset W$ for all $t \in (t_0 - \epsilon, t_0)$ and for some $\epsilon > 0$. Decreasing t and by compactness, let $t_1 < t_0$ be the first time where $\Sigma_{t_1}^*$ touches $\Sigma_{t_1}^-$. If $p \in \Sigma_{t_1}^* \cap \Sigma_{t_1}^-$, then the surfaces $\Sigma_{t_1}^*$ and $\Sigma_{t_1}^-$ are H-surfaces with the same orientations at p. Moreover, $\Sigma_{t_1}^*$ lies in one side of $\Sigma_{t_1}^-$ around p. Then, the maximum principle implies $\Sigma_{t_1}^* = \Sigma_{t_1}^-$ in an open set around p [**20**]. By connectedness, $\Sigma_{t_1}^* = \Sigma_{t_1}^-$. This shows that Π_{t_1} is a plane of symmetry of Σ. Since the fixed orientation was arbitrary, then Σ is a sphere.

Now consider a compact embedded cmc surface spanning a circle $\mathbb{S}^1$. Suppose that $\mathbb{S}^1 \subset P$ is centered at the origin. If we repeat the reflection argument, a first problem we face is the existence of the domain W. For example, if we simply attach to Σ the planar disk D bounded by $\mathbb{S}^1$, the existence of W is not assured because the surface may cross D.

A situation where the reflection method applies is when Σ lies in one side of P. This hypothesis is not only because now $\Sigma \cup D$ encloses a 3-domain W, but also because Σ does not intersect the plane P; see Figure 2.1 left. By repeating the argument as in the closed case, we now fix an arbitrary horizontal direction, say, the x-direction. The time t_1 can occur is when $p \notin \mathbb{S}^1$ or $p \in \mathbb{S}^1$. In the first case, Π_{t_1} is a plane of symmetry, in particular, of symmetry of $\mathbb{S}^1$. This implies $t_1 = 0$. The second case implies $t_1 = 0$. Let us follow this situation. If Σ_0^* is tangent to Σ_0^-, then the maximum principle applies proving that Π_0 is a plane of symmetry. On the contrary, Σ_0^* is contained in W and it does not touch Σ_0^-. We see that this situation is not possible. For this we repeat the argument but coming from $t = -\infty$. Then there is a time $t_2 < 0$ such that the reflection of $\Sigma_{t_2}^+$ about Π_{t_2} tangentially touches $\Sigma_{t_2}^-$. By the maximum principle, the plane Π_{t_2} would be a plane of symmetry of Σ. This is a contradiction because $t_2 < 0$ and Π_{t_2} cannot be a plane of symmetry of $\mathbb{S}^1$. Summarizing, we have proved that for any horizontal direction, the plane orthogonal to this direction and containing the z-axis is a plane of symmetry of Σ. This proves that Σ is rotationally symmetric about the z-axis, hence a spherical cap.

Thus Problem 2.1 can be affirmatively answered if one proves that a compact embedded cmc surface spanning a circle is contained in one side of P.

PROBLEM 2.2 (3*). Under what conditions on the boundary is a compact embedded cmc surface of $\mathbb{R}^3$ spanning a planar closed curve C contained in one side of the boundary plane? And if C is convex?

A partial answer to Problem 2.2 was obtained by Koiso [**31**] assuming that $\Sigma \cap (P \setminus \overline{D}) = \emptyset$. In fact, in [**31**] it is not necessary that H is constant but only $H \neq 0$ on Σ.

The existence of the domain W under other circumstances does not solve Problem 2.1. In Figure 2.1 right, the surface Σ together with D defines W but $\text{int}(\Sigma)$ intersects the plane P. Then the reflection process may find a touching point p between Σ_t^* and $\partial\Sigma$, where Σ_t^* and Σ_t^- are not tangent: the maximum principle cannot be applied.

Let us observe that the surface of Figure 2.1 right is star-shaped with respect to the point q, that is, any ray from q meets the surface once at most. If Σ is compact, the star-shaped property implies that Σ is embedded. This is equivalent to say that Σ is a radial graph on some domain of a sphere centered at q. In the nineteenth century, Jellet proved that a closed star-shaped cmc surface is a sphere [**28**]. This motivates the boundary version of this result.

PROBLEM 2.3 (5*). Is a compact star-shaped cmc surface of $\mathbb{R}^3$ spanning a circle a planar disk or a spherical cap?

Although a positive answer is a partial answer to Problem 2.1, I rate it as a 5* problem because it is the analog of a classical result. The proof in the closed case

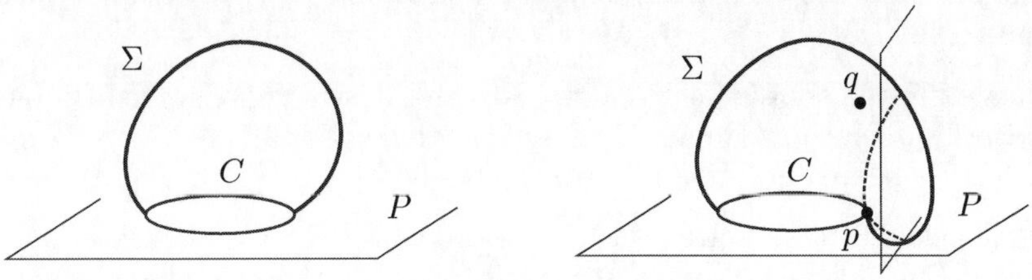

FIGURE 2.1. Left: a compact embedded surface over the boundary plane P. Right: an embedded surface where the Alexandrov reflection method fails at p. Notice that the surface Σ is star-shaped from the point q.

is the following. Assume that Σ is star-shaped from the origin of $\mathbb{R}^3$. If x is the position vector of Σ, the support function $h = \langle N, x\rangle$ has sign, say $h < 0$ with the inward orientation (and thus, $H > 0$). Moreover, we have

$$\begin{aligned}\Delta|x|^2 &= 4 + 4Hh \\ \Delta h &= -2H - |\sigma|^2 h,\end{aligned} \tag{2.1}$$

where Δ is the Beltrami–Laplace operator and σ is the second fundamental form of Σ. The first equation holds for any surface but the second one holds because H is constant. Integrating and using that Σ is closed, we have $A + H\int_\Sigma h = 0$ and $2HA + \int_\Sigma |\sigma|^2 h = 0$, where $A = \text{area}(\Sigma)$. Combining both equations and using $|\sigma|^2 \geq 2H^2$, we have

$$0 = 2HA + \int_\Sigma |\sigma|^2 h \leq 2HA + 2H^2 \int_\Sigma h = 2H(A + H\int_\Sigma h) = 0.$$

Thus $|\sigma|^2 = 2H^2$ on Σ, concluding that the surface is totally umbilical, hence a sphere. In the boundary case, the troubles appear in the boundary terms coming from the divergence theorem for both equations (2.1).

Returning to the difficulty to find the domain W, it was proved in [**14**] that if $\partial\Sigma = C$ is convex and Σ is transverse to the boundary plane P along C, then Σ lies in one side of P. Thus in Problem 2.1 it would remain to study that it is not possible that along $\mathbb{S}^1$, the surface have parts in both sides of P. In the same paper, the authors propose the following problem.

PROBLEM 2.4 (4*). *Let C be a simple closed convex curve in a plane P. Let Σ be a compact embedded cmc surface in $\mathbb{R}^3$ with $\partial\Sigma = C$. Assuming that Σ lies in one side of P, is Σ a topological disk?*

A partial affirmative answer is when Σ is a graph on D in a neighborhood of C. Under this hypothesis, Σ is a graph on D. Here $D \subset P$ is the domain bounded by C [**34**]. In this result is not necessary that C is convex. The proof is an application of the Alexandrov method by means of parallel planes to P. Returning to Problem 2.4 we propose Problem 2.5.

PROBLEM 2.5 (4*). *Under the same hypothesis of Problem 2.4, if Σ is a graph on $P \setminus \overline{D}$ around C, is Σ a bigraph? that is, is there a planar domain Ω, $\overline{D} \subset \Omega$,* such that $\Sigma = \Sigma_1 \cup \Sigma_2$, where Σ_1 is a graph on D and Σ_2 is an annulus which it is a graph on $\Omega \setminus \overline{D}$?

Related to the above two problems, Ros and Rosenberg answered Problem 2.4 affirmatively when the volume of Σ is sufficiently large [**49**].

A further related question is when the boundary $\partial\Sigma$ is formed of two curves.

PROBLEM 2.6 (4*). Let C_1 and C_2 be two convex curves in parallel planes. Is a compact embedded cmc surface spanning $C_1 \cup C_2$ a topological annulus?

If $H = 0$, this problem was proposed by Meeks [**45**] and it is still open. Problem 2.6 has an interesting version when the boundary is formed by two coaxial circles.

PROBLEM 2.7 (4*). Let C_1 and C_2 be two coaxial circles in parallel planes. Is a compact embedded cmc surface spanning $C_1 \cup C_2$ a surface of revolution?

In order to apply the Alexandrov reflection method, if the surface is contained in the slab determined by the two boundary planes, then the surface is rotational [**35**]. However, and in contrast to Problem 2.2, one cannot expect that the surface is included in the slab: there are embedded pieces of nodoids spanning two coaxial circles with parts outside of the slab.

We now continue with Conjecture 2.

PROBLEM 2.8 (5*). Is a topological cmc disk spanning a circle a planar disk or a spherical cap?

In the closed case, the proof is based in the properties of the Hopf differential form. Assume that $\Psi = \Psi(z)$ is a conformal parametrization of the surface, where $z = u + iv$. Now we denote by $\langle , \rangle$ the metric in $\mathbb{R}^3$ as well as in $\mathbb{C}^3$. Then $2\langle \Psi_z, \Psi_{\bar{z}} \rangle = E > 0$ and $4HE = \langle N, \Psi_{uu} + \Psi_{vv} \rangle$. Hopf defined the quadratic form $Q(dz)^2$ where $Q = \langle N, \Psi_{zz} \rangle = \frac{1}{4}(e - g + 2if)$, being $\{e, f, g\}$ the coefficients of σ with respect to Ψ. Note that the zeroes of Q are the umbilical points of Σ. Using the Codazzi equations, one gets $Q_{\bar{z}} = EH_z/2$ [**25**]. Consequently, Q is holomorphic if and only if H is constant. Moreover, either the umbilical points of the surface are isolated or the surface is totally umbilical. Suppose now that Ψ is an immersion of a topological sphere. Then we can assume that the immersion is $\Psi\colon \overline{\mathbb{C}} \to \mathbb{R}^3$. Because $Q\colon \overline{\mathbb{C}} \to \mathbb{C}$ is holomorphic, then $Q \equiv 0$. Thus Ψ is totally umbilical, hence it is a sphere. In the boundary version, if we have a cmc topological disk Ψ, the zeroes of Q coincide again with the umbilical points of Ψ. However, we cannot assure $\Psi \equiv 0$ on the surface because the existence of boundary of Σ.

A result related to Problem 2.8 is [**42**]. For immersed disks Σ in $\mathbb{R}^3$ Barbosa and do Carmo proved the isoperimetric inequality $L^2 \geq 4\pi A(1 - \frac{k_0 A}{4\pi})$, where L is the length of $\partial\Sigma$, A is the area of Σ and $k_0 \in \mathbb{R}$ is an upper bound of K [**7**]. The equality holds if and only if Σ is a totally geodesic disk with curvature $K \equiv k_0$. If Σ has constant mean curvature H, we can choose $k_0 = H^2$. In case that $\partial\Sigma$ is a circle $\mathbb{S}^1(r)$ of radius r, this inequality implies that either $A \leq A_-$ or $A \geq A_+$, where

$$A_- = \frac{2\pi}{H^2}(1 - \sqrt{1 - r^2H^2}), \quad A_+ = \frac{2\pi}{H^2}(1 - \sqrt{1 + r^2H^2}).$$

The values A_- and A_+ are the areas of the small and the big H-spherical caps with boundary $\mathbb{S}^1(r)$, respectively (Figure 1.1). In [**42**] it is proved that if Σ is a topological H-disk with $A \leq A_-$, then Σ is the small spherical cap.

The following problem is Conjecture 3.

PROBLEM 2.9. Is a stable compact cmc surface spanning a circle a circle a planar disk or a spherical cap?

Stability means that the second derivative of the area functional is nonnegative for any volume preserving variation of the surface. This is equivalent to

$$-\int_{\Sigma} u(\Delta u + |\sigma|^2 u) \geq 0 \tag{2.2}$$

for all $u \in C^2(\Sigma)$ such that

$$\int_{\Sigma} u = 0, \quad u_{|\partial\Sigma} = 0. \tag{2.3}$$

The mean zero integral $\int_{\Sigma} u = 0$ is a consequence of the fact that all admissible variations of Σ preserve the volume of the surface. The boundary condition $u_{|\partial\Sigma} = 0$ says that the variation fixes the boundary of the surface. In the closed case the condition $u_{|\partial\Sigma} = 0$ makes no sense and it is dropped. In such a case, Barbosa and do Carmo consider the function $u = 1 + H\langle N, x\rangle$ in (2.2) as from (2.1), and we know that $\int_{\Sigma} u = 0$. After some computations, and using the second identity of (2.1), the stability inquality (2.2) becomes

$$-\int_{\Sigma}(|\sigma|^2 - 2H^2) \geq 0.$$

Thus $|\sigma|^2 = 2H^2$, hence Σ is totally umbilical, and thus it is a sphere.

Coming back to the boundary case, the extra condition $u_{|\partial\Sigma} = 0$ adds difficulties to find a suitable test function to put in (2.2). Notice that the above function $u = 1 + H\langle N, x\rangle$ does not satisfy $\int_{\Sigma} u = 0$ neither $u_{|\partial\Sigma} = 0$. In [**3**] it was proved that if Σ is a stable topological cmc disk, then it is a planar disk or a spherical cap. This gives a partial answer to Conjecture 3. The proof uses the Hopf differential 2-form $Q(dz)^2$ and the behavior of the first and second Dirichlet eigenfunctions of the Jacobi equation.

To finish this section, we return to Kapouleas's examples [**29**]. These examples have self-intersections, genus $g > 2$, and they are included in one of the half-spaces determined by the boundary plane. The methods employed in [**29**] are the same that for the construction of closed cmc surfaces. There are a great number of pictures of closed cmc surfaces; see, for example, [**56–58**]. However, up to today, we do not have any numerical pictures of the Kapouleas's examples with circular boundary.

PROBLEM 2.10 (3*). Give numerical pictures of the Kapouleas's examples with circular boundary.

As far as we know, graphics of cmc surfaces with circular boundary a circle appear in [**12**], where the authors solve the Björling problem for H-surfaces. In [**12**, Figure 1], the authors take a circle $\mathbb{S}^1$ in the Björling data and the graphic shows the corresponding Björling solution. Of course, the result is local in the sense that the surface is defined in a neighborhood of $\mathbb{S}^1$: it is unknown if the surface closes to be compact.

Related to the three conjectures, an easier approach is assuming two of the three hypothesis. Recall that only if the surface is a topological disk and stable, it is known that the surface is a planar disk or a spherical cap [**3**].

PROBLEM 2.11 (4*). Is a compact stable embedded cmc surface of $\mathbb{R}^3$ spanning a circle a planar disk or a spherical cap?

PROBLEM 2.12 (4*). Is a compact embedded cmc disk of $\mathbb{R}^3$ spanning a circle a planar disk or a spherical cap?

The proposed problems in this section can be formulated in space forms. Notice that examples of compact cmc surfaces spanning a circle are geodesic spheres of totally umbilical surfaces. The hyperbolic space $\mathbb{H}^3$ is special because the behavior of the H-surfaces in $\mathbb{H}^3$ depends on the value of H in relation with the value 1, the absolute value of the sectional curvature of $\mathbb{H}^3$. If $|H| = 1$, then 1-surfaces are similar to minimal surfaces of $\mathbb{R}^3$. If $|H| < 1$, then H-surfaces have not a counterpart in $\mathbb{R}^3$. In this case, the three conjectures are true. Even more, any compact H-surface in $\mathbb{H}^3$ spanning a circle is umbilical without any further assumption on the surface [**9, 36**]. However, if $|H| > 1$, then the properties of H-surfaces are similar to cmc surfaces of $\mathbb{R}^3$: the same results that hold in $\mathbb{R}^3$ are valid in $\mathbb{H}^3$; the same open questions in $\mathbb{R}^3$ are open in $\mathbb{H}^3$.

In Lorentzian space forms, the mean curvature is defined for surfaces with non-degenerate induced metric. However the behavior of spacelike surfaces and timelike surfaces is very different. If the ambient space is the Lorentz–Minkowski space, it was proved that the only immersed compact spacelike cmc surfaces spanning a circle are planar disks and hyperbolic caps: no further assumptions are needed [**4**].

3. Part II: Arbitrary boundary curve

In this section we will consider that the boundary of a compact cmc surface is not necessarily a circle. Let $C \subset \mathbb{R}^3$ be a closed curve. If Σ is a compact H-surface bounded by C, then the value of H is not arbitrary and it depends on the geometry of C. This is due to the following argument ([**43**]; also in [**22, 32, 33**] for particular cases). The 1-form $v \mapsto (Hp + N) \times v$, $v \in T_p\Sigma$, is closed because H is constant. The divergence theorem yields $H \int_{\partial\Sigma} \alpha \times \alpha' + \int_{\partial\Sigma} \nu = 0$, where α is a parametrization by arc-length of $\partial\Sigma$ and $\nu = N \times \alpha'$ is the unit conormal vector along $\partial\Sigma$. If $\vec{v} \in \mathbb{R}^3$ is a unit vector, then

$$H \int_{\partial\Sigma} \langle \alpha \times \alpha', \vec{v} \rangle + \int_{\partial\Sigma} \langle \nu, \vec{v} \rangle = 0. \tag{3.1}$$

This identity is know as the flux formula of cmc surfaces because $\int_{\partial\Sigma} \langle \nu, \vec{v} \rangle$ is the flux of the vector field $\vec{v}$ along Σ. For example, on a compact minimal surface, the the sum of the flux along all the boundary components must vanish.

The first integral $\int_{\partial\Sigma} \langle \alpha \times \alpha', \vec{v} \rangle$ is twice the algebraic area $A_{\vec{v}}$ of the projection of C on a plane Π orthogonal to $\vec{v}$. Then $2|H||A_{\vec{v}}| \leq \int_{\partial\Sigma} |\langle \nu, \vec{v} \rangle| \leq L$, where L is the length of C. If $H \neq 0$, we conclude

$$|H| \leq \frac{L}{2|A_{\vec{v}}|}.$$

This implies that H is not arbitrary and that the value of H only depends on C. For example, suppose that C is a simple closed curve contained in a plane P and let D be the bounded domain by C in P. Then $\max\{|A_{\vec{v}}| : |\vec{v}| = 1\}$ is attained when $\vec{v}$ is orthogonal to P and its value is the area of D. Therefore $|H| \leq \frac{L}{2|D|}$. If $C = \mathbb{S}^1(r)$, then $|H| \leq \frac{1}{r}$. The case $\mathbb{S}^1(r)$ as boundary curve is special because for each $-\frac{1}{r} \leq H \leq \frac{1}{r}$ there exists an H-surface spanning $\mathbb{S}^1(r)$: planar disks and spherical caps. This phenomenon is unknown when C is an arbitrary curve.

PROBLEM 3.1 (3*). Let C be a simple planar closed curve. Is there an H-surface bounded by C for each H of the interval $[0, \frac{L}{2|D|}]$? Particularize to the case that C is a convex curve or C is an ellipse.

If J_C is the subset of $\mathbb{R}$ formed by the numbers H such that there is an H-surface bounded by C, Problem 3.1 asks the relation between J_C and the interval $[0, \frac{L}{2|D|}]$, where we know that $J_C \subset [0, \frac{L}{2|D|}]$. It is also known that for values close to 0, there exist H-graphs on D bounded by C (independently whether C is convex or not). So, there is $\epsilon > 0$ such that $[0, \epsilon) \subset J_C$. Nothing is known about the distribution of values H in the set J_C. The last value $\frac{L}{2|D|}$ is of interest. If $C = \mathbb{S}^1(r)$, this value is $1/r$. It was proved that hemispheres are the only $1/r$-surfaces bounded by $\mathbb{S}^1(r)$: not further hypotheses are needed [**13**]. For a simple planar closed curve, if $H = \frac{L}{2|D|}$, the flux formula (3.1) says that Σ is orthogonal to P along C. In particular, C is a line of curvature of Σ by the Joachimsthal theorem. With the extra hypothesis that C is convex and Σ is embedded, we know from [**14**] that Σ lies in one side of P. Then the Alexandrov reflection method can be employed obtaining that C is a circle and Σ is a spherical cap. This result also holds assuming that the angle between Σ and P is constant (but not necessarily $\pi/2$). In the general case, if Σ is an immersed H-disk and C is a simple closed curve, then $|H| = \frac{L}{2|D|}$ implies that Σ is a hemisphere and C is a circle [**37**].

PROBLEM 3.2 (3*). Let C be a simple planar closed curve. Is there a H-surface bounded by C where $H = \frac{L}{2|D|}$? What can be said about the geometry of this surface?

A method to construct cmc surfaces is by solving the Plateau problem: given a boundary curve C and $H \in \mathbb{R}$, find an immersed disk of mean curvature H and boundary C. The techniques of functional analysis look for a critical point of the functional $X \mapsto \int_\Omega \frac{1}{2}|\nabla X|^2 + \frac{2H}{3} \int_\Omega \langle X, X_u \times X_v \rangle$, where $X = X(u,v)\colon \Omega \to \mathbb{R}^3$ is a conformal map from the unit disk Ω and $X_{|\partial\Omega}$ a parametrization of C. Critical points of this functional are H-surfaces. Hildebrandt proved that if C lies included in a ball $\mathcal{B}_R$ of radius $R > 0$, then for any H such that $|H| < 1/R$, there is an H-surface bounded by C. Moreover, this surface is included in the closure of $\mathcal{B}_R$ [**23**]. Later, Gulliver and Spruck proved a similar result replacing $\mathcal{B}_R$ by a solid cylinder $\mathcal{C}_R$ of radius R [**21**]. It is natural to ask if it is possible to get a similar result bounding only one spatial coordinate. Notice that given a circle $\mathbb{S}^1(r)$, H-spherical caps bounded by $\mathbb{S}^1(r)$ lies in a slab of width $2/|H|$.

PROBLEM 3.3 (3*). Let $H \neq 0$ and let C be a closed curve contained in a slab $\mathcal{S}_w \subset \mathbb{R}^3$ of width w such that $|H| < 2/w$. Does an H-topological disk exist spanning C and included in $\mathcal{S}_w$?

It is crucial in the results of [**21, 23**] that the boundary of the sphere or of the cylinder has positive mean curvature: this is not the case of a slab. Related with the results of Hildebrandt, Gulliver and Spruck, and Barbosa considered the particular case that C is a circle. He proved that an H-surface bounded by a circle $\mathbb{S}^1(r)$ and included in a solid cylinder of radius $1/|H|$ must be a spherical cap [**5, 6**]: here there is no further hypothesis on the surface. Thus we have the following problem.

PROBLEM 3.4 (3*). Let $H \neq 0$ and let Σ be a compact H-surface spanning a circle $\mathbb{S}^1(r)$. If Σ included in the slab $\mathcal{S}_w$, with $|H| < \frac{2}{w}$, is Σ a spherical cap? If we replace $\frac{2}{w}$ by $w\frac{1}{w}$, is Σ a small spherical cap?

The results proved in [**5**, **6**] also hold for closed cmc surfaces. However, in the case that the surface is included in a slab, we can propose:

PROBLEM 3.5 (3*). Is the sphere of radius $1/H$ the only closed H-surface included in the closure of a slab of width $2/|H|$?

Notice that small spherical caps bounded by $\mathbb{S}^1(r)$ are included in the solid cylinder, where $D \subset P$ is the domain bounded by $\mathbb{S}^1(r)$.

PROBLEM 3.6 (3*). Let $C \subset P$ be a simple planar convex curve. Let Σ be a compact cmc surface spanning C. If Σ is included in $D \times \mathbb{R}$, is Σ a graph on D?

If Σ is embedded, then Σ is a graph [**34**]. On the other hand, if there exists an H-graph Σ^* on D, where H is the mean curvature of Σ, then $\Sigma = \Sigma^*$ up to reflections about P [**37**]. This is a consequence of the flux formula (3.1) and the maximum principle. In particular, this is the situation when H is sufficiently small because the existence of H-graphs on D with boundary C is assured when H varies in an interval $[0, \epsilon)$ for some $\epsilon > 0$. The above two results does not use the convexity of C.

In the same direction of the results in [**21**, **23**], Wente proved that given a real number $V > 0$, there is a topological disk Σ bounded by C and with volume V [**54**]. This surface is a critical point of the Dirichlet energy $\int_\Omega \frac{1}{2}|\nabla X|^2$ in a suitable family of conformal immersions $X\colon \Omega \to \mathbb{R}^3$ with boundary C and fixed volume V. In particular, the mean curvature of Σ must be constant. In case that C is a convex planar curve, it was proved in [**42**] that a cmc surface bounded by C is a graph if the volume of Σ is sufficiently small. Recall here the result of Ros and Rosenberg for large volume [**49**], mentioned in Section 2.

The following problem was first formulated by Rosenberg [**24**, p. 21].

PROBLEM 3.7 (4*). Let C be a non-circular closed convex planar curve. Prove that there is $V_C > 0$ depending only on C such that, if Σ is a compact cmc embedded disk spanning C and with volume $V > 0$, it holds:

(1) If $V \leq V_C$, then Σ lies in one side of P.
(2) If $V = V_C$, then Σ is tangent to P at the points of minimum curvature of C.
(3) If $V > V_C$, then Σ has points in both sides of P.

Hoffman used the Brakke's Surface Evolver to show numerical graphics that support that (3) is true [**24**]. Notice that C is not a circle in Problem 3.7.

When the volume is sufficiently small and motivated by the result of [**42**], we can ask the following problem.

PROBLEM 3.8 (4*). Let $C_1, C_2 \subset \mathbb{R}^2$ be two simple planar convex curves with disjoint interiors. Does $V_0 > 0$ exist such that if Σ is a cmc surface with boundary $C_1 \cup C_2$ and volume $V < V_0$, then Σ is not connected.

4. The Dirichlet problem for the cmc equation

A method to find H-surfaces spanning a given curve is by solving the Dirichlet problem

$$\begin{cases} \operatorname{div} \frac{Du}{\sqrt{1+|Du|^2}} & = 2H, \quad \text{in } \Omega \\ u & = \varphi, \quad \text{in } \partial\Omega, \end{cases} \tag{4.1}$$

where $u \in C^2(\Omega) \cap C^0(\overline{\Omega})$. Here Ω is a domain of $\mathbb{R}^2$, $H \in \mathbb{R}$ and $\varphi : \partial\Omega \to \mathbb{R}$ is a continuous function. The graph $\Sigma = \{(x, y, u(x, y)) : (x, y) \in \Omega\}$ is an H-surface whose boundary C is the graph of φ. Notice that integrating the first equation of (4.1) and using the divergence theorem, we obtain the flux formula (3.1). By the maximum principle, the solution of (4.1), if exists, is unique. The solvability of (4.1) requires to establish *a priori* C^1 estimates for the solutions of (4.1). This means finding estimates $|u|$ in Ω and gradient estimates $|Du|$ along $\partial\Omega$. In general, results in the literature employ suitable H-surfaces as barriers to get these estimates. A result of Serrin shows that (2.2) has a solution for *any* function φ if and only if Ω is convex and the curvature κ of $\partial\Omega$ satisfies $\kappa > 2|H|$ [**51**].

From now on we assume $\varphi = 0$ (planar boundary curve). Then it is natural to relax the Serrin's condition on H. For example, if Ω is a convex domain, a first result asserts that if $\kappa > |H|$ then (4.1) is solvable. Here the condition $\kappa > |H|$ allows the use of suitable small H-spherical caps as barriers in establishing the C^1-estimates [**37**]. In [**43**] it was proved an estimate of the height of an H-surface Σ with planar boundary, namely,

$$h \leq \frac{2\pi}{|H|} A, \tag{4.2}$$

where h is the height of Σ with respect to the boundary plane and A is the area of Σ. Thanks to (4.2), the Dirichlet problem (4.1) was solved under a variety of assumptions, namely, $2|\Omega|H^2 < \pi$, $LH^2 < \sqrt{3}\pi$ and $\operatorname{diam}(\Omega) < 1/|H|$ [**37–39, 42**]; see also [**40, 47, 48**] for other results of existence. However, if Ω is a round disk, the above inequalities are not sharp such as small spherical caps show.

PROBLEM 4.1 (4*). Let Ω be a bounded convex domain. Solve the Dirichlet problem (4.1) for $\varphi = 0$ under each one of the following hypothesis:

(1) $|\Omega|H^2 < \pi$.
(2) $LH^2 < 2\pi$.
(3) $\operatorname{diam}(\Omega) < 2/|H|$.

If Ω is a nonbounded convex domain, it was proved in [**38**] that (4.1) has a solution for $\varphi = 0$ if and only if Ω is included in a strip of width $1/|H|$. Notice that a half-cylinder of radius $\frac{1}{2|H|}$ is a graph on a strip of width $1/|H|$. Interesting questions on cmc graphs on convex domains are about the convexity of the solutions or of its level curves. Here we are assuming planar boundary curve ($\varphi = 0$ on $\partial\Omega$). It is known that the level curves are convex if the mean curvature H or the volume V of the graph are sufficiently small [**44, 50**]. However, Wang gave an example of a cmc graph on a convex domain whose level curves are not all convex [**53**].

PROBLEM 4.2 (3*). Suppose that Ω is a bounded convex domain. Let u be a solution (4.1) for $\varphi = 0$ on $\partial\Omega$. Give a control of the distribution of the convex level curves in terms of the value H. Extend the Wang's example, if possible, when Ω is not bounded.

Following (4.1) and $\varphi = 0$ along $\partial\Omega$, it is natural to ask if a solution of the Dirichlet problem is convex in the sense that the Gauss curvature K of the graph is positive everywhere. Notice that there exist pieces of unduloids obtained by cutting-off an unduloid for some particular planes parallel to the rotation axis that are graphs on convex domains but the surface has parts with negative Gauss curvature. See related results in [**18**, **26**, **41**].

PROBLEM 4.3 (4*). Let Ω be a bounded convex domain. Give conditions that assure the convexity of a cmc graph on Ω and with planar boundary

References

[1] U. Abresch, *Constant mean curvature tori in terms of elliptic functions*, J. Reine Angew. Math. **374** (1987), 169–192, DOI 10.1515/crll.1987.374.169. MR876223

[2] A. D. Aleksandrov, *Uniqueness theorems for surfaces in the large. V*, Amer. Math. Soc. Transl. (2) **21** (1962), 412–416. MR150710

[3] L. J. Alías, R. López, and B. Palmer, *Stable constant mean curvature surfaces with circular boundary*, Proc. Amer. Math. Soc. **127** (1999), no. 4, 1195–1200, DOI 10.1090/S0002-9939-99-04950-3. MR1618705

[4] L. J. Alías, R. López, and J. A. Pastor, *Compact spacelike surfaces with constant mean curvature in the Lorentz-Minkowski* 3*-space*, Tohoku Math. J. (2) **50** (1998), no. 4, 491–501, DOI 10.2748/tmj/1178224893. MR1653414

[5] J. L. M. Barbosa, *Hypersurfaces of constant mean curvature on* $\mathbf{R}^{n+1}$ *bounded by a Euclidean sphere*, Geometry and topology of submanifolds, II (Avignon, 1988), World Sci. Publ., Teaneck, NJ, 1990, pp. 1–9. MR1068731

[6] J. L. M. Barbosa, *Constant mean curvature surfaces bounded by a plane curve*, Mat. Contemp. **1** (1991), 3–15. VIIth School on Differential Geometry (Portuguese) (Campinas, 1990). MR1304297

[7] J. L. Barbosa and M. do Carmo, *A proof of a general isoperimetric inequality for surfaces*, Math. Z. **162** (1978), no. 3, 245–261, DOI 10.1007/BF01186367. MR508841

[8] J. L. Barbosa and M. do Carmo, *Stability of hypersurfaces with constant mean curvature*, Math. Z. **185** (1984), no. 3, 339–353, DOI 10.1007/BF01215045. MR731682

[9] J. L. M. Barbosa and R. Sa Earp, *Prescribed mean curvature hypersurfaces in* $H^{n+1}(-1)$ *with convex planar boundary. I*, Geom. Dedicata **71** (1998), no. 1, 61–74, DOI 10.1023/A:1005046021921. MR1624726

[10] H. Berestycki, L. A. Caffarelli, and L. Nirenberg, *Monotonicity for elliptic equations in unbounded Lipschitz domains*, Comm. Pure Appl. Math. **50** (1997), no. 11, 1089–1111, DOI 10.1002/(SICI)1097-0312(199711)50:11¡1089::AID-CPA2¿3.0.CO;2-6. MR1470317

[11] A. I. Bobenko, *All constant mean curvature tori in* $\mathbf{R}^3$, S^3, H^3 *in terms of theta-functions*, Math. Ann. **290** (1991), no. 2, 209–245, DOI 10.1007/BF01459243. MR1109632

[12] D. Brander and J. F. Dorfmeister, *The Björling problem for non-minimal constant mean curvature surfaces*, Comm. Anal. Geom. **18** (2010), no. 1, 171–194, DOI 10.4310/CAG.2010.v18.n1.a7. MR2660462

[13] F. Brito and R. Sa Earp, *Geometric configurations of constant mean curvature surfaces with planar boundary*, An. Acad. Brasil. Ciênc. **63** (1991), no. 1, 5–19. MR1115489

[14] R. Earp, F. Brito, W. H. Meeks III, and H. Rosenberg, *Structure theorems for constant mean curvature surfaces bounded by a planar curve*, Indiana Univ. Math. J. **40** (1991), no. 1, 333–343, DOI 10.1512/iumj.1991.40.40017. MR1101235

[15] J. E. Brothers (ed.), *Some open problems in geometric measure theory and its applications suggested by participants of the 1984 AMS summer institute*, Geometric measure theory and the calculus of variations (Arcata, Calif., 1984), Proc. Sympos. Pure Math., vol. 44, Amer. Math. Soc., Providence, RI, 1986, pp. 441–464, DOI 10.1090/pspum/044/840292. MR840292

[16] M. Del Pino, F. Pacard, and J. Wei, *Serrin's overdetermined problem and constant mean curvature surfaces*, Duke Math. J. **164** (2015), no. 14, 2643–2722, DOI 10.1215/00127094-3146710. MR3417183

[17] C. Delaunay, *Sur la surface de révolution dont la courbure moyenne est constante*, J. Math. Pures Appl. **6**, 309–315 (1841).

[18] H. Deng, H. Liu, and L. Tian, *Critical points of solutions for the mean curvature equation in strictly convex and nonconvex domains*, Israel J. Math. **233** (2019), no. 1, 311–332, DOI 10.1007/s11856-019-1906-2. MR4013976

[19] B. Gidas, W. M. Ni, and L. Nirenberg, *Symmetry and related properties via the maximum principle*, Comm. Math. Phys. **68** (1979), no. 3, 209–243. MR544879

[20] D. Gilbarg and N. S. Trudinger, *Elliptic partial differential equations of second order*, Classics in Mathematics, Springer-Verlag, Berlin, 2001. Reprint of the 1998 edition. MR1814364

[21] R. Gulliver and J. Spruck, *The Plateau problem for surfaces of prescribed mean curvature in a cylinder*, Invent. Math. **13** (1971), 169–178, DOI 10.1007/BF01390099. MR298515

[22] E. Heinz, *On the nonexistence of a surface of constant mean curvature with finite area and prescribed rectifiable boundary*, Arch. Rational Mech. Anal. **35** (1969), 249–252, DOI 10.1007/BF00248159. MR246237

[23] S. Hildebrandt, *On the Plateau problem for surfaces of constant mean curvature*, Comm. Pure Appl. Math. **23** (1970), 97–114, DOI 10.1002/cpa.3160230105. MR256276

[24] D. Hoffman, *Comment utiliser un ordinateur pour trouver de nouvelles surfaces minimales et des bulles de savon*, Surfaces minimales et solutions de problèmes variationnels, SMF Journ. Annu., vol. 1993, Soc. Math. France, Paris, 1993, pp. 29. MR1483381

[25] H. Hopf, *Differential geometry in the large*, Lecture Notes in Mathematics, vol. 1000, Springer-Verlag, Berlin, 1983. Notes taken by Peter Lax and John Gray; With a preface by S. S. Chern, DOI 10.1007/978-3-662-21563-0. MR707850

[26] W.-H. Huang and C.-C. Lin, *Negatively curved sets on surfaces of constant mean curvature in* $\mathbf{R}^3$ *are large*, Arch. Rational Mech. Anal. **141** (1998), no. 2, 105–116, DOI 10.1007/s002050050074. MR1615516

[27] C. Isenberg, *The science of soap films and soap bubbles*, Tieto Ltd., Clevedon, 1978. With a foreword by George Porter. MR620322

[28] J. H. Jellet, *Sur la surface dont la courbure moyenne est constant*, J. Math. Pures Appl. **18**, 163–167 (1853).

[29] N. Kapouleas, *Compact constant mean curvature surfaces in Euclidean three-space*, J. Differential Geom. **33** (1991), no. 3, 683–715. MR1100207

[30] N. Kapouleas, *Constant mean curvature surfaces constructed by fusing Wente tori*, Proc. Nat. Acad. Sci. U.S.A. **89** (1992), no. 12, 5695–5698, DOI 10.1073/pnas.89.12.5695. MR1165926

[31] M. Koiso, *Symmetry of hypersurfaces of constant mean curvature with symmetric boundary*, Math. Z. **191** (1986), no. 4, 567–574, DOI 10.1007/BF01162346. MR832814

[32] N. Korevaar and R. Kusner, *The global structure of constant mean curvature surfaces*, Invent. Math. **114** (1993), no. 2, 311–332, DOI 10.1007/BF01232673. MR1240641

[33] R. Kusner, *Bubbles, conservation laws, and balanced diagrams*, Geometric analysis and computer graphics (Berkeley, CA, 1988), Math. Sci. Res. Inst. Publ., vol. 17, Springer, New York, 1991, pp. 103–108, DOI 10.1007/978-1-4613-9711-3_11. MR1081331

[34] R. López, *A note on H-surfaces with boundary*, J. Geom. **60** (1997), no. 1-2, 80–84, DOI 10.1007/BF01252219. MR1477073

[35] R. López, *Surfaces of constant mean curvature bounded by two planar curves*, Ann. Global Anal. Geom. **15** (1997), no. 3, 201–210, DOI 10.1023/A:1006545410182. MR1456507

[36] R. López, *Constant mean curvature surfaces with boundary in hyperbolic space*, Monatsh. Math. **127** (1999), no. 2, 155–169, DOI 10.1007/s006050050030. MR1669875

[37] R. López, *Constant mean curvature surfaces with boundary in Euclidean three-space*, Tsukuba J. Math. **23** (1999), no. 1, 27–36, DOI 10.21099/tkbjm/1496163774. MR1693145

[38] R. López, *Constant mean curvature graphs on unbounded convex domains*, J. Differential Equations **171** (2001), no. 1, 54–62, DOI 10.1006/jdeq.2000.3830. MR1816793

[39] R. López, *An existence theorem of constant mean curvature graphs in Euclidean space*, Glasg. Math. J. **44** (2002), no. 3, 455–461, DOI 10.1017/S0017089502030100. MR1956553

[40] R. López, *Constant mean curvature graphs in a strip of* $\mathbb{R}^2$, Pacific J. Math. **206** (2002), no. 2, 359–373, DOI 10.2140/pjm.2002.206.359. MR1926782

[41] R. López, *Gradient estimates for the constant mean curvature equation in hyperbolic space*, Proc. Roy. Soc. Edinburgh Sect. A **150** (2020), no. 6, 3216–3230, DOI 10.1017/prm.2019.71. MR4190109

[42] R. López and S. Montiel, *Constant mean curvature discs with bounded area*, Proc. Amer. Math. Soc. **123** (1995), no. 5, 1555–1558, DOI 10.2307/2161147. MR1286001

[43] R. López and S. Montiel, *Constant mean curvature surfaces with planar boundary*, Duke Math. J. **85** (1996), no. 3, 583–604, DOI 10.1215/S0012-7094-96-08522-1. MR1422358

[44] J. McCuan, *Continua of H-graphs: convexity and isoperimetric stability*, Calc. Var. Partial Differential Equations **9** (1999), no. 4, 297–325, DOI 10.1007/s005260050143. MR1731469

[45] W. H. Meeks III, *Lectures on Plateau's Problem*, Instituto de Matematica Pura e Aplicada (IMPA), Rio de Janeiro, Brazil, 1978.

[46] U. Pinkall and I. Sterling, *On the classification of constant mean curvature tori*, Ann. of Math. (2) **130** (1989), no. 2, 407–451, DOI 10.2307/1971425. MR1014929

[47] J. Ripoll, *Some characterization, uniqueness and existence results for Euclidean graphs of constant mean curvature with planar boundary*, Pacific J. Math. **198** (2001), no. 1, 175–196, DOI 10.2140/pjm.2001.198.175. MR1831977

[48] J. Ripoll, *Some existence results and gradient estimates of solutions of the Dirichlet problem for the constant mean-curvature equation in convex domains*, J. Differential Equations **181** (2002), no. 1, 230–241, DOI 10.1006/jdeq.2001.4050. MR1900466

[49] A. Ros and H. Rosenberg, *Constant mean curvature surfaces in a half-space of* $\mathbf{R}^3$ *with boundary in the boundary of the half-space*, J. Differential Geom. **44** (1996), no. 4, 807–817. MR1438193

[50] S. Sakaguchi, *Uniqueness of critical point of the solution to the prescribed constant mean curvature equation over convex domain in* $\mathbf{R}^2$, Recent topics in nonlinear PDE, IV (Kyoto, 1988), North-Holland Math. Stud., vol. 160, North-Holland, Amsterdam, 1989, pp. 129–151, DOI 10.1016/S0304-0208(08)70510-3. MR1041379

[51] J. Serrin, *The problem of Dirichlet for quasilinear elliptic differential equations with many independent variables*, Philos. Trans. Roy. Soc. London Ser. A **264** (1969), 413–496, DOI 10.1098/rsta.1969.0033. MR282058

[52] J. Serrin, *A symmetry problem in potential theory*, Arch. Rational Mech. Anal. **43** (1971), 304–318, DOI 10.1007/BF00250468. MR333220

[53] X.-J. Wang, *Counterexample to the convexity of level sets of solutions to the mean curvature equation*, J. Eur. Math. Soc. (JEMS) **16** (2014), no. 6, 1173–1182, DOI 10.4171/JEMS/457. MR3226739

[54] H. C. Wente, *A general existence theorem for surfaces of constant mean curvature*, Math. Z. **120** (1971), 277–288, DOI 10.1007/BF01117500. MR282300

[55] A. M. de Souza Doering, *The depth conjecture: a counterexample*, J. Algebra **77** (1982), no. 2, 443–448, DOI 10.1016/0021-8693(82)90265-4. MR673127

[56] The Center for Geometry, Analysis, Numerics & Graphics (GANG). `http://www.gang.umass.edu/gallery/cmc/`

[57] GeometrieWerkstatt Surface Gallery. `https://www.math.uni-tuebingen.de/user/nick/gallery/`

[58] m:iv research network. `https://www.miv-gallery.net/`

Departamento de Geometría y Topología, Universidad de Granada, 18071 Granada, Spain

Email address: `rcamino@ugr.es`

Contemporary Mathematics
Volume **821**, 2025
https://doi.org/10.1090/conm/821/16440

On holomorphic conformal structures associated with lattice polarized K3 surfaces

Andreas Malmendier and Michael T. Schultz

Abstract. We discuss the connection between Picard–Fuchs equations for certain families of lattice polarized K3 surfaces and the construction of integrable holomorphic conformal structures on their period domains. We then compute an explicit example of a locally conformally flat holomorphic metric associated with generic Jacobian Kummer surfaces, which allows for a novel description of the local variation of complex structure.

1. Introduction

Among complex manifolds, the existence of holomorphic conformal structures are quite restrictive [**24**], as opposed to the well known conformal structures in the real Riemannian or semi-Riemannian case [**23**]. In this article, we will explore holomorphic conformal structures that are integrable, or equivalently, locally conformally flat. Quadric hypersurfaces $Q^n \subset \mathbb{P}^{n+1}$ naturally possess such an integrable holomorphic conformal structure and serve as the model spaces of (holomorphic) conformal geometry; see Kuiper [**25**] and Kobayashi [**22**, IV. Theorem 6.2]. As we will explain, the study of integrable holomorphic conformal structures, well known in the field of differential geometry, finds another powerful application in the field of algebraic geometry, namely, within the study of algebraic families of certain algebraic varieties.

Such a situation arises when a proper smooth holomorphic map $f\colon \mathfrak{X} \to M$ is given between complex algebraic varieties, where $\mathfrak{X}, M$, and the general fiber $X_t = f^{-1}(t)$ for $t \in M$ are smooth and connected, and a distinguished class of projective embeddings $\mathfrak{X} \subset \mathbb{P}^N$ is given; see [**17**]. Fix a generic point $t = 0 \in M$. Then we can think of this situation as a family $\{X_t\}_{t\in M}$ of generically smooth, compact, complex projective algebraic varieties with the algebraic parameter space M. However, in general, M is not complete, and X_t acquires singularities as t tends to ∞ or, more generally, towards a more complicated discriminant locus contained in M with $t \neq 0$. The fundamental problem is to describe the variation of the complex structure of the compact algebraic manifold X_0.

2020 *Mathematics Subject Classification.* Primary 14D07, 14J28, 53C18.

Key words and phrases. K3 surfaces, lattice polarizations, period map, Picard-Fuchs equations, holomorphic conformal structures.

The first author acknowledges support from the Simons Foundation through grant no. 202367.

In this article, we will investigate algebraic families of varieties arising as lattice polarized K3 surfaces. A family of lattice polarized K3 surfaces can be constructed for every even, primitive sublattice L of rank $\rho = 20 - n > 0$ of the K3 lattice $\Lambda_{\mathrm{K3}} \cong 3H \oplus 2E_8(-1)$ of signature $(3, 19)$. These L-polarized K3 surfaces are classified, up to isomorphism, by a coarse moduli space $\mathfrak{M}_L$, which is a quasi-projective variety of dimension n. Given an algebraic family of L-polarized K3 surfaces $f_L : \mathfrak{X} \to M$ over an n-dimensional open subspace M, disjoint from the discriminant locus, one can compute a holonomic system E of rank $n + 2$ in n variables, whose solutions are the period integrals of X_t. This linear system of partial differential equations is called the *Picard–Fuchs system* for the family f_L. This system is a fundamental object of study, encoding crucial information about the geometry of the K3 surface as the complex structure varies with $t \in M$. In fact, a connection on a vector bundle over the base space M can now be constructed where the fibers are the direct sum of the Hodge groups $H^{p,q}(X_t)$. As such a family is locally topologically trivial, cohomology classes can be moved from one fiber in the family to nearby fibers, providing flat local sections of the bundle. In turn, from these flat sections, the existence of such a connection can be inferred, called the Gauss–Manin connection [**16**].

The goal of this article is to explain the great utility of the framework of integrable holomorphic conformal structures in the study of Picard–Fuchs equations for families of lattice polarized K3 surfaces. While this connection has been known for quite sometime, it does not appear to have been used much since the pioneering work of Sasaki and Yoshida [**36**]. We will demonstrate how the Picard–Fuchs system for a family $f_L : \mathfrak{X} \to M$ of L-polarized K3 surfaces imparts M with an integrable holomorphic conformal structure. In fact, the period map for f_L determines a (multivalued) inverse of the canonical projection $\mathcal{D}_L \to \mathfrak{M}_L$ where $\mathcal{D}_L$ denotes the period domain. The latter can be identified with a relatively open subset of a quadric hypersurface in $\mathbb{P}^{n+1}$, and thus admits a natural integrable conformal structure. While the holomorphic conformal structure is sufficiently interesting in its own right, we will see that in this context it efficiently encodes much more information—essentially, the entire local variation of Hodge (or equivalently in this case, complex) structure.

This article is structured as follows: after reviewing some aspects of lattice polarized K3 and Kummer surfaces and the geometry of their period domains, we provide a new example of an integrable holomorphic conformal structure associated with Jacobian Kummer surfaces, allowing for a novel description of their local variation of complex structure.

2. Lattice polarized K3 & Kummer surfaces

Let us start by reviewing a few classical lattice theory facts. We shall use the following standard notations: $L_1 \oplus L_2$ is orthogonal direct sum of the two lattices L_1 and L_2, $nL = L^{\oplus n}$ is the n-fold orthogonal direct sum of L, $L(\lambda)$ is obtained from the lattice L by multiplication of its form by $\lambda \in \mathbb{Z}$, $\langle R \rangle$ is a lattice with the matrix R in some basis; A_n, D_m, and E_k are the positive definite root lattices for the corresponding root systems; H is the unique even unimodular hyperbolic rank-two lattice. Given a lattice L, one has the discriminant group $A(L) = L^\vee / L$ and its associated discriminant form, denoted q_L. For a lattice L and a field $\mathbb{K}$ of characteristic zero, let $L_{\mathbb{K}} = L \otimes_{\mathbb{Z}} \mathbb{K}$. We will continue to denote by $(\cdot,\cdot)$, by a slight misuse of notation, the $\mathbb{K}$-extension of the bilinear form on L to $L_{\mathbb{K}}$.

A K3 surface X is a smooth compact complex surface with trivial canonical bundle $\omega_X := \bigwedge^2 \Omega^1_X \cong \mathcal{O}_X$ and $h^1(X, \mathcal{O}_X) = 0$. Here, Ω^1_X is the holomorphic cotangent bundle of X. It is known that every K3 surface is Kähler. The second cohomology group $H^2(X,\mathbb{Z})$, equipped with the pairing induced by the cup product $(\cdot,\cdot)$, forms a lattice isometric to the even rank-22 K3 lattice $\Lambda_{K3} \cong 3H \oplus 2E_8(-1)$ of signature $(3,19)$. Up to isometry, Λ_{K3} is the only such even lattice with the given signature. The isometry is called the marking of the K3 surface X. Thus, we have

$$\Lambda_{K3} \otimes \mathbb{C} \ \cong\ H^2(X,\mathbb{C}) = H^{2,0}(X) \oplus H^{1,1}(X) \oplus H^{0,2}(X)$$

by the Hodge decomposition. The Néron–Severi lattice is given by

$$\mathrm{NS}(X) := H^{1,1}(X) \cap H^2(X,\mathbb{Z}). \tag{2.1}$$

Via the first Chern class map it is isomorphic to the Picard lattice $\mathrm{Pic}(X)$. This lattice is even and has signature $(1,\rho_X - 1)$, where $1 \le \rho_X \le 20$ is the Picard number of X. The transcendental lattice T_X is defined to be the smallest sublattice of $H^2(X,\mathbb{Z})$ whose complexification contains a generator of $H^{2,0}(X)$. In the non-degenerate case, it follows that T_X is the orthogonal complement of $\mathrm{NS}(X)$ in $H^2(X,\mathbb{Z})$ and has signature $(2, 20-\rho_X)$. Conversely, for every primitive sublattice L (or N) of Λ_{K3} of signature $(1,\rho_X - 1)$ (or $(2, 20-\rho_X)$), there is an algebraic K3 surface X together with a marking such that $L \cong \mathrm{NS}(X)$ (or $N \cong T_X$). Here, the term algebraic surface refers to a complex algebraic variety of dimension two.

A (pseudo-)polarized K3 surface of degree $2k$ with $k > 0$ is a pair (X,h) consisting of a K3 surface X and a primitive (pseudo-)ample class $h \in \mathrm{NS}(X)$ with $(h,h) = 2k$. Moreover, two (pseudo-)polarized K3 surfaces (X,h) and (X',h') of degree $2k$ are equivalent if there exists an analytic isomorphism $\alpha\colon X \to X'$ such that $\alpha^* h' = h$. If $\mathrm{NS}(X)$ contains a pseudo-ample class, then it follows that X is projective, hence every pseudo-polarized K3 surface is projective [**21**].

There are important classes of examples of K3 surfaces that arise as the result of a pseudo-polarization. The first is the class of double sextic surfaces; they are pseudo-polarized K3 surfaces (X,h) of degree 2 with an effective divisor D on X such that $[D] = h \in \mathrm{NS}(X)$: in the generic case, the linear system $|D|$ then defines a map from X to $\mathbb{P}^2$ that is generically 2:1. Accordingly, X is birational to the double cover of $\mathbb{P}^2$ branched on a sextic curve. A second class of examples are quartic hypersurfaces. They are pseudo-polarized K3 surfaces (X,h) of degree 4 with an effective divisor D on X such that $[D] = h$: in the generic case, $|D|$ defines a birational morphism onto a quartic hypersurface in $\mathbb{P}^3$. Quartic hypersurfaces, for example the Fermat quartic or the Hudson quartic, constitute some of the famous examples of complex projective K3 surfaces. For degree 4, there is also a second situation that can arise, called a double quadric surface. Here, $|D|$ defines a generically 2:1 map from X onto a quadric hypersurface in $\mathbb{P}^3$ that is isomorphic to $\mathbb{P}^1 \times \mathbb{P}^1$. Accordingly, X is birational to the double cover of $\mathbb{P}^1 \times \mathbb{P}^1$ branched on a curve of bi-degree $(4,4)$.

Let X be a smooth complex projective K3 surface. A lattice polarization on X is defined as a primitive lattice embedding $\jmath\colon L \hookrightarrow \mathrm{NS}(X)$, where $\jmath(L)$ contains a pseudo-ample class. Here, L is an even indefinite lattice of signature $(1,\rho_L - 1)$, with $1 \le \rho_L \le 20$. Two K3 surfaces $(X,\jmath)$ and $(X',\jmath')$ are said to be isomorphic under L-polarization if there exists an analytic isomorphism $\alpha\colon X \to X'$ and a lattice isometry $\beta \in O(L)$ such that $\alpha^* \circ \jmath' = \jmath \circ \beta$, where α^* is the induced morphism at the cohomology level.[1] L-polarized K3 surfaces are classified up to isomorphism by a coarse moduli space $\mathfrak{M}_L$, which is a quasi-projective variety of dimension $20 - \rho_L$, see §3.1. A general L-polarized K3 surface $(X, \jmath)$ satisfies $\jmath(L) = \mathrm{NS}(X)$.

2.1. Kummer surfaces. We also consider special K3 surfaces known as Kummer surfaces; they are associated with abelian surfaces. First, we recall that a complex 2-dimensional torus T^2 is a surface isomorphic to the quotient of $\mathbb{C}^2$ by a real non-degenerate lattice G of rank 4. The group $H^2(T^2, \mathbb{Z})$, equipped with the pairing induced by the cup product, is an even unimodular lattice of signature $(3,3)$. Up to isometry, there exists only one lattice with these properties, namely the lattice $3H$. An abelian surface A is an algebraic torus. Moreover, for every primitive sublattice S (or T) of $3H$ of signature $(1, \rho - 1)$ (or $(2, 4 - \rho)$) with $1 \le \rho \le 4$, there is an abelian surface A with a marking such that $S \cong \mathrm{NS}(A)$ (or $T \cong T_A$). Every abelian surface admits an involution $\imath$: the involution $\imath$ maps every element $a \in A$ to its inverse $-a \in A$ with respect to the group law. The fixed points of this map are the sixteen points of order two on A. The Kummer surface Y associated with A is the K3 surface obtained as the minimal resolution of the quotient $A/\langle \imath \rangle$. Let $\widetilde{A}$ be the blow up of A in the sixteen fixed points of the involution $\imath$ and let $\pi \colon \widetilde{A} \to Y$ be the 2:1 cover between $\widetilde{A}$ and Y. One can easily check that $\pi_* H^2(\widetilde{A}, \mathbb{Z})^\imath \cong 3H(2)$.

By construction, the Kummer surface Y admits an even set of sixteen rational curves, coming from the blow up of the order-2 points on the abelian surface, with classes $K_1, \dots, K_{16}$. The term even refers to the fact that the classes add up to an even element in $\mathrm{NS}(Y)$, that is, $K_1 + \dots + K_{16} \in 2\mathrm{NS}(Y)$. An even set made up of disjoint rational curves on any K3 surface must contain either eight or sixteen curves [**34**]. The classes $K_1, \dots, K_{16}$ on Y generate a minimal primitive sublattice K of $\mathrm{NS}(Y)$, and hence a sublattice of Λ_{K3}. The lattice K is the orthogonal complement of $\pi_* H^2(\widetilde{A}, \mathbb{Z})^\imath$ in Λ_{K3}, whence a negative definite even lattice of rank 16 and determinant of the discriminant form 2^6. It was described explicitly in [**34**].

A polarization of an abelian surface $A \cong \mathbb{C}^2/G$, over the complex numbers, is a positive definite hermitian form H on $\mathbb{C}^2$, satisfying $\operatorname{Im} H(\Lambda, \Lambda) \subset \mathbb{Z}$. Every such hermitian form determines a line bundle $\mathcal{L} \to A$ in the Néron–Severi lattice $\mathrm{NS}(A)$. A general fact from linear algebra asserts that one can always choose a basis of G such that $E = \operatorname{Im} H$ is given by the matrix $\left(\begin{smallmatrix} 0 & M \\ -M & 0 \end{smallmatrix}\right)$ with $M = \left(\begin{smallmatrix} d_1 & 0 \\ 0 & d_2 \end{smallmatrix}\right)$ where $d_1, d_2 \in \mathbb{N}$, $d_1, d_2 \ge 0$, and d_1 divides d_2. The pair (d_1, d_2) gives the type of the polarization; see [**2**]. A polarization of type $(1,1)$ is also called a principal polarization.

For a principally polarized abelian variety A the line bundle $\mathcal{L}$ defining its principal polarization is ample and satisfies $h^0(\mathcal{L}) = 1$. There exists an effective divisor Θ such that $\mathcal{L} = \mathcal{O}_A(\Theta)$, uniquely defined only up to translations. The divisor $\Theta \in \mathrm{NS}(A)$ is called a theta divisor associated with the polarization. Moreover, it is known that a principally polarized abelian surface is either the Jacobian of a smooth curve C of genus 2 with canonical theta divisor [**2**, §11.1], or the product of two complex elliptic curves, with the product polarization [**33**].

For the principally polarized Abelian surfaces $A = \operatorname{Jac} C$ where C is a smooth curve of genus 2 with $T_A = 2H + \langle -2 \rangle$, the general Kummer surface associated with

[1] Our definition of isomorphic lattice polarizations coincides with the one used by Vinberg, and is slightly more general than the one used in [**12**, Section 1].

A satisfies

$$\text{NS}(Y) = H \oplus D_8(-1) \oplus D_4(-1) \oplus A_3(-1) \cong H \oplus D_7(-1) \oplus 2D_4(-1). \tag{2.2}$$

This follows from the results in **[26]**. We also refer to the K3 surface Y as a *generic* Jacobian Kummer surface. A similar statement holds for Kummer surfaces associated with the product of two non-isogenous elliptic curves such that $T_A = 2H$, a specialization of the Kummer surface where the lattice polarization extends. See **[27]**.

2.2. Double sextic surfaces. The Jacobian Kummer surfaces Y can be described explicitly as double sextic surfaces as follows: let C be given in affine coordinates (ξ, η) by the Rosenhain normal form

$$C\colon \quad \eta^2 = \xi\,(\xi-1)\,(\xi-\lambda_1)\,(\xi-\lambda_2)\,(\xi-\lambda_3). \tag{2.3}$$

We denote the hyperelliptic involution on C by $\imath_C$. An ordered tuple $(\lambda_1, \lambda_2, \lambda_3)$—where the λ_i are pairwise distinct and different from $(\lambda_4, \lambda_5, \lambda_6) = (0, 1, \infty)$—determines a point in the moduli space of genus-2 curves with level-two structure.[2]

For the symmetric product $C^{(2)}$, the quotient $C^{(2)}/\langle \imath_C \times \imath_C \rangle$ is realized as a variety in terms of variables $t = \xi^{(1)}\xi^{(2)}$, $x = \xi^{(1)} + \xi^{(2)}$, and $y = \eta^{(1)}\eta^{(2)}$, satisfying the affine equation

$$y^2 = t\big(t - x + 1\big) \prod_{i=1}^{3} \big(\lambda_i^2\, t - \lambda_i\, x + 1\big). \tag{2.4}$$

The affine variety completes to a hypersurface in $\mathbb{P}(1,1,1,3)$ which is birational to the Kummer surface $\text{Kum}(\text{Jac}\, C)$, i.e., we have

$$Y \cong \text{Kum}(\text{Jac}\, C). \tag{2.5}$$

We introduce the following six lines $\ell_1, \dots, \ell_6$ in $\mathbb{P}^2 = \mathbb{P}(z_1, z_2, z_3)$:

$$\ell_1 = z_1\,,\ \ell_2 = z_2\,,\ \ell_3 = z_1 + z_2 - z_3\,,\ \ell_{3+i} = z_1 + \lambda_i^2 z_2 - \lambda_i z_3 \ (\text{for } i = 1, 2, 3)\,, \tag{2.6}$$

which are tangent to a common conic $z_3^2 - 4 z_1 z_2 = 0$. Then, Equation (2.4) is the double cover of $\mathbb{P}^2$ branched on $\ell_1, \dots, \ell_6$ in the affine chart $z_1 = 1, z_2 = t, z_3 = x$.

Generalizing further, we consider the minimal resolution of a double cover of the projective plane $\mathbb{P}^2$ branched along the union of six lines in general position. We denote the six-line configuration by $\boldsymbol{\ell} = \{\ell_1, \dots, \ell_6\}$. Using the weighted homogeneous coordinates $[z_1 : z_2 : z_3 : y] \in \mathbb{P}(1,1,1,3)$, the family of double sextics is given by

$$y^2 = \prod_{i=1}^{6} \big(a_{i1} z_1 + a_{i2} z_2 + a_{i3} z_3\big), \tag{2.7}$$

where the lines $\ell_i = \{[z_1 : z_2 : z_3] \mid a_{i1} z_1 + a_{i2} z_2 + a_{i3} z_3 = 0\} \subset \mathbb{P}^2$ for parameters $a_{ij} \in \mathbb{C}$, $i = 1, \dots, 6$, $j = 1, 2, 3$ are assumed to be in general position. Let $(a_{ij}) \in \text{Mat}(3, 6; \mathbb{C})$ be the matrix whose entries are the coefficients encoding the six-line configuration $\boldsymbol{\ell}$. The general member of the family in Equation (2.7) is a smooth K3 surface of Picard number 16. It follows from **[9]** that Equation (2.7) defines a family

[2] A full level-2 structure is basis of rational 2-torsion points on the principally polarized abelian surface $\text{Jac}\, C$.

of L-polarized K3 surfaces, where L has rank 16 with the following isomorphic presentations:

$$L \cong H \oplus E_8(-1) \oplus 6A_1(-1) \cong H \oplus D_8(-1) \oplus D_4(-1) \oplus 2A_1(-1). \tag{2.8}$$

Let $\mathcal{D}$ be the configuration space of six lines $\boldsymbol{\ell}$ whose minimal resolution is a K3 surface. Note that isomorphic K3 surfaces are obtained if we act on an element (a_{ij}) by matrices induced by automorphisms of $\mathbb{P}^2$ on the left and overall scale changes of each line $\ell_i \in \boldsymbol{\ell}$ on the right. Thus, we are led to consider the four-dimensional quotient space

$$\mathcal{M} = \mathrm{SL}(3,\mathbb{C})\backslash\mathcal{D}/(\mathbb{C}^*)^6. \tag{2.9}$$

The family of double-sextics in Equation (2.7) has been studied in the literature, famously by Matsumoto [**29**], and Matsumoto et al. [**30–32**]. In [**30**], the authors used the action of $\mathrm{SL}(3,\mathbb{C})$ and $(\mathbb{C}^*)^6$ to bring the matrix (a_{ij}) into the form

$$\begin{pmatrix} 1 & 0 & 0 & 1 & 1 & 1 \\ 0 & 1 & 0 & 1 & x_1 & x_2 \\ 0 & 0 & 1 & 1 & x_3 & x_4 \end{pmatrix}, \tag{2.10}$$

so that the associated K3 surfaces are the minimal resolution of the double sextic surfaces

$$\mathcal{X}\colon \quad y^2 = z_1 z_2 z_3 \big(z_1 + z_2 + z_3\big)\big(z_1 + x_1 z_2 + x_3 z_3\big)\big(z_1 + x_2 z_2 + x_4 z_3\big). \tag{2.11}$$

A six-line configuration $\boldsymbol{\ell}$ is tangent to a conic if and only if a certain projective invariant R vanishes; see [**11**]. For a six-line configuration given by Equation (2.10), one finds

$$R = -x_1 x_2 x_3 + x_1 x_2 x_4 + x_1 x_3 x_4 - x_2 x_3 x_4 - x_1 x_4 + x_2 x_3. \tag{2.12}$$

The subvariety determined by $R = 0$ is also refereed to as the *Kummer sublocus* or the set of Kummer configurations of six lines. For a general member Y of the Kummer locus the Néron-Severi lattice is given by Equation (2.2). Conversely, a K3 surface $\mathcal{X}$ with $\mathrm{NS}(\mathcal{X}) \cong L$ with L in Equation (2.2) is a generic Jacobian Kummer surface and isomorphic to a double sextic surface branched on a Kummer configuration of six lines.

3. Geometry of period domains for lattice polarized K3 surfaces

In this section we review aspects of the period map for lattice polarized K3 surfaces. The corresponding periods are the solutions of the Picard-Fuchs system. We then review Sasaki and Yoshida's [**36**] differential geometric interpretation of such equations, which equips the moduli space with a canonical integrable holomorphic conformal structure.

3.1. Period domains and the period map. Let us start by reviewing how the transcendental lattice of a general member in a family of lattice polarized K3 surfaces determines a local deformation of complex structure.

Let $L \overset{\jmath}{\hookrightarrow} \Lambda_{\mathrm{K3}}$ be a primitive sublattice of rank ρ of the K3 lattice, and let X be a general member in the family of L-polarized K3 surfaces, that is also equipped with a marking as in §2. Let $N = L^{\perp}$ be the orthogonal complement in the K3 lattice.. Assume that N is an even lattice of signature $(2,n)$ with $n \in \mathbb{N}$ so that

$$n + 2 = \mathrm{rank}(N) = \mathrm{rank}(\Lambda_{\mathrm{K3}}) - \mathrm{rank}(L) = 22 - \rho .$$

Since $H^{2,0}(X) \perp L_{\mathbb{C}}$, the line $H^{2,0}(X) \subset N_{\mathbb{C}}$ is contained in the complexified transcendental lattice. If $\pi : N_{\mathbb{C}} - \{0\} \to \mathbb{P}(N_{\mathbb{C}})$ is the canonical projection, it follows from the Torelli theorem of Burns & Rapoport for algebraic K3 surfaces [**5**] that the marked K3 surface X determines a unique point in $\mathbb{P}(N_{\mathbb{C}})$, called the period point, given by

$$p_X = \pi\big(H^{2,0}(X)\big) \in \mathbb{P}(N_{\mathbb{C}}) = \mathbb{P}^{n+1} . \tag{3.1}$$

Geometrically, the period point $p_X \in \mathbb{P}^{n+1}$ can be described as follows. Choosing a global nonvanishing section of the canonical bundle ω_X is equivalent to a nonvanishing holomorphic 2-form Ω with $[\Omega] \in H^{2,0}(X)$. Using Poincaré duality, we have $H_2(X,\mathbb{Z}) \cong N \oplus L$. We fix an ordered basis $\{\gamma_0, \dots, \gamma_{n+1}, \gamma_{n+2}, \dots, \gamma_{21}\} \subset H_2(X,\mathbb{Z})$ of 2-cycles so that $N \cong \mathrm{span}_{\mathbb{Z}}\{\gamma_0, \dots, \gamma_{n+1}\}$ and $L \cong \mathrm{span}_{\mathbb{Z}}\{\gamma_{n+2}, \dots, \gamma_{21}\}$. It follows that the integral of Ω over any algebraic cycle $\gamma \in L$ vanishes, i.e.,

$$\oint_\gamma \Omega = 0 ,$$

and the only nontrivial period integrals arise by integrating over transcendental cycles. In this way, the period point can also be obtained as

$$p_X := \left[\oint_{\gamma_0} \Omega : \cdots : \oint_{\gamma_{n+1}} \Omega \right] \in \mathbb{P}^{n+1} . \tag{3.2}$$

Since $[\omega \wedge \omega] \in H^{4,0}(X) = \{0\}$ we have $(\Omega, \Omega) = 0$, and it follows that p_X lies on an n-dimensional quadric hypersurface in $\mathbb{P}^{n+1}$ which we will denote by $Q^n \subset \mathbb{P}^{n+1}$. The hypersurface Q^n is defined as the projectivized locus of isotropic vectors in $N_{\mathbb{C}}$ with respect to the quadratic form $(\cdot, \cdot)$. Moreover, since

$$\big(\Omega, \overline{\Omega}\big) > 0 , \tag{3.3}$$

it follows that $p_X \in \mathcal{D}_L$, where $\mathcal{D}_L$ is the relatively open subset of Q^n defined by Equation (3.3). Hence, $\mathcal{D}_L$ is a quasi-projective subvariety of Q^n, called the period domain. The period domain inherits more geometric structure than that of a quasi-projective variety. In fact, identifying $H^{2,0}(X) \subset N_{\mathbb{C}}$ with a positive definite oriented 2-plane in $N_{\mathbb{R}}$, it follows that $\mathcal{D}_L$ can be identified with the symmetric homogeneous space $\mathcal{D}_L \cong \mathrm{O}(2,n)/(\mathrm{SO}(2) \times \mathrm{O}(n))$ of positive oriented 2-planes in $N_{\mathbb{R}}$. We refer the interested reader to Aspinwall [**1**, §2.5] for an overview and a discussion of the connection with the moduli space of Kähler-Einstein metrics on a smooth K3 surface. Then $\mathcal{D}_L$ has two connected components which are exchanged under the induced involution by complex conjugation on forms. Each component is isomorphic to a bounded Hermitian symmetric domain of type IV_n. The coarse moduli space $\mathfrak{M}_L$ introduced in §2 then takes the form of $\mathfrak{M}_L = \mathcal{D}_L/\Gamma_L$, where Γ_L is the kernel of the natural homomorphism $\mathrm{O}(L,\mathbb{Z}) \to A(L)$; see Dolgachev [**12**, §3] for a detailed discussion.

Next, we consider the family of marked L-polarized K3 surfaces $f_L : \mathfrak{X} \to M$ over a complex quasi-projective variety M where f_L is a proper smooth holomorphic map and each fiber over $t \in M$ is an L-polarized K3 surface X_t, that is also equipped with a basis $\{\gamma_i(t)\}_{i=0}^{21} \subset H_2(X_t,\mathbb{Z})$ of locally constant 2-cycles on which to integrate a nondegenerate holomorphic 2-form $\Omega_t \in H^{2,0}(X_t)$; see Dolgachev [**12**, Remark 3.4] for more details. The holomorphic map

$$\begin{aligned} p: \quad & M \to \mathcal{D}_L \subset Q^n \\ & t \mapsto p_{X_t} = \left[\oint_{\gamma_0(t)} \Omega_t : \cdots : \oint_{\gamma_{n+1}(t)} \Omega_t \right] \in \mathcal{D}_L , \end{aligned} \tag{3.4}$$

is called the period map of the family $f_L : \mathfrak{X} \to M$. The family is called maximal if the period mapping is surjective. In this case we must that the family is not isotrivial and that $\dim(M) = n$, the latter by the local Torelli theorem of Burns & Rapoport [**5**] which guarantees that the period mapping is locally an isomorphism. In this way, the period map of a family encodes the local variation of the complex structure in the K3 fibers of the family f_L.

By the global Torelli theorem [**5**], the period map also characterizes the variation of Hodge structure, that is, how Hodge groups $H^{2,0}(X_t), H^{1,1}(X_t), H^{0,2}(X_t)$ vary within $H^2(X_{t_0}, \mathbb{C})$ (the latter can be described in purely topological terms) as $t \in M$ varies. Moreover, for $M = \mathfrak{M}_L$ the solutions of the Picard-Fuchs system determine a (multivalued) inverse of the canonical projection $\mathcal{D}_L \to \mathfrak{M}_L$ and provide a uniformization of the moduli space.

3.2. Linear PDEs modeled after quadric hypersurfaces. Inspired at least partially by the discussion above, Sasaki and Yoshida [**36**] sought to develop a *geometric* theory of linear PDEs in n variables of rank $n+2$, whose solutions provide a map of the equation manifold M into a quadric hypersurface $Q^n \subset \mathbb{P}^{n+1}$. Such a theory had already been developed classically by Wilczynski [**38**] for $n = 1, 2$, and Sasaki and Yoshida successfully generalized Wilczynski's results for $n \geq 3$ by casting the equations in a canonical differential geometric form. We focus here on the case $n \geq 3$, which has a markedly different flavor than $n = 1, 2$. Sasaki and Yoshida brought Wilczynski's results into the modern era for $n = 2$ in [**35**]. See Doran [**13, 14**] for aspects related to the $n = 1$ case, and Clingher, Doran, and the first author [**7**] for some aspects related to the $n = 2$ case. The second author [**37**, §6.2.3] connected some of these results with a particular $n = 3$ case relevant to §2.2 that is briefly discussed at the end of §4.2.

Let M be a complex n-dimensional manifold with $n \geq 3$, equipped with a locally finite open covering by local holomorphic coordinate charts. Let E be a second order holonomic system on X in a single unknown of rank $n+2$, as defined by Sasaki and Yoshida. In a local coordinate chart $(t^1, \dots, t^n)$, the system E is then given by a second order system of linear homogeneous differential equations with polynomial coefficients (satisfying a suitable dimension condition in terms of D-modules theory) for a single unknown function $w = w(t^1, \dots, t^n)$ of rank $n+2$, whose solutions $w_0, \dots, w_{n+1}$ are unique up to scale. Therefore, it is natural to consider the following map determined by the solutions of the holonomic system:

$$
\begin{aligned}
p : \quad M &\to \mathbb{P}^{n+1}, \\
t &\mapsto p(t) = [\, w_0(t) : \cdots : w_{n+1}(t) \,] \in \mathbb{P}^{n+1}.
\end{aligned}
\tag{3.5}
$$

Sasaki and Yoshida asked the following question: what is the geometry of $p(M) \subset \mathbb{P}^{n+1}$, and, conversely, what effect does imposing the existence of a geometric structure on $p(M)$ have on the solutions of E? Here, we will always assume that $p(M) \subseteq V \subset \mathbb{P}^{n+1}$, where V is a smooth, closed complex hypersurface, $p(M)$ is relatively open in V, and p is locally an isomorphism. By a celebrated result of Chow [**6**], V is then a projective *subvariety* of $\mathbb{P}^{n+1}$, and is given by the smooth locus of some algebraic equations. Hence, $p(M) \subseteq V$ is a complex quasi-projective variety, so locally near a point $t \in M$, the solutions $w_0, \dots, w_{n+1}$ satisfy the same algebraic relations that also define $V \subset \mathbb{P}^{n+1}$ around $p(t) \in V$.

The simplest nontrivial algebraic relationship that can hold between solutions $w_0, \ldots, w_{n+1}$ is a quadratic relation. For example, the period integrals of lattice polarized K3 surfaces in §3.1 always satisfy such quadratic relations. In general, for a second order holonomic system E on X, we say that E satisfies the *quadric condition* if its solutions satisfy a quadratic relation. We can then take $V = Q^n \subset \mathbb{P}^{n+1}$ to be a quadratic hypersurface. Among Sasaki and Yoshida's results are necessary and sufficient conditions [**36**, Proposition 1.1] to ensure that $p(M) \subseteq Q^n$. These conditions can be stated entirely in differential geometric terms on M. In fact, the relevant geometric structure to consider turns out to be the one of a *holomorphic conformal structure* on the holomorphic tangent bundle $T_{\mathbb{C}}M$. The existence of such a structure for a complex manifold is quite restrictive [**24**]. This is quite different from the situation for real (semi-)Riemannian manifolds that, as it is well known, always admit natural conformal structures.

We follow Sasaki and Yoshida's presentation [**36**, §1.1]: let $h = (h_{ij}) \in S^2(\mathbb{C}^n)$ be the matrix representation of a fixed nondegenerate symmetric bilinear form on $\mathbb{C}^n$ with respect to some fixed basis, and $S^2(\mathbb{C}^n)$ denote the collection of all such forms. The conformal group $\mathrm{CO}(h)$ is defined to be

$$\mathrm{CO}(h) = \left\{ \lambda a \mid a \in \mathrm{GL}(n,\mathbb{C}), aha^T = h, \lambda \in \mathbb{C}^* \right\},$$

which is the direct product of the multiplicative group $\mathbb{C}^\times$ and the complex orthogonal group $\mathrm{O}(h,\mathbb{C})$.

We say that M, a complex n-dimensional manifold with $n \geq 3$, admits a holomorphic conformal structure (HCS) if the structure group of the holomorphic tangent bundle reduces as a holomorphic bundle to the conformal group. This is equivalent to the condition that the complex frame bundle $F(M) \to M$ admits a holomorphic principal subbundle $P \to M$ with structure group $\mathrm{CO}(h)$ for some nondegenerate bilinear form h. As the bundle is topologically trivial, it is naturally associated with a holomorphic section $g\colon M \to F(M)/\mathrm{CO}(h)$, and hence, any HCS on M corresponds to a holomorphic, nondegenerate symmetric covariant tensor g of rank-2, called the conformal metric. In a local coordinate chart, we write g as

$$g = g_{ij}\, dt^i \odot dt^j \in S^2(M)\,. \tag{3.6}$$

Here, $S^2(M)$ is the holomorphic bundle of symmetric rank-2 covariant tensors on $T_{\mathbb{C}}M$, $\odot$ is the symmetric tensor product, and we employ the summation convention.

We call a HCS $g \in S^2(M)$ integrable if g is locally conformally flat. It is then well that integrability of a metric can be detected from the vanishing of certain curvature tensors associated with g, namely the Weyl tensor $W(g)$ if $n \geq 4$, and the Cotton tensor $C(g)$ if $n = 3$. Recall that these tensors are defined as follows: let Γ^j_{ik} be the Christoffel symbols of g, i.e.,

$$\Gamma^j_{ik} = \frac{1}{2} g^{jl} \left(\partial_k g_{il} + \partial_i g_{kl} - \partial_l g_{ik} \right). \tag{3.7}$$

The Christoffel symbols define a unique torsion free symmetric connection ∇ on $T_{\mathbb{C}}M$ associated with g. The Riemannian curvature tensor $R^l{}_{ijk}$ is defined as

$$R^l{}_{ijk} = \partial_j \Gamma^l_{ki} - \partial_k \Gamma^l_{ji} + \Gamma^l_{jr}\Gamma^r_{ki} - \Gamma^l_{kr}\Gamma^r_{ji}\,. \tag{3.8}$$

Moreover, the Ricci curvature R_{ij} and the scalar curvature R are given by the following contractions of Equation (3.8) with g:

$$R_{ij} = R^l{}_{ilj}, \quad R = g^{ij} R_{ij}\,. \tag{3.9}$$

Finally, the Weyl tensor $W(g)$ is given by

$$W^j{}_{ikl} = R^j{}_{ikl} + S_{ik}\delta^j_l - S_{il}\delta^j_k + g_{ik}g^{jm}S_{ml} - g_{il}g^{jm}S_{mk}, \tag{3.10}$$

and the Cotton tensor $C(g)$ is

$$C_{ijk} = (\nabla_i S(g))_{jk} - (\nabla_j S(g))_{ik}\,, \tag{3.11}$$

where $S(g)$ is the Schouten tensor

$$S_{ik} = \frac{1}{n-2}\left(R_{ik} - \frac{R}{2(n-1)}g_{ik}\right).$$

Every quadric hypersurfaces $Q^n \subset \mathbb{P}^{n+1}$ naturally possesses an integrable HCS. By definition, Q^n is the projectivized locus of isotropic vectors in $\mathbb{C}^{n+2}$ of a nondegenerate symmetric bilinear form $\tilde{h} \in S^2(\mathbb{C}^{n+2})$. Under a suitable automorphism of $\mathbb{C}^{n+2}$, $\tilde{h}$ can be brought into the form

$$\tilde{h} = \begin{pmatrix} 0 & 0 & -1 \\ 0 & h & 0 \\ -1 & 0 & 0 \end{pmatrix},$$

where $h \in S^2(\mathbb{C}^n)$ is nondegenerate. Identifying $\mathbb{P}^{n+1} = \mathbb{P}(X^0, \dots, X^{n+1})$, Q^n is then given by the homogeneous equation

$$Q^n: \quad -2X^0X^{n+1} + h_{ij}X^iX^j = 0\,. \tag{3.12}$$

One checks that we always have $p_0 = [0 : \cdots : 1] \in Q^n$. The stabilizer group $\mathrm{O}(\tilde{h}) = \{g \in \mathrm{GL}(n+2, \mathbb{C}) \mid g\tilde{h}g^T = \tilde{h}\}$ acts transitively on Q^n, with an isotropy group H at p_0, given by matrices of the form

$$\begin{pmatrix} \lambda & 0 & 0 \\ b & a & 0 \\ \mu & c^T & \nu \end{pmatrix} \quad \text{with } \lambda\nu = 1,\ aha^T = h,\ b = \lambda ahc,\ \mu = \lambda c^T hc/2, \tag{3.13}$$

where $\lambda, \mu, \nu \in \mathbb{C}^\times$, $b, c \in \mathbb{C}^n$, $a \in \mathrm{GL}(n, \mathbb{C})$. This is the standard presentation of the quadric hypersurface $Q^n = \mathrm{O}(\tilde{h})/H$ as a homogeneous space and shows the reduction of structure group for the frame bundle $F(M)$ to H. Furthermore, the linear isotropy representation at $p_0 \in Q^n$ has a kernel N, given by matrices of the form

$$\begin{pmatrix} \pm 1 & 0 & 0 \\ b & \pm\mathbb{I}_n & 0 \\ \mu & c^T & \pm 1 \end{pmatrix}. \tag{3.14}$$

It follows that $H/N \cong \mathrm{CO}(h)$. Hence, there is a further reduction of the structure group of the frame bundle to $\mathrm{CO}(h)$.

The corresponding HCS is called the canonical conformal structure of the quadric hypersurface Q^n, and it can be explicitly constructed as follows: we consider the constant tensor $\varphi = -2dX^0 \odot dX^{n+1} + h_{ij}dX^i \odot dX^j$ on $\mathbb{C}^{n+2} - \{0\}$, and σ a local section of the tautological bundle $\mathcal{O}_{\mathbb{P}^{n+1}}(-1)$ with projection $\pi\colon \mathbb{C}^{n+2} - \{0\} \to \mathbb{P}^{n+1}$. We then obtain a conformal metric on $Q^n \subset \mathbb{P}^{n+1}$ by writing $\mathsf{g} = \sigma^*\varphi|_{Q^n}$. Although g depends on σ, changing the section results in rescaling by a nonvanishing holomorphic function, and hence the conformal class of g is well defined. It follows easily from the construction that g is also locally conformally flat.

To summarize, for a holonomic system E of rank n+2 over M with $n = \dim M \geq 3$, satisfying the quadric condition, the parameter space M can be equipped with an integrable HCS $g \in S^2(M)$ by setting $g = p^*\mathrm{g}$, for p given in Equation (3.5). Furthermore, the converse holds as well, as the following result shows.

THEOREM 3.1 (Theorem 2.5, [**36**]). *Let M be an $n \geq 3$-dimensional complex manifold with an integrable HCS $g \in S^2(M)$. The system of linear differential equations, given by*

$$g_{ij}\left(w_{kl} - \Gamma^p_{kl} w_p + \frac{1}{n-2} R_{kl} w\right) = g_{kl}\left(w_{ij} - \Gamma^p_{ij} w_p + \frac{1}{n-2} R_{ij} w\right), \tag{3.15}$$

is a holonomic system of rank $n+2$ satisfying the quadric condition. Here, subscripts on w indicate partial derivatives, and Γ^i_{jk}, R_{ij} are respectively the Christoffel symbols and Ricci curvature of g, given in Equations (3.7), (3.9).

It follows then from the discussion in §3.1 that the Picard-Fuchs system of any maximal family of L-polarized K3 surfaces with $\mathrm{rank}(N) \geq 5 = 3 + 2$ satisfies the quadric condition and can be represented in the form of Equation (3.15) for an integrable HCS on M determined by the pairing $(\cdot, \cdot)$ on $N_{\mathbb{C}}$. In particular, the period domain $\mathcal{D}_L \subset Q^n$ naturally carries an integrable HCS via the inclusion map, and fundamental properties of the period map in Equation (3.4) from §3.1, such as the local Torelli theorem, impart M with such an integrable HCS $g \in S^2(M)$ by pullback. Hence, the existence of such a $g \in S^2(M)$ completely characterizes the local variation of complex structure for such families. Let us see an example.

4. Holomorphic conformal geometry of the Kummer sublocus

In this section we compute the HCS of the Kummer sublocus of the parameter space of double sextic K3 surfaces from Equation (2.12) introduced in §2.2. We then utilize Sasaki and Yoshida's geometric theory from §3.2 to compute the Picard–Fuchs system of the family of generic Jacobian Kummer surfaces from §2.1.

4.1. Computation of the HCS. We use a different, birational presentation for the family $\mathcal{X}$ in Equation (2.11), which is similar to the one studied by Hoyt and Schwartz [**20**]. In fact, a top-dimensional open subspace of $\mathcal{M}$ in Equation (2.9) is given by elements $[(a_{ij})] \in \mathcal{M}$ of the form

$$\begin{pmatrix} 1 & 1 & 1 & 0 & 0 & -d \\ 0 & 1 & 0 & -1 & -1 & -1 \\ 0 & 0 & 1 & a & b & c \end{pmatrix}.$$

Setting $z_1 = x, z_2 = -t, z_3 = -1$ in Equation (2.7) we obtain a family of double sextic surfaces, given by the equation

$$X: \quad y^2 = x(x-1)(x-t)(t-a)(t-b)(t-c-dx), \tag{4.1}$$

over the four-dimensional parameter space

$$M = \left\{(a,b,c,d) \in \mathbb{C}^4 \mid a \neq b, \ (c,d) \neq (a,0),(b,0),(0,1)\right\}. \tag{4.2}$$

This family of projective K3 surfaces was studied by the authors in [**28**, §3]. As in §3.1, we can construct a family of marked L-polarized K3 surfaces $f_L : \mathfrak{X} \to M$ where the general fiber, obtained for generic parameters $(a,b,c,d) \in M$, is a smooth K3 surface X in Equation (4.2) with $\mathrm{NS}(X) \cong L$ where L was given in Equation (2.8), and we have set $\Omega_{a,b,c,d} = dt \wedge dx/y$.

It follows from [**28**, Proposition 3.5] that there is birational map $\psi\colon X \dashrightarrow \mathcal{X}$; in the affine coordinate system $z_1 = -1$, it is given by

$$\psi: \quad z_2 = \frac{x_3 t - 1}{x_3 t - x_1}, \quad z_3 = \frac{x(1-x_1)}{x_3 t - x_1}, \quad \tilde{z}_4 = \frac{x_3 (x_1 - 1)^2 y}{(x_3 t - x_1)^3}, \tag{4.3}$$

induces a birational map $\Psi : M \dashrightarrow \mathcal{M}$ between M and $\mathcal{M}$, given by

$$\Psi: \quad x_1 = \frac{a}{b}, \quad x_2 = \frac{a-c}{b-c}, \quad x_3 = \frac{1}{b}, \quad x_4 = \frac{d}{b-c}. \tag{4.4}$$

Matsumoto, Sasaki, and Yoshida [**32**, Equation (0.8.4)] found an integrable holomorphic conformal structure $G = G_{ij}\, dx_i \odot dx_j \in S^2(\mathcal{M})$ on $\mathcal{M}$: they used the GKZ formalism [**15**] for the period integrals of the family over $\mathcal{M}$ whose fibers are the projective K3 surfaces, determined by Equation (2.11). The components $G_{ij} = G_{ji}$ of the HCS are as follows:

$$G_{12} = \frac{x_4 - x_3}{x_1 - x_2}, \quad G_{13} = \frac{x_4 - x_2}{x_1 - x_3}, \quad G_{14} = 1,$$

$$G_{23} = 1, \quad G_{24} = \frac{x_3 - x_1}{x_2 - x_4}, \quad G_{34} = \frac{x_2 - x_1}{x_3 - x_4},$$

$$\begin{aligned} G_{11} &= \frac{x_2 x_3 - x_4}{x_1(1-x_1)} - \frac{x_3(x_4 - x_2)}{x_1(x_1 - x_3)} - \frac{x_2(x_4 - x_3)}{x_1(x_1 - x_2)}, \\ G_{22} &= \frac{x_1 x_4 - x_3}{x_2(1-x_2)} - \frac{x_1(x_3 - x_4)}{x_2(x_2 - x_1)} - \frac{x_4(x_3 - x_1)}{x_2(x_2 - x_4)}, \\ G_{33} &= \frac{x_1 x_4 - x_2}{x_3(1-x_3)} - \frac{x_1(x_2 - x_4)}{x_3(x_3 - x_1)} - \frac{x_4(x_2 - x_1)}{x_3(x_3 - x_4)}, \\ G_{44} &= \frac{x_2 x_3 - x_1}{x_4(1-x_4)} - \frac{x_3(x_1 - x_2)}{x_4(x_4 - x_3)} - \frac{x_2(x_1 - x_3)}{x_4(x_4 - x_2)}. \end{aligned} \tag{4.5}$$

Then G is nondegenerate on M, and integrability is verified checking that the vanishing of the Weyl tensor, $W(G) = 0$ from Equation (3.10). We then obtain an integrable holomorphic conformal structure g on M by pullback via Ψ, i.e., we set $g = \Psi^* G \in S^2(M)$. The integrability of g follows again by verifying the vanishing of the Weyl tensor, i.e., $W(g) = 0$.

The Kummer sublocus M' within M can be computed as the vanishing locus of the projective invariant R in Equation (2.12). Notice that for $(a,b,c,d) \in M$ we have $a \neq b$; it follows from [**28**, Proposition 3.12] that the Kummer locus in M is given by

$$(ab - c)d - (b-c)(a-c) = 0. \tag{4.6}$$

It was shown in [**28**, Proposition 3.14] that the lattice polarization of $f_L : \mathfrak{X} \to M$ upon restriction to M' extends to the lattice

$$L' \cong H \oplus D_8(-1) \oplus D_4(-1) \oplus A_3(-1) \cong H \oplus D_7(-1) \oplus 2D_4(-1).$$

We then construct a family of marked L'-polarized K3 surfaces $f_{L'} : \mathfrak{X}' \to M'$ where the fiber is the double sextic surface, given by the equation

$$X': \quad y^2 = x(x-1)(x-t)(t-a)(t-b)\big((ab-c)(t-c) - (b-c)(a-c)x\big), \tag{4.7}$$

and the family varies over the three-dimensional parameter space

$$M' = \big\{(a,b,c) \in \mathbb{C}^3 \mid a \neq b,\ a \neq c,\ b \neq c,\ c \neq ab\big\}. \tag{4.8}$$

The general fiber, obtained for generic parameters $(a,b,c) \in M'$, is a smooth K3 surface X' with $\mathrm{NS}(X') \cong L'$ of Picard number 17. A choice of holomorphic 2-form generating the period integrals on X' is $\Omega'_{a,b,c} = dt \wedge dx/y$. As explained in §2, $f_{L'}$ is a family of Jacobian Kummer surfaces.

Let $\imath\colon M' \hookrightarrow M$ be the inclusion map determined by Equation (4.6). A conformal metric on M' is then given by

$$g' = \Big(-2bc^2(b-c)(b-1)(a-1)\Big)^{-1} \imath^* g \in S^2(M'). \tag{4.9}$$

The components $g'_{ij} = g'_{ji}$, with $i,j \in \{a,b,c\}$ are as follows:

$$\begin{aligned} g'_{aa} &= 1, \quad g'_{ab} = 0 = g'_{cc}, \quad g'_{ac} = -\frac{a}{2c}, \\ g'_{bb} &= -\frac{a(a-c)(a-1)}{b(b-c)(b-1)}, \quad g'_{bc} = \frac{a(a-c)(a-1)}{2c(b-c)(b-1)}. \end{aligned} \tag{4.10}$$

Here we have chosen the additional holomorphic conformal scaling factor in front of $\imath^* g$ to simplify the metric. It follows that g' defines a nondegenerate holomorphic conformal structure on M'. First, one checks that the determinant of the metric is nonzero everywhere; this can be seen from Equation (4.8). Second, one verifies the integrability of g' by checking that the Cotton tensor $C(g') = 0$ vanishes, using Equation (3.11).

4.2. Picard–Fuchs equations for generic Kummer surfaces. We can now use the results of Theorem 3.1 with $n = 3$ and $g' \in S^2(M')$ from Equation (4.10) to characterize explicitly the Picard–Fuchs system for the family of Jacobian Kummer surfaces $f_{L'}\colon \mathfrak{X}' \to M'$. To the best of our knowledge, such a system has not been determined before.

THEOREM 4.1. *The system of linear differential equations, given by*

$$g'_{ij}\left(w_{kl} - \Gamma^r_{kl} w_r + R_{kl} w\right) = g'_{kl}\left(w_{ij} - \Gamma^r_{ij} w_r + R_{ij} w\right) \quad \text{for } i,j,k,l,r \in \{a,b,c\}, \tag{4.11}$$

is a rank-5 holonomic system in the variables a,b,c. It is the Picard–Fuchs system annihilating the period integrals for the family $f_{L'}\colon \mathfrak{X}' \to M'$, and any other representation of the Picard–Fuchs system is projectively equivalent to Equation (4.11). *Here, $g' \in S^2(M')$ is the integrable holomorphic conformal structure in Equation* (4.10), *and Γ^r_{ij}, R_{ij} are the Christoffel connection and Ricci curvature of g', respectively.*

The Picard–Fuchs system in Equation (4.11) necessarily satisfies the quadric condition by Theorem 3.1 since the HCS is integrable. Such quadratic relationships, when attached to Picard–Fuchs equations, always contain intricate geometric and arithmetic information about the K3 surface itself. Sasaki and Yoshida [**36**, Equation 3.1] determined a closely related system of rank-5, acting as the uniformizing equations for the Siegel modular threefold $\mathcal{A}_2[2] = \mathbb{H}_2/\Gamma(2)$. Here, $\Gamma(2) \subset \mathrm{Sp}(4,\mathbb{Z})$ is the Siegel modular group of level-two. It follows from [**32**, §0.19] that this system is the Picard–Fuchs system for Kummer surfaces associated with the Jacobian of smooth genus-2 curves C with level-2 structure as in Equation (2.3). Along with Hara [**18**], they showed that their system decomposes as the exterior product of two copies of Lauricella's F_D system, each of which represents precisely the Picard–Fuchs system of such a genus-2 curve C.

In [**37**, Theorem 6.2.89], the second author showed that the rank-5 uniformizing system for $\mathcal{A}_2[2]$ is closely related to the Picard–Fuchs system of the subfamily $X'' = X_{a,b,c,0}$ of Equation (4.1). This family is also known as the twisted Legendre pencil, and it was originally studied by Hoyt [**19**]. The family does *not* represent Kummer surfaces of a principally polarized abelian surface, since Equation (4.6) does not hold. Rather, Equation (4.1) with $d = 0$ represents a six line configuration in $\mathbb{P}^2$ where three lines intersect in a single point [**8**], and X'' gives the relative Jacobian fibration of a Kummer surface of a $(1,2)$-polarized abelian surface; see [**10**]. Then under a change of variables determined by Braeger, Sung, and the first author [**4**, Eq. (3.11)], the second author showed [**4**] that the Picard–Fuchs system for X'' also decomposes into the exterior product of Lauricella F_D systems.

4.3. Future directions. At present it is not known how a multivariate system like the one in Equation (3.15) decomposes in general when imposing the quadric condition. It would be interesting to determine precisely how the Picard–Fuchs system in Equation (4.11) decomposes under the quadric condition. One might also ask how the Picard–Fuchs system decomposes further when a Kummer configuration of six lines in the projective plane attains certain special curves; essential information of the principally polarized abelian surface A are reflected in these configurations of six lines, used to construct $\mathrm{Kum}(A)$ as a double sextic surface. For example, a geometric characterizations of the Humbert surfaces, within the moduli space $\mathcal{A}_2$ of principally polarized abelian surfaces, in terms of the presence of certain curves on the associated Kummer plane was given in [**3**]. These questions will be discussed in forthcoming works by the authors.

References

[1] P. S. Aspinwall, *K3 surfaces and string duality*, Fields, strings and duality (Boulder, CO, 1996), World Sci. Publ., River Edge, NJ, 1997, pp. 421–540. MR1479699 ↑125

[2] C. Birkenhake and H. Lange, *Complex abelian varieties*, 2nd ed., Grundlehren der mathematischen Wissenschaften [Fundamental Principles of Mathematical Sciences], vol. 302, Springer-Verlag, Berlin, 2004, DOI 10.1007/978-3-662-06307-1. MR2062673 ↑122

[3] C. Birkenhake and H. Wilhelm, *Humbert surfaces and the Kummer plane*, Trans. Amer. Math. Soc. **355** (2003), no. 5, 1819–1841, DOI 10.1090/S0002-9947-03-03238-0. MR1953527 ↑132

[4] N. Braeger, A. Malmendier, and Y. Sung, *Kummer sandwiches and Greene-Plesser construction*, J. Geom. Phys. **154** (2020), 103718, 18, DOI 10.1016/j.geomphys.2020.103718. MR4099481 ↑132

[5] D. Burns Jr. and M. Rapoport, *On the Torelli problem for kählerian $K-3$ surfaces*, Ann. Sci. École Norm. Sup. (4) **8** (1975), no. 2, 235–273. MR447635 ↑125, 126

[6] W.-L. Chow, *On compact complex analytic varieties*, Amer. J. Math. **71** (1949), 893–914, DOI 10.2307/2372375. MR33093 ↑126

[7] A. Clingher, C. F. Doran, and A. Malmendier, *Special function identities from superelliptic Kummer varieties*, Asian J. Math. **21** (2017), no. 5, 909–951, DOI 10.4310/AJM.2017.v21.n5.a6. MR3767270 ↑126

[8] A. Clingher and A. Malmendier, *On the geometry of (1,2)-polarized Kummer surfaces*, Apr. 2017. arXiv:1704.04884 [hep-th]. 132

[9] A. Clingher, A. Malmendier, and T. Shaska, *Six line configurations and string dualities*, Comm. Math. Phys. **371** (2019), no. 1, 159–196, DOI 10.1007/s00220-019-03372-0. MR4015343 ↑123

[10] A. Clingher, A. Malmendier, and T. Shaska, *On isogenies among certain Abelian surfaces*, Michigan Math. J. **71** (2022), no. 2, 227–269, DOI 10.1307/mmj/20195790. MR4484238 ↑132

[11] I. Dolgachev and D. Ortland, *Point sets in projective spaces and theta functions* (English, with French summary), Astérisque **165** (1988), 210 pp. (1989). MR1007155 ↑124

[12] I. V. Dolgachev, *Mirror symmetry for lattice polarized $K3$ surfaces*, J. Math. Sci. **81** (1996), no. 3, 2599–2630, DOI 10.1007/BF02362332. Algebraic geometry, 4. MR1420220 ↑122, 125

[13] C. F. Doran, *Picard-Fuchs uniformization and modularity of the mirror map*, Comm. Math. Phys. **212** (2000), no. 3, 625–647, DOI 10.1007/s002200000228. MR1779161 ↑126

[14] C. F. Doran, *Picard-Fuchs uniformization: modularity of the mirror map and mirror-moonshine*, The arithmetic and geometry of algebraic cycles (Banff, AB, 1998), CRM Proc. Lecture Notes, vol. 24, Amer. Math. Soc., Providence, RI, 2000, pp. 257–281, DOI 10.1090/crmp/024/13. MR1738862 ↑126

[15] I. M. Gel′fand, M. M. Kapranov, and A. V. Zelevinsky, *Generalized Euler integrals and A-hypergeometric functions*, Adv. Math. **84** (1990), no. 2, 255–271, DOI 10.1016/0001-8708(90)90048-R. MR1080980 ↑130

[16] P. A. Griffiths, *Periods of integrals on algebraic manifolds. I. Construction and properties of the modular varieties*, Amer. J. Math. **90** (1968), 568–626, DOI 10.2307/2373545. MR229641 ↑120

[17] P. A. Griffiths, *Periods of integrals on algebraic manifolds: Summary of main results and discussion of open problems*, Bull. Amer. Math. Soc. **76** (1970), 228–296, DOI 10.1090/S0002-9904-1970-12444-2. MR258824 ↑119

[18] M. Hara, T. Sasaki, and M. Yoshida, *Tensor products of linear differential equations—a study of exterior products of hypergeometric equations*, Funkcial. Ekvac. **32** (1989), no. 3, 453–477. MR1040172 ↑131

[19] W. L. Hoyt, *On twisted Legendre equations and Kummer surfaces*, Theta functions—Bowdoin 1987, Part 1 (Brunswick, ME, 1987), Proc. Sympos. Pure Math., vol. 49, Amer. Math. Soc., Providence, RI, 1989, pp. 695–707, DOI 10.1090/pspum/049.1/1013162. MR1013162 ↑132

[20] W. L. Hoyt and C. F. Schwartz, *Yoshida surfaces with Picard number $\rho \geq 17$*, Proceedings on Moonshine and related topics (Montréal, QC, 1999), CRM Proc. Lecture Notes, vol. 30, Amer. Math. Soc., Providence, RI, 2001, pp. 71–78, DOI 10.1090/crmp/030/06. MR1877757 ↑129

[21] D. Huybrechts, *Lectures on K3 surfaces*, Cambridge Studies in Advanced Mathematics, vol. 158, Cambridge University Press, Cambridge, 2016, DOI 10.1017/CBO9781316594193. MR3586372 ↑121

[22] S. Kobayashi, *Transformation groups in differential geometry*, Ergebnisse der Mathematik und ihrer Grenzgebiete [Results in Mathematics and Related Areas], Band 70, Springer-Verlag, New York-Heidelberg, 1972. MR355886 ↑119

[23] S. Kobayashi, *Transformation groups in differential geometry*, Classics in Mathematics, Springer-Verlag, Berlin, 1995. Reprint of the 1972 edition. MR1336823 ↑119

[24] S. Kobayashi and T. Ochiai, *Holomorphic structures modeled after hyperquadrics*, Tohoku Math. J. (2) **34** (1982), no. 4, 587–629, DOI 10.2748/tmj/1178229159. MR685426 ↑119, 127

[25] N. H. Kuiper, *On conformally-flat spaces in the large*, Ann. of Math. (2) **50** (1949), 916–924, DOI 10.2307/1969587. MR31310 ↑119

[26] A. Kumar, *Elliptic fibrations on a generic Jacobian Kummer surface*, J. Algebraic Geom. **23** (2014), no. 4, 599–667, DOI 10.1090/S1056-3911-2014-00620-2. MR3263663 ↑123

[27] M. Kuwata and T. Shioda, *Elliptic parameters and defining equations for elliptic fibrations on a Kummer surface*, Algebraic geometry in East Asia—Hanoi 2005, Adv. Stud. Pure Math., vol. 50, Math. Soc. Japan, Tokyo, 2008, pp. 177–215, DOI 10.2969/aspm/05010177. MR2409557 ↑123

[28] A. Malmendier and M. T. Schultz, *On the mixed-twist construction and monodromy of associated Picard-Fuchs systems*, Commun. Number Theory Phys. **16** (2022), no. 3, 459–513. MR4494119 ↑129, 130

[29] K. Matsumoto, *Theta functions on the classical bounded symmetric domain of type* $\mathrm{I}_{2,2}$, Proc. Japan Acad. Ser. A Math. Sci. **67** (1991), no. 1, 1–5. MR1103969 ↑124

[30] K. Matsumoto, T. Sasaki, and M. Yoshida, *The period map of a 4-parameter family of $K3$ surfaces and the Aomoto-Gel′fand hypergeometric function of type* $(3,6)$, Proc. Japan Acad. Ser. A Math. Sci. **64** (1988), no. 8, 307–310. MR973860 ↑124

[31] K. Matsumoto, T. Sasaki, and M. Yoshida, *The Aomoto-Gel′fand hypergeometric function and period mappings* (Japanese), Sūgaku **41** (1989), no. 3, 258–263. MR1073363 ↑124

[32] K. Matsumoto, T. Sasaki, and M. Yoshida, *The monodromy of the period map of a 4-parameter family of $K3$ surfaces and the hypergeometric function of type* $(3,6)$, Internat. J. Math. **3** (1992), no. 1, 164, DOI 10.1142/S0129167X92000023. MR1136204 ↑124, 130, 131

[33] D. Mumford, *Abelian varieties*, Tata Institute of Fundamental Research Studies in Mathematics, vol. 5, Published for the Tata Institute of Fundamental Research, Bombay; by Hindustan Book Agency, New Delhi, 2008. With appendices by C. P. Ramanujam and Yuri Manin; Corrected reprint of the second (1974) edition. MR2514037 ↑122

[34] V. V. Nikulin, *Kummer surfaces* (Russian), Izv. Akad. Nauk SSSR Ser. Mat. **39** (1975), no. 2, 278–293, 471. MR429917 ↑122

[35] T. Sasaki and M. Yoshida, *Linear differential equations in two variables of rank four. I*, Math. Ann. **282** (1988), no. 1, 69–93, DOI 10.1007/BF01457013. MR960834 ↑126

[36] T. Sasaki and M. Yoshida, *Linear differential equations modeled after hyperquadrics*, Tohoku Math. J. (2) **41** (1989), no. 2, 321–348, DOI 10.2748/tmj/1178227829. MR996019 ↑120, 124, 126, 127, 129, 131

[37] M. T. Schultz. *On the Geometry of the Moduli Space of Certain Lattice Polarized K3 Surfaces and Their Picard-Fuchs Operators.* PhD thesis, Utah State University, 2021. 126, 132

[38] E. J. Wilczynski, *Projective differential geometry of curves and ruled surfaces*, Chelsea Publishing Co., New York, 1962. MR131232 ↑126

Department of Mathematics & Statistics, Utah State University, Logan, Utah 84322
Email address: andreas.malmendier@usu.edu

Department of Mathematics, Virginia Tech, Blacksburg, Virginia 24060
Email address: michaelschultz@vt.edu

Contemporary Mathematics
Volume **821**, 2025
https://doi.org/10.1090/conm/821/16441

Triharmonic curves in the 3-dimensional Sol space

S. Montaldo and A. Ratto

ABSTRACT. The main aim of this paper is to study triharmonic curves in the 3-dimensional homogeneous space Sol. In the first part of the paper we shall obtain a complete classification of proper triharmonic curves with constant geodesic curvature and torsion. In the final section we shall show that these triharmonic curves form a constant angle with a suitable Killing field of constant length along the curve.

1. Introduction

Harmonic maps are the critical points of the *energy functional*

$$E(\varphi)=\frac{1}{2}\int_M |d\varphi|^2\,dV\,, \tag{1.1}$$

where $\varphi : M \to N$ is a smooth map between two Riemannian manifolds (M,g) and (N,h). We refer to [**7**,**8**] for an introduction to this important topic.

The study of *polyharmonic maps of order* r, shortly r-*harmonic maps*, was first proposed by Eells and Lemaire in [**7**]. These maps are defined as the critical points of the functionals

$$E_r(\varphi)=\frac{1}{2}\int_M |(d+d^*)^r\varphi|^2\,dV$$

which represent a version of order r of the classical energy (1.1). We refer to [**3**] for a detailed discussion on the definition of the functionals $E_r(\varphi)$.

In this context, the most widely studied case is $r=2$, where we have the so-called *bienergy* functional and its critical points are known as *biharmonic maps.* At present, a very ample literature on biharmonic maps is available (see [**18**] and references therein).

More recently, several authors worked intensively on the case $r \geq 3$. For instance, we refer to [**3**, **5**, **6**, **9**, **11**–**14**, **16**, **17**] for background and results.

In this paper we shall focus on the study of 3-*harmonic curves* in the 3-dimensional solvable Lie group Sol.

2020 *Mathematics Subject Classification.* Primary 58E20; Secondary 53A04, 53C30.

Key words and phrases. Triharmonic curves, 3-dimensional Sol space, helices, Killing fields.

The authors are members of the Italian National Group G.N.S.A.G.A. of INdAM. The work was partially supported by the Project ISI-HOMOS funded by Fondazione di Sardegna and by PNRR e.INS Ecosystem of Innovation for Next Generation Sardinia (CUP F53C22000430001, codice MUR ECS00000038).

More generally, let $\gamma : I \subset \mathbb{R} \to N$ denote a smooth curve parametrized by the arc length s and let T denote its unit tangent vector. Then γ is r-harmonic if and only if (see [**2**, **11**, **15**])

$$\nabla_T^{2r-1}T + \sum_{\ell=0}^{r-2}(-1)^\ell R\left(\nabla_T^{2r-3-\ell}T, \nabla_T^\ell T\right)T = 0\,, \tag{1.2}$$

where

$$\nabla_T^0 T = T,\ \nabla_T^k T = \nabla_T\left(\nabla_T^{k-1}T\right)$$

and R is the Riemannian curvature operator of N.

We point out that any geodesic is trivially r-harmonic for all $r \geq 1$. Therefore, we say that γ is a *proper* r-harmonic curve if it is r-harmonic and *not* a geodesic.

The Chen–Maeta conjecture states that any r-harmonic submanifolds of $\mathbb{R}^n$ is minimal. In this direction, Maeta (see [**11**]) proved that the conjecture is true for curves. Note that, in order to avoid trivialities arising from re-parametrization, it is convenient to restrict the attention to curves parametrized by the arc length, and we shall adopt this assumption throughout the whole paper.

By contrast, when the target is an n-dimensional sphere, examples of proper r-harmonic curves do exist. We refer to the paper of Branding [**2**] and references therein for an illustration of the case of r-harmonic curves into space forms.

An investigation of triharmonic curves in 3-dimensional homogeneous spaces with a 4-dimensional group of isometries was carried out in [**15**].

As a successive step, in this paper we undertake the study of triharmonic curves in the 3-dimensional homogeneous geometry with isometries group of dimension 3, i.e., the 3-dimensional solvable Lie group Sol. A study of triharmonic curves in the 3-dimensional solvable Lie group Sol was initiated in [**19**], where the author gives the explicit relevant differential equations.

It is natural to start this investigation by considering a special class of geometrically significant curves, the so-called helices.

Classically, in the Euclidean 3-dimensional space $\mathbb{R}^3$, a *general helix* is a curve characterized by the property that its tangent vector makes a constant angle with a fixed straight line which represents the axis of the helix.

A famous result, stated by M. A. Lancret in 1802 and first proved by B. de Saint Venant in 1845, is: *A necessary and sufficient condition that a curve be a general helix is that the ratio of curvature to torsion be constant.*

Helices arise in nanosprings, carbon nanotubes, and the double helix shape is commonly associated with the structure of DNA.

In a Riemannian geometric context, the definition of a general helix is more delicate and some new interesting phenomena appear (see [**1**]). For instance, for general helices according to [**1**], in the 3-dimensional sphere $\mathbb{S}^3$ the ratio of the curvature and the torsion is not necessarily constant.

In the first part of the paper we follow the approach and the notation of [**15**] and say that a Frenet curve $\gamma(s)$ is a *helix* if its geodesic curvature $\kappa(s)$ and its torsion $\tau(s)$ are constant. In Theorem 3.2 we shall obtain a *complete classification of proper* 3*-harmonic helices in the* 3*-dimensional Sol space.*

Our paper is organized as follows. In Section 2 we recall some basic facts about curves and the geometry of Sol. The main aim of this section is to make this paper reasonably self-contained.

In Section 3 we shall state our main result. Proofs will be given in Section 4.

Finally, in Section 5 we shall follow [**1**] and consider a more general definition of a helix. This approach involves Killing fields which have constant length along γ. In this context, we shall show that the proper 3-harmonic curve determined in Theorem 3.2 is an integral curve of one of these Killing fields.

2. Preliminaries

We adopt the following notation and sign convention for the Riemannian curvature tensor field R:

$$R(X,Y)Z = \nabla_X \nabla_Y Z - \nabla_Y \nabla_X Z - \nabla_{[X,Y]} Z\,, \quad R(X,Y,Z,W) = \langle R(X,Y)W, Z\rangle\,,$$

where X, Y, Z, W are vector fields on N.

Let Sol $= (\mathbb{R}^3, g_{Sol})$ denote the 3-dimensional Riemannian manifold given by $\mathbb{R}^3$ endowed with the metric tensor $g_{Sol} = e^{2z}\mathrm{d}x^2 + e^{-2z}\mathrm{d}y^2 + \mathrm{d}z^2$ with respect to the standard Cartesian coordinates (x, y, z).

The space Sol is a solvable Lie group and the metric g_{Sol} is left-invariant with respect to the group operation

$$(x,y,z).(x',y',z') = \left(e^z x' + x, e^{-z} y' + y, z' + z\right).$$

The group of isometries of Sol has dimension 3 and the connected component of the identity is generated by the following three families:

$$(2.1)\quad (x,y,z) \to (x+c, y, z),\ \ (x,y,z) \to (x, y+c, z),\ \ (x,y,z) \to (e^{-c}x, e^c y, z+c).$$

For future use, we point out that

$$(2.2)\qquad (x,y,z) \to (-x,y,z) \text{ and } (x,y,z) \to (x,-y,z)$$

are also isometries of Sol. Moreover, one can easily check that a global orthonormal frame field on Sol is

$$E_1 = e^{-z}\frac{\partial}{\partial x},\ E_2 = e^{z}\frac{\partial}{\partial y},\ E_3 = \frac{\partial}{\partial z}.$$

With respect to this orthonormal frame field, the Lie brackets and the Levi-Civita connection can be easily computed as:

$$[E_1, E_2] = 0,\ [E_2, E_3] = -E_2,\ [E_1, E_3] = E_1,$$

$$(2.3)\qquad \begin{array}{lll} \nabla_{E_1} E_1 = -E_3, & \nabla_{E_1} E_2 = 0, & \nabla_{E_1} E_3 = E_1, \\ \nabla_{E_2} E_1 = 0, & \nabla_{E_2} E_2 = E_3, & \nabla_{E_2} E_3 = -E_2, \\ \nabla_{E_3} E_1 = 0, & \nabla_{E_3} E_2 = 0, & \nabla_{E_3} E_3 = 0. \end{array}$$

A further computation gives

$$(2.4)\qquad \begin{array}{lll} R(E_1,E_2)E_1 = -E_2, & R(E_1,E_3)E_1 = E_3, & R(E_1,E_2)E_2 = E_1, \\ R(E_2,E_3)E_2 = E_3, & R(E_1,E_3)E_3 = -E_1, & R(E_2,E_3)E_3 = -E_2, \end{array}$$

and from this it is easy to deduce

$$(2.5)\qquad R_{1212} = 1,\ R_{1313} = -1,\ R_{2323} = -1.$$

Now, let $\gamma(s) = (x(s), y(s), z(s))$ be a *Frenet curve* in Sol parametrized by arc length. Then,

$$(2.6)\qquad T = x' e^{z} E_1 + y' e^{-z} E_2 + z' E_3 = (T_1, T_2, T_3)\,,$$

and we can associate to γ the classical Frenet equations

$$\begin{aligned} \nabla_T T &= \kappa\, N, \\ \nabla_T N &= -\kappa\, T + \tau\, B, \\ \nabla_T B &= -\tau\, N, \end{aligned} \tag{2.7}$$

where $\{T, N, B\}$ is a global orthonormal frame field along γ called the *Frenet frame field* and the functions $\kappa = \kappa(s) > 0$, $\tau = \tau(s)$ are called the *geodesic curvature* and the *torsion* of γ, respectively.

Next, let $V = V_1(s)E_1 + V_2(s)E_2 + V_3(s)E_3 = (V_1(s), V_2(s), V_3(s))$ be any vector field along γ. Then a computation using (2.3) and (2.6) shows that

$$\nabla_T V = \left(V_1{}' + T_1 V_3, V_2{}' - T_2 V_3, V_3{}' - T_1 V_1 + T_2 V_2\right). \tag{2.8}$$

For our purposes, it will be useful to know the explicit expression of $\kappa(s)$ and $\tau(s)$ for a Frenet curve $\gamma(s) = (x(s), y(s), z(s))$. Observing that

$$\kappa^2(s) = \langle \nabla_T T, \nabla_T T\rangle \quad \text{and} \quad \tau(s) = -\langle \nabla_T B, N\rangle,$$

a long, straightforward computation using (2.8) gives

$$\kappa^2(s) = e^{2z}\left(x'' + 2x'z'\right)^2 + e^{-2z}\left(y'' - 2y'z'\right)^2 + \left(z'' - e^{2z}x'^2 + e^{-2z}y'^2\right)^2 \tag{2.9}$$

and

$$\tau(s) = \frac{\mathcal{A}}{\mathcal{B}}, \tag{2.10}$$

where

$$\begin{aligned} \mathcal{A} = {} & 2e^{8z}x'^5y' + 2x'y'^5 + e^{6z}x'(3x''^2y' + x'^2(y^{(3)} - 6y''z' + 8y'(2z'^2 - z'')) \\ & + x'(y'(14x''z' - x^{(3)}) - 3x''y'')) + e^{2z}y'(x'(3y''^2 + 8y'^2(z'' + 2z'^2) \\ & - y'(y^{(3)} + 14y''z')) + y'(y'(x^{(3)} + 6x''z') - 3x''y'')) \\ & + e^{4z}(-4x'^3y'^3 + x'(-y^{(3)}z'' + 2y^{(3)}z'^2 + z^{(3)}y'' - 8y''z'^3 + y'(6z''^2 + 8z'^4 - 4z^{(3)}z')) \\ & + y'(x^{(3)}z'' + 2x^{(3)}z'^2 - z^{(3)}x'' + 8x''z'^3) + z'(y^{(3)}x'' - y''(x^{(3)} + 6x''z'))) \end{aligned}$$

and

$$\begin{aligned} \mathcal{B} = {} & e^{4z}(z''^2 - 2x'^2y'^2) + e^{8z}x'^4 + e^{6z}((x'' + 2x'z')^2 - 2x'^2z'') \\ & + y'^4 + e^{2z}(2y'^2z'' + (y'' - 2y'z')^2). \end{aligned}$$

As a special case of (1.2), we see that γ is triharmonic if it satisfies

$$\nabla^5_T T + R\left(\nabla^3_T T, T\right)T - R\left(\nabla^2_T T, \nabla_T T\right)T = 0. \tag{2.11}$$

In the following proposition we compute explicitly the left side of (2.11).

PROPOSITION 2.1. *Let $\gamma(s)$ be a Frenet curve in Sol parametrized by arc length. Then γ is proper triharmonic if and only if the following three equations hold:*

$$5\left[2\kappa^3\kappa' + \kappa\left(\tau^2\kappa' - \kappa^{(3)}\right) - 2\kappa'\kappa'' + \kappa^2\tau\tau'\right] = 0; \tag{2.12}$$

(2.13)

$$\begin{aligned} & -2\kappa^2\left(B_3T_3\tau + 5\kappa''\right) + \kappa\left[-2B_3N_3\tau' - 15\kappa'^2 + \left(1 - 2B_3{}^2\right)\tau^2 - 4\tau\tau'' - 3\tau'^2 + \tau^4\right] \\ & -4\tau\kappa'\left(B_3N_3 + 3\tau'\right) + \kappa^{(4)} - 2\kappa''\left(1 - B_3{}^2 + 3\tau^2\right) + \kappa'' + 2\kappa^3\left(1 - 2B_3{}^2 + \tau^2\right) + \kappa^5 = 0; \end{aligned}$$

(2.14)

$$-2\kappa''(B_3N_3-3\tau')+\kappa\big(2\tau^2(B_3N_3-3\tau')+(2{N_3}^2-1)\tau'+\tau^{(3)}\big)+\kappa^3(4B_3N_3-\tau') \\ +\kappa^2\tau(2N_3T_3-9\kappa')+4\kappa'\tau''-4\tau^3\kappa'+2\tau\big(2\kappa^{(3)}+(2{N_3}^2-1)\kappa'\big)=0\,.$$

The left sides of (2.12), (2.13) and (2.14) are precisely the component of the 3-tension field with respect to the Frenet frame field. The proof of Proposition 2.1 will be carried out in Section 4.

3. Main theorem

A simple look at Proposition 2.1, together with the general formulae (2.9) and (2.10), suggest that a complete study of proper triharmonic curves in Sol may be a rather difficult task. Therefore, as a starting point, we restrict our investigation to a suitable family of geometrically significant curves.

DEFINITION 3.1. We say that a Frenet curve $\gamma(s)$ in Sol is a *Frenet helix* (shortly, a *helix*) if its geodesic curvature $\kappa(s)$ and its torsion $\tau(s)$ are both constant, say $\kappa(s)=a>0$, $\tau(s)=b$.

We prove a complete classification of proper triharmonic Frenet helices in Sol.

THEOREM 3.2. *There exist a proper triharmonic Frenet helix γ in Sol if and only if $\kappa(s)=a=1/2$ and $\tau(s)=b=\pm 1/2$. Moreover, up to isometries of Sol, the parametrization of such proper triharmonic helix when $b=1/2$ is*

$$\gamma(s)=\frac{1}{\sqrt{2}}\left(-e^{-\frac{s}{\sqrt{2}}},\, e^{\frac{s}{\sqrt{2}}},\, s\right). \tag{3.1}$$

REMARK 3.3. In [**4**] the authors proved that there is no proper biharmonic curve in Sol. By contrast, Theorem 3.2 shows that proper triharmonic curves in Sol do exist.

4. Proofs

PROOF OF PROPOSITION 2.1. We want to compute the 3-tension field, i.e., the left side of (2.11) expressing it with respect to the Frenet frame field $\{T,N,B\}$.

We shall find that, when $r=3$, the left-hand side of (2.12), (2.13) and (2.14) are precisely the components of the left-hand side of (1.2) along T, N, and B, respectively.

We shall use again the notation $(v_1,v_2,v_3)=v_1T+v_2N+v_3B$. First, a simple computation gives:

$$\begin{aligned}
&\nabla_T T=(0,\kappa,0),\\
&\nabla_T^2 T=(-\kappa^2,\kappa',\kappa\tau),\\
&\nabla_T^3 T=(-3\kappa\kappa',-\kappa^3-\kappa\tau^2+\kappa'',2\kappa'\tau+\kappa\tau'),\\
&\nabla_T^5 T=\Big(5\big(2\kappa^3\kappa'+\kappa(\tau^2\kappa'-\kappa^{(3)})-2\kappa'\kappa''+\kappa^2\tau\tau'\big),\\
&\qquad \kappa^{(4)}-6\tau^2\kappa''-10\kappa^2\kappa''-12\tau\kappa'\tau'+\kappa\left(\tau^4-3\left(5\kappa'^2+\tau'^2\right)-4\tau\tau''\right)\\
&\qquad +2\kappa^3\tau^2+\kappa^5,\\
&\qquad 4\kappa^{(3)}\tau+6\kappa''\tau'+4\kappa'\tau''-9\kappa^2\tau\kappa'-4\tau^3\kappa'-\kappa^3\tau'+\kappa\big(\tau^{(3)}-6\tau^2\tau'\big)\Big).
\end{aligned} \tag{4.1}$$

Next, using (2.4) and (2.5) we compute

$$\begin{array}{rcl} R(T,N,T,N) & = & -1+2B_3^2, \\ R(T,N,T,B) & = & -2N_3B_3, \\ R(B,T,B,T) & = & -1+2N_3^2, \\ R(B,N,N,T) & = & 2T_3B_3, \\ R(B,N,B,T) & = & -2T_3N_3. \end{array} \tag{4.2}$$

Finally, a computation using (4.1) and (4.2) into (2.11) leads us to complete the proof of this proposition. □

PROOF OF THEOREM 3.2. First, we establish a technical Lemma.

LEMMA 4.1. *Let γ be a proper triharmonic Frenet helix in Sol. Then*

(i) *$N_3 \equiv 0$ along γ.*
(ii) *Both T_3 and B_3 are constant along γ.*

PROOF OF LEMMA 4.1.

(i) We use Proposition 2.1. Direct substitution shows that (2.12) always holds when $\kappa(s) \equiv a$ and $\tau(s) \equiv b$. As for equations (2.13) and (2.14), they become

$$a\Big[a^4 - 2abB_3T_3 + 2a^2(1+b^2-2B_3^2) + b^2(1+b^2-2B_3^2)\Big] = 0 \tag{4.3}$$

and

$$2aN_3\big(2a^2B_3 + b^2B_3 + abT_3\big) = 0, \tag{4.4}$$

respectively. Now, let us assume that $N_3(s_0) \neq 0$ for some s_0. Then, near s_0, we must have

$$B_3 = -\frac{abT_3}{2a^2+b^2}.$$

Inserting this condition into the left side of (4.3) and simplifying we find

$$a\Big[a^4 + b^2 + b^4 + 2a^2(1+b^2)\Big] = 0,$$

which is clearly impossible since $a > 0$. Then $N_3 \equiv 0$.

(ii) Since $\{T, N, B\}$ is an orthonormal frame field along γ we deduce that $T_3^2 + N_3^2 + B_3^2 = 1$. Thus, since $N_3 = 0$, we must have

$$T_3 = \pm\sqrt{1-B_3^2}.$$

Replacing this into (4.3) we obtain

$$a^4 + 2a^2(1+b^2-2B_3^2) + b^2(1+b^2-2B_3^2) \mp 2abB_3\sqrt{1-B_3^2} = 0.$$

Then it is easy to deduce that

$$4a^2b^2B_3^2(-1+B_3^2) + \Big(a^4 + 2a^2(1+b^2-2B_3^2) + b^2(1+b^2-2B_3^2)\Big)^2 = 0$$

along γ. It follows that B_3 is a root of a not identically zero polynomial with constant coefficients. Therefore, B_3 is constant along γ and so is T_3. □

REMARK 4.2. Part (i) of Lemma 4.1 was obtained by similar methods in [**19**], but to help the reader we have inserted here its proof.

As a consequence of Lemma 4.1, we can assume that $N_3 = 0$ along γ. According to (4.3), we have to understand when there exists a Frenet helix $\gamma(s) = (x(s), y(s), z(s))$ such that

$$T_3^2 + B_3^2 = 1 \tag{4.5}$$

and

$$a^4 + 2a^2(1 + b^2 - 2B_3^2) + b^2(1 + b^2 - 2B_3^2) - 2abB_3\sqrt{1 - B_3^2} = 0 \tag{4.6}$$

with $T_3 = \sqrt{1 - B_3^2}$ constant along γ, or

$$a^4 + 2a^2(1 + b^2 - 2B_3^2) + b^2(1 + b^2 - 2B_3^2) + 2abB_3\sqrt{1 - B_3^2} = 0 \tag{4.7}$$

with $T_3 = -\sqrt{1 - B_3^2}$.

A routine computation shows that for any helix γ we have:

$$\begin{aligned} &\text{(i)} \quad T_3 = \; z', \\ &\text{(ii)} \quad N_3 = \; -\frac{1}{a}\big(e^{2z}x'^2 - e^{-2z}y'^2 - z''\big), \\ &\text{(iii)} \quad B_3 = \; \frac{1}{a}\big(- y'x'' + x'(-4y'z' + y'')\big). \end{aligned} \tag{4.8}$$

Now, since $N_3 = 0$ and T_3 is constant, we deduce from (4.8)(ii) that either

$$z(s) = z_1(s) = \frac{1}{2}\log\left(\frac{y'}{x'}\right) \tag{4.9}$$

or

$$z(s) = z_2(s) = \frac{1}{2}\log\left(-\frac{y'}{x'}\right) \tag{4.10}$$

with the obvious sign restrictions on x', y'. Let us first assume that $z(s) = z_1(s)$. Then

$$T_3 = z_1' = \frac{1}{2}\left(-\frac{x''}{x'} + \frac{y''}{y'}\right) = c_1 \qquad (|c_1| \leq 1).$$

Integration of this condition yields

$$y' = c_2 e^{2c_1 s} x' \quad (c_2 > 0)\,. \tag{4.11}$$

Next, using (2.6), (4.9) and imposing $\langle T, T\rangle = 1$ we deduce

$$c_1^2 + 2c_2 e^{2c_1 s} x'^2 = 1$$

along γ. Integration of this equality tells us that either

$$x(s) = x_1(s) = -\frac{\sqrt{(1 - c_1^2)}}{c_1\sqrt{2c_2}} e^{-c_1 s} + c_x \tag{4.12}$$

or

$$x(s) = \tilde{x}_1(s) = \frac{\sqrt{(1 - c_1^2)}}{c_1\sqrt{2c_2}} e^{-c_1 s} + c_x. \tag{4.13}$$

First, let us assume that we are in the case (4.12). Then direct substitution into (4.11) and integration enables us to deduce the explicit expression for $y(s)$ and, using (4.9), for $z(s)$ as well. The final output is:

$$y(s) = y_1(s) = \frac{\sqrt{(1-c_1^2)c_2}}{c_1\sqrt{2}} e^{c_1 s} + c_y\,, \quad z_1(s) = \frac{1}{2}\log\left(c_2 e^{2c_1 s}\right).$$

Now, using the explicit expression $(x_1(s), y_1(s), z_1(s))$ into (2.9) and (2.10) we find that we must have:

$$a^2 = c_1^2 - c_1^4\,, \quad b = 1 - c_1^2.$$

Finally, using these values for a, b into (4.6), (4.7), together with $B_3 = \pm\sqrt{1-c_1^2}$, it is straightforward to check that the only acceptable values are $c_1 = \pm 1/\sqrt{2}$. Both of these values for c_1 give $a = 1/2$ and $b = 1/2$. If we now choose $x = \tilde{x}_1$, the analysis is similar. We find

$$y(s) = \tilde{y}_1(s) = -\frac{\sqrt{(1-c_1^2)c_2}}{c_1\sqrt{2}} e^{c_1 s} + c_y\,, \quad z_1(s) = \frac{1}{2}\log\left(c_2 e^{2c_1 s}\right)$$

and the conclusion is as before. By way of summary, the only possible solutions associated to the case $z(s) = z_1(s)$ are

$$\gamma(s) = (x_1, y_1, z_1)$$

where

$$\begin{aligned}
&x_1 = -\frac{e^{-\frac{s}{\sqrt{2}}}}{\sqrt{2c_2}} + c_x\,, \quad y_1 = \frac{e^{\frac{s}{\sqrt{2}}}\sqrt{c_2}}{\sqrt{2}} + c_y\,, \quad z_1(s) = \tfrac{1}{2}\log\left(c_2 e^{\sqrt{2}s}\right), \text{ or}\\
&x_1 = \frac{e^{\frac{s}{\sqrt{2}}}}{\sqrt{2c_2}} + c_x\,, \quad y_1 = -\frac{e^{\frac{-s}{\sqrt{2}}}\sqrt{c_2}}{\sqrt{2}} + c_y\,, \quad z_1(s) = \tfrac{1}{2}\log\left(c_2 e^{-\sqrt{2}s}\right), \text{ or}\\
&x_1 = \frac{e^{-\frac{s}{\sqrt{2}}}}{\sqrt{2c_2}} + c_x\,, \quad y_1 = -\frac{e^{\frac{s}{\sqrt{2}}}\sqrt{c_2}}{\sqrt{2}} + c_y\,, \quad z_1(s) = \tfrac{1}{2}\log\left(c_2 e^{\sqrt{2}s}\right), \text{ or}\\
&x_1 = -\frac{e^{\frac{s}{\sqrt{2}}}}{\sqrt{2c_2}} + c_x\,, \quad y_1 = \frac{e^{\frac{-s}{\sqrt{2}}}\sqrt{c_2}}{\sqrt{2}} + c_y\,, \quad z_1(s) = \tfrac{1}{2}\log\left(c_2 e^{-\sqrt{2}s}\right),
\end{aligned} \tag{4.14}$$

where c_1, c_2, c_x, c_y are integration constants with $c_2 > 0$. Taking into account (2.1) and (2.2), we conclude that, up to isometries, the only proper triharmonic curve associated to the case $z(s) = z_1(s)$ is (3.1), which is the first parametrization of (4.14) with $c_x = 0, c_y = 0, c_2 = 1$.

Next, if we repeat the same analysis with $z(s) = z_2(s)$ as in (4.10), we find that

$$c_1^2 - c_1^4 = a^2\,, \quad b = -1 + c_1^2.$$

Again, the only acceptable values are $c_1 = \pm 1/\sqrt{2}$. Both these values for c_1 give $a = 1/2$ and $b = -1/2$. The parametrization of γ in this case, up to isometries, is

$$\gamma(s) = \frac{1}{\sqrt{2}}\left(e^{-\frac{s}{\sqrt{2}}}, e^{\frac{s}{\sqrt{2}}}\,,\ s\right)$$

which, up to the isometry $(x, y, z) \to (-x, y, z)$, coincides with (3.1) and so the proof is completed. □

5. General helices

As we pointed out in the introduction, the most suitable definition of a general helix in a Riemannian geometric context is a delicate matter and may depend on the specific situation. Definition 3.1 above seemed to us the most appropriate for a first approach to the study of triharmonic curves in Sol.

In this section we study the connection between the proper triharmonic Frenet helices provided by Theorem 3.2 and the notion of a *general helix* according to the paper [**1**] of Barros. In this article the author, along the lines of previous work by Langer and Singer (see [**10**]), proposes the following definition.

DEFINITION 5.1. A smooth, regular curve $\gamma(s)$ into a 3-dimensional space form will be called a *general helix* if there exists a Killing vector field $\mathcal{V}(s)$ with constant length along γ and such that the angle between $\mathcal{V}$ and the unit tangent T is a non-zero constant along γ. We will say that $\mathcal{V}$ is an axis of the general helix.

Using this definition, Barros proved a version of Lancret Theorem for both $\mathbb{H}^3$ and $\mathbb{S}^3$. The spherical case is the most interesting because of the appearance of general helices whose geodesic curvature and torsion are related by a rule of the type $\tau(s) = \kappa(s) \pm 1$ (see [**1**]).

Now, we discuss this notion of a general helix in our context.

First, we recall that a base for the Killing fields on Sol is $\{\mathcal{V}_1, \mathcal{V}_2, \mathcal{V}_3\}$, where

$$\mathcal{V}_1 = e^z E_1\,, \quad \mathcal{V}_2 = e^{-z} E_2\,, \quad \mathcal{V}_3 = -xe^z E_1 + ye^z E_2 + E_3\,. \tag{5.1}$$

Our result is Proposition 5.2.

PROPOSITION 5.2.

(i) *There exists no proper triharmonic curve $\gamma(s)$ in Sol such that both $|\mathcal{V}_1|$ and β are constant along γ, where β denotes the angle between $\mathcal{V}_1$ and the unit tangent vector field T. The same is true for $\mathcal{V}_2$.*

(ii) *Let $\gamma(s)$ be a proper triharmonic curve in Sol as in Theorem* 3.2. *Then $|\mathcal{V}_3| = \sqrt{2}$ and $\beta = 0$ along γ, where β is the angle between $\mathcal{V}_3$ and T. Then γ is an integral curve of $\mathcal{V}_3/\sqrt{2}$.*

PROOF.

(i) We argue by contradiction. We observe that $|\mathcal{V}_1|^2 = e^{2z(s)}$. Therefore, we must have $z(s) = c$ along γ. Then explicit integration of the conditions $\langle T, T\rangle = 1$ and $\langle T, \mathcal{V}_1\rangle = |\mathcal{V}_1| \cos\beta$ tells us that $\gamma(s)$ must be of the type

$$\gamma(s) = (\cos\beta e^{-c}s + c_x, \sin\beta e^c s + c_y, c)\,.$$

Then a computation using (2.9) and (2.10) shows that, for this type of curve, $\kappa(s) = \sqrt{\cos^2(2\beta)}$, $\tau(s) = \sin(2\beta)$. Moreover, equations (2.12) and (2.14) are verified. But (2.13) becomes equivalent to

$$\sqrt{\cos^2(2\beta)}(5 + \cos(4\beta)) = 0\,.$$

This is a contradiction since $\cos(2\beta) = 0$ is not acceptable because γ is not a geodesic. The case of $\mathcal{V}_2$ is analogous.

(ii) We use the parametrization (3.1) of γ. Computing we find that along γ

$$|\mathcal{V}_3|^2 = x^2e^{2z} + y^2e^{-2z} + 1 = 2\,; \quad |\mathcal{V}_3|\cos\beta = \langle T, \mathcal{V}_3\rangle = -xx'e^{2z} + yy'e^{-2z} + z' = \sqrt{2}$$

so that $|\mathcal{V}_3| = \sqrt{2}$ and $\beta = 0$. Then γ is an integral curve of $\mathcal{V}_3/\sqrt{2}$, as required to end the proof. □

References

[1] M. Barros, *General helices and a theorem of Lancret*, Proc. Amer. Math. Soc. **125** (1997), no. 5, 1503–1509, DOI 10.1090/S0002-9939-97-03692-7. MR1363411
[2] V. Branding, *On polyharmonic helices in space forms*, Arch. Math. (Basel) **120** (2023), no. 2, 213–225, DOI 10.1007/s00013-022-01809-y. MR4542687
[3] V. Branding, S. Montaldo, C. Oniciuc, and A. Ratto, *Higher order energy functionals*, Adv. Math. **370** (2020), 107236, 60, DOI 10.1016/j.aim.2020.107236. MR4106647
[4] R. Caddeo, S. Montaldo, C. Oniciuc, and P. Piu, *The Euler-Lagrange method for biharmonic curves*, Mediterr. J. Math. **3** (2006), no. 3-4, 449–465, DOI 10.1007/s00009-006-0090-x. MR2274737
[5] H. Chen and Z. Guan, *Triharmonic CMC hypersurfaces in space forms with at most 3 distinct principal curvatures*, Results Math. **77** (2022), no. 4, Paper No. 155, 17, DOI 10.1007/s00025-022-01698-1. MR4444191
[6] H. Chen and Z. Guan, *Triharmonic CMC hypersurfaces in* $\mathbb{R}^5(c)$, Manuscripta Math. **172** (2023), no. 1-2, 209–220, DOI 10.1007/s00229-022-01409-8. MR4632927
[7] J. Eells and L. Lemaire, *Selected topics in harmonic maps*, CBMS Regional Conference Series in Mathematics, vol. 50, Published for the Conference Board of the Mathematical Sciences, Washington, DC; by the American Mathematical Society, Providence, RI, 1983, DOI 10.1090/cbms/050. MR703510
[8] J. Eells and L. Lemaire, *Two reports on harmonic maps*, World Scientific Publishing Co., Inc., River Edge, NJ, 1995, DOI 10.1142/9789812832030. MR1363513
[9] Y. Fu and D. Yang, *On triharmonic hypersurfaces in space forms*, J. Geom. Anal. **33** (2023), no. 8, Paper No. 263, 19, DOI 10.1007/s12220-023-01325-7. MR4598081
[10] J. Langer and D. A. Singer, *The total squared curvature of closed curves*, J. Differential Geom. **20** (1984), no. 1, 1–22. MR772124
[11] S. Maeta, *k-harmonic maps into a Riemannian manifold with constant sectional curvature*, Proc. Amer. Math. Soc. **140** (2012), no. 5, 1835–1847, DOI 10.1090/S0002-9939-2011-11049-9. MR2869168
[12] S. Maeta, *Construction of triharmonic maps*, Houston J. Math. **41** (2015), no. 2, 433–444. MR3403738
[13] S. Maeta, N. Nakauchi, and H. Urakawa, *Triharmonic isometric immersions into a manifold of non-positively constant curvature*, Monatsh. Math. **177** (2015), no. 4, 551–567, DOI 10.1007/s00605-014-0713-4. MR3371364
[14] S. Montaldo, C. Oniciuc, and A. Ratto, *Polyharmonic hypersurfaces into space forms*, Israel J. Math. **249** (2022), no. 1, 343–374, DOI 10.1007/s11856-022-2315-5. MR4462636
[15] S. Montaldo and A. Pámpano, *Triharmonic curves in 3-dimensional homogeneous spaces*, Mediterr. J. Math. **18** (2021), no. 5, Paper No. 198, 17, DOI 10.1007/s00009-021-01837-y. MR4308322
[16] S. Montaldo and A. Ratto, *New examples of r-harmonic immersions into the sphere*, J. Math. Anal. Appl. **458** (2018), no. 1, 849–859, DOI 10.1016/j.jmaa.2017.09.046. MR3711937
[17] S. Montaldo and A. Ratto, *Proper r-harmonic submanifolds into ellipsoids and rotation hypersurfaces*, Nonlinear Anal. **172** (2018), 59–72, DOI 10.1016/j.na.2018.03.002. MR3790367
[18] Y.-L. Ou and B.-Y. Chen, *Biharmonic submanifolds and biharmonic maps in Riemannian geometry*, World Scientific Publishing Co. Pte. Ltd., Hackensack, NJ, 2020. MR4265170
[19] B. Senoussi, *Triharmonic curves in* SOL_3 *space*, Palest. J. Math. **11** (2022), no. 2, 264–273. MR4424860

Università degli Studi di Cagliari, Dipartimento di Matematica e Informatica, Via Ospedale 72, 09124 Cagliari, Italia
Email address: montaldo@unica.it

Università degli Studi di Cagliari, Dipartimento di Matematica e Informatica, Via Ospedale 72, 09124 Cagliari, Italia
Email address: rattoa@unica.it

Contemporary Mathematics
Volume **821**, 2025
https://doi.org/10.1090/conm/821/16442

A note on Weingarten surfaces foliated by helices

Ana Irina Nistor

ABSTRACT. This paper represents the starting point of studying a class of surfaces obtained by using a family of isometries acting on a helix. We classify these surfaces under the properties as flatness or minimality. We obtain a nonexistence result for Special Weingarten surfaces. Finally, we classify the Weingarten surfaces foliated by helices.

1. Introduction

The first reference and maybe the most important one in the study of the helicoidal surfaces is represented by the Darboux Leçons [**Dar**]. According to [**DD**], the helicoidal surfaces with axis Oz and pitch h are invariant under a one-parameter group of rigid motions in $\mathbb{R}^3$ and, moreover, they represent a natural generalization of rotation surfaces. In this approach there are many beautiful results obtained in the literature, namely in the study of minimal, flat, constant mean curvature or constant Gaussian curvature; see, e.g., [**BK**], [**HR**], [**R**], and references therein.

In the present paper we define the surfaces foliated by helices, parametrized as $X : M^2 \to \mathbb{R}^3$, $X(u,v) = A(u)\gamma_0(v) + c(u)$, where $\gamma_0 : I \subset \mathbb{R} \to \mathbb{R}^3$ is a circular helix parametrized by arc-length as $\gamma_0(v) = (\cos v,\ \sin v, v)$,

$$A(u) = \begin{pmatrix} 0 & 0 & b(u) \\ a(u) & 0 & 0 \\ 0 & a(u) & 0 \end{pmatrix} \in Gl(3,\mathbb{R})$$

and $c(u) \in \mathbb{R}^3$. The idea of this definition comes from the following fact. Taking any helix, for example the one with Ox axis, its parametrization can be written as

$${}^t\Big(bv, a\cos v, a\sin v\Big) = \begin{pmatrix} 0 & 0 & b \\ a & 0 & 0 \\ 0 & a & 0 \end{pmatrix} \begin{pmatrix} \cos v \\ \sin v \\ v \end{pmatrix},$$

or the one with Oz axis

$${}^t\Big(a\cos v, a\sin v, bv\Big) = \begin{pmatrix} a & 0 & 0 \\ 0 & a & 0 \\ 0 & 0 & b \end{pmatrix} \begin{pmatrix} \cos v \\ \sin v \\ v \end{pmatrix},$$

2020 *Mathematics Subject Classification.* Primary 53B25, 53A05.

The author is grateful to professor Marian Ioan Munteanu for suggesting this topic and to professor Rafael López for useful hints during the preparation of the manuscript. The author would also like to thank the referees for the careful reading of the manuscript and helpful suggestions.

where $a,\ b \in \mathbb{R}$, $a^2 + b^2 \neq 0$. We notice that with a convenient matrix $A \in Gl(3, \mathbb{R})$ and $\gamma_0(v)$ we can generate every circular helix in $\mathbb{R}^3$. For this reason, we choose for example the matrix A giving the helix withe the axis Ox and we let the constants a and b to vary, becoming functions $a(u)$ and $b(u)$, respectively. Since

$$A(u) = \begin{pmatrix} 0 & 0 & b(u) \\ a(u) & 0 & 0 \\ 0 & a(u) & 0 \end{pmatrix}$$

belongs to $Gl(3, \mathbb{R})$, the functions $a(u)$ and $b(u)$ cannot vanish.

Next, we obtain two main results, according to the choice of $c(u)$, as follows.

THEOREM 1.1. *Let $X : M^2 \to \mathbb{R}^3$, be a surface foliated by helices, parametrized as $X(u, v) = (b(u)v,\ a(u)\cos v,\ a(u)\sin v) + c_0$. Then:*

- *The constant mean curvature surfaces are the helicoid and the cylinder.*
- *The constant Gaussian curvature surfaces are the flat surfaces.*
- *There are no proper Special Weingarten surfaces.*
- *The only Weingarten surfaces of this type are the flat surfaces.*

THEOREM 1.2. *Let $X : M^2 \to \mathbb{R}^3$, be a surface foliated by helices, parametrized as $X(u, v) = (u + b(u)v,\ a(u)\cos v,\ a(u)\sin v)$. Then:*

- *The constant mean curvature surfaces are the cylinder and the helicoidal surfaces.*
- *The constant Gaussian curvature surfaces are obtained when $a(u)$ and $b(v)$ are purely affine functions.*
- *There are no proper Special Weingarten surfaces.*
- *The Weingarten surfaces of this type are obtained for $a(u) = Au + B$ and $b(u) = \rho_0(Au + B)$, where $A, B, \rho_0 \in \mathbb{R}$, and they are flat surfaces.*

2. Proof of the theorems

Let U be a (complete) vector field in $\mathbb{R}^3$ and let us denote by $\{\varphi_t\}_t \in \mathbb{R}$ its one-paremeter group. Let $\gamma : I \subset \mathbb{R} \to \mathbb{R}^3$ be a circular helix parametrized by $\gamma(v) = (a\cos v,\ a\sin v, bv)$, $a, b \in \mathbb{R} \setminus \{0\}$.

We define a surface parametrized by

$$X : D \subset \mathbb{R}^2 \to \mathbb{R}^3, \quad X(u, v) = \varphi_u(\gamma(v)).$$

If U is a Killing vector field, then $\{\varphi_t\}$ is a family of isometries of $\mathbb{R}^3$. In the following we will discuss two cases: when U is a translational Killing field along the helix's axis and when it is a rotational Killing field with respect to the helix's axis.

• $U = (0, 0, 1) \equiv \dfrac{\partial}{\partial z}$, the one-parameter group is $\varphi_t(x, y, z) = (x, y, z + t)$, and the surface is a circular right cylinder, $x^2 + y^2 = a^2$ parametrized as

$$X(u, v) = (a\cos v, a\sin v, bv + u). \tag{2.1}$$

• $U = (y, -x, 0) = y\dfrac{\partial}{\partial x} - x\dfrac{\partial}{\partial y}$, the one-parameter group is $\varphi_t(x, y, z) = (x\cos t + y\sin t, y\cos t - x\sin t, z)$, and the surface is again the right circular cylinder, $x^2 + y^2 = a^2$, this time parametrized as

$$\begin{aligned} X(u, v) &= (a\cos v\cos u + a\sin v\sin u, a\sin v\cos u - a\cos v\sin u, bv) \\ &= (a\cos(v - u), a\sin(v - u), bv). \end{aligned} \tag{2.2}$$

Let us notice that the above two surfaces are part of a wider class of surfaces, parametrized as:

$$X(u, v) = A(u)\gamma(v) + c(u), \tag{2.3}$$

where $A(u) \in O(3)$ and $c(u) \in \mathbb{R}^3$, $\forall u$. Subsequently, in the first case we have $A = I_3$ and $c(u) = (0, 0, u)$, while in the second case

$$A(u) = \begin{pmatrix} \cos u & \sin u & 0 \\ -\sin u & \cos u & 0 \\ 0 & 0 & 1 \end{pmatrix}$$

and $c(u) = (0, 0, 0)$.

We would like to point out that equation (2.3) is not surprising, since the general form of an isometry in $\mathbb{R}^3$ is $f(\boldsymbol{x}) = A\boldsymbol{x} + b$, where $\boldsymbol{x} = (x, y, z) \in \mathbb{R}^3$, $A \in O(3)$ and $b \in \mathbb{R}^3$ and the two vector fields we used are Killing. Hence, the above surfaces are obtained using a family of isometries acting on the helix γ, that is $X(u, v) = f_u(\gamma(v))$. As further questions, we can discuss two aspects: *What happens if*

(1) *U is not Killing*, and
(2) *A is only a nonsingular matrix, not necessarily an orthogonal one, i.e., $A(u) \in Gl(3; \mathbb{R})$, $\forall u$?*

- Finally, if

$$U = \frac{\partial}{\partial x}$$

we also get a cylindrical surface (hence flat), but not a right cylinder. The same happens when

$$U = \frac{\partial}{\partial y}.$$

In both cases $A = I_3$ and $c(u)$ is a straight line orthogonal to the helix's axis. In Figure 1 we plotted such a surface obtained starting from the circular helix

$$\gamma(v) = \left(\frac{1}{\sqrt{2}} \cos v,\ \frac{1}{\sqrt{2}} \sin v,\ \frac{1}{\sqrt{2}} v\right).$$

In the next two subsections, we prove Theorem 1.1 and Theorem 1.2, each being divided into four problems which we solve.

Let us fix the notations according to the local classical differential geometry theory on surfaces (see, e.g., [**Lop**]). Denoting $X = X(u, v)$ a local parametrization of a surface in $\mathbb{R}^3$, the unit normal vector field is

$$N = \frac{X_u \times X_v}{\|X_u \times X_v\|},$$

where

$$X_u = \frac{\partial X}{\partial u},\quad X_v = \frac{\partial X}{\partial v},$$

$\times$ denotes the cross product in $\mathbb{R}^3$ and $\|\cdot\| = \sqrt{\langle \cdot, \cdot \rangle}$ stands for the Euclidean norm defined using the scalar product $\langle \cdot, \cdot \rangle$.

The first fundamental form is given by $I = \langle dX, dX \rangle = Edu^2 + 2Fdudv + Gdv^2$, where the coefficients are $E = \langle X_u, X_u \rangle$, $F = \langle X_u, X_v \rangle$ and $G = \langle X_v, X_v \rangle$.

The second fundamental form is $II = -\langle dN, dX \rangle = edu^2 + 2fdudv + gdv^2$, where the coefficients are $e = \langle N, X_{uu} \rangle$, $f = \langle N, X_{uv} \rangle$, $g = \langle N, X_{vv} \rangle$.

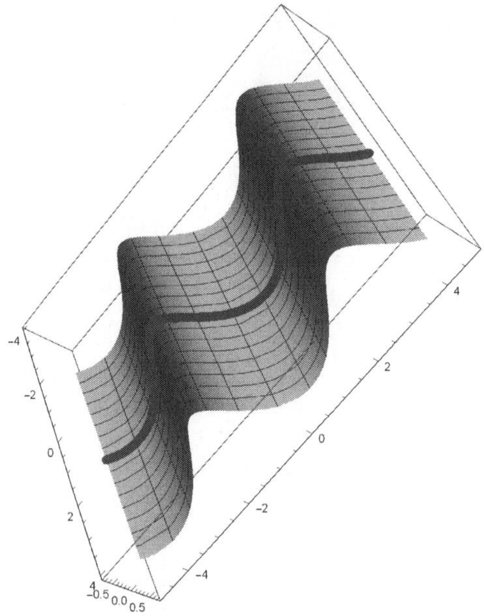

FIGURE 1. Flat surface for translational $U = \dfrac{\partial}{\partial x}$.

Further on, the classical formulas for the mean curvature H and the Gaussian curvature K are

$$H = \frac{eG - 2fF + gE}{2(EG - F^2)}, \quad K = \frac{eg - f^2}{EG - F^2}.$$

In order to proceed with the computations, denoting by $[\cdot,\cdot,\cdot]$ the determinant of order 3 in $\mathbb{R}^3$ and by $W := EG - F^2$, we have:

$$H = \frac{G[X_u, X_v, X_{uu}] - 2F[X_u, X_v, X_{uv}] + E[X_u, X_v, X_{vv}]}{2W^{3/2}} := \frac{H_1}{2W^{3/2}}, \tag{2.4}$$

$$K = \frac{[X_u, X_v, X_{uu}][X_u, X_v, X_{vv}] - [X_u, X_v, X_{uv}]^2}{W^2} := \frac{K_1}{W^2}. \tag{2.5}$$

2.1. Proof of Theorem 1.1. The case when $c(u) = c_0 = (x_0,\ y_0,\ z_0)$ is a constant vector. The surface is parametrized by

$$X(u,v) = (b(u)v,\ a(u)\cos v,\ a(u)\sin v) + c_0, \tag{2.6}$$

when

$$A(u) = \begin{pmatrix} 0 & 0 & b(u) \\ a(u) & 0 & 0 \\ 0 & a(u) & 0 \end{pmatrix} \in Gl(3, \mathbb{R}).$$

Let us point out first two particular situations.

REMARK 2.1. If $a(u) = a_0 \in \mathbb{R} \setminus \{0\}$, then (2.6) becomes

$$X(u,v) = (b(u)v,\ a_0\cos v,\ a_0\sin v) + c_0,$$

which is a flat and constant mean curvature surface, $H = \pm 1/2a_0$, hence a cylinder $(y - y_0)^2 + (z - z_0)^2 = a_0^2$.

REMARK 2.2. If $b(u) = b_0 \in \mathbb{R} \setminus \{0\}$, then (2.6) becomes

$$X(u,v) = (b_0 v,\ a(u)\cos v,\ a(u)\sin v) + c_0,$$

which is a minimal surface with the Gaussian curvature

$$K = -\frac{b_0^2}{(b_0^2 + a(u)^2)^2},$$

hence it is the helicoid.

We investigate four problems for each class of surfaces we study.

2.1.1. *Problem 1.* Find the surfaces of constant mean curvature $H = H_0$. We have the equation:

$$H_1^2 - 4H_0 W^3 = 0. \tag{2.7}$$

The left-hand part is a 6th degree polynomial in v and the coefficient of v^6 is:

$$a(u)^4 b'(u)^6 \left(1 - 4H_0 a(u)^2\right).$$

It must vanish so, either $b'(u) = 0$, hence $b(u) = b_0 \neq 0$ or $a^2(u) = \dfrac{1}{4H_0}$. These cases were already discussed previously.

2.1.2. *Problem 2.* Find the surfaces of constant Gaussian curvature $K = K_0$. We have the equation:

$$K_1 - K_0 W^2 = 0. \tag{2.8}$$

The left-hand part is a 4th degree polynomial in v and the coefficient of v^4 is:

$$-K_0 a(u)^4 b'(u)^4.$$

We can discuss the following two cases:

• $b(u) = b_0 \neq 0$ and subsequently $a(u) \neq a_0$, else $W = 0$. The equation becomes:

$$-a'(u)^4 \left(b_0^2 + b_0^4 K_0 + 2b_0^2 K_0 a(u) + K_0 a(u)^4\right) = 0,$$

and hence, the term inside the parenthesis must vanish,

$$b_0^2 + b_0^4 K_0 + 2b_0^2 K_0 a(u) + K_0 a(u)^4 = 0. \tag{2.9}$$

Taking a first derivative with respect to u, we get $2b_0^2 K_0 a'(u) + 4K_0 a(u)^3 a'(u) = 0$. Hence, $K_0 = 0$, i.e., the surface is flat.

• $K_0 = 0$ implies $K_1 = 0$. Now K_1 is a second degree polynomial in v, having the coefficient of v^2 given by

$$a(u)^3 b'(u) \left(a'(u)b''(u) - a''(u)b'(u)\right).$$

As $b'(u) \neq 0$ (obviously $a(u) \neq 0$), we must have

$$a'(u)b''(u) = a''(u)b'(u).$$

It follows that

$$b(u) = \rho_0 a(u) + \rho_1, \quad \rho_0, \rho_1 \in \mathbb{R}. \tag{2.10}$$

The constant term of K_1 is:

$$a(u)^3 b'(u) \left(a'(u)b(u) - a(u)b'(u)\right).$$

Now, using (2.10) in (2.9), we get $(\rho_0 a(u) + \rho_1)a'(u) - \rho_0 a(u)a'(u) = 0$, which yields $\rho_1 = 0$ and the surface is parametrized by $X(u,v) = a(u)\left(\rho_0 v, \cos v, \sin v\right)$, namely it is a cone over the helix having Ox axis and pitch ρ_0.

2.1.3. *Problem 3.* Find the Special Weingarten (SW) surfaces, i.e., those surfaces for which $\lambda H + \mu K = \zeta$, where λ, μ, ζ are real numbers.

Using the expressions of K and H, the condition for the surface of being SW becomes:

$$\lambda^2 H_1^2 W - 4(\zeta W^2 - \mu K_1)^2 = 0.$$

The left-hand part is a 8th degree polynomial in v and the coefficient of v^8 is:

$$a(u)^6 b'(u)^8 \left(\lambda^2 - 4\zeta^2 a(u)^2\right).$$

We distinguish the following situations:

(a) $b(u) = b_0$ and the surface is the helicoid;

(b) $\zeta \neq 0$, $a(u) = a_0 = \pm\dfrac{\lambda}{2\zeta}$ and the surface is a cylinder;

(c) $\zeta = 0$, $\lambda = 0$ yields $K = 0$, case already studied.

Hence, there are no proper SW surfaces.

2.1.4. *Problem 4.* Let us study Weingarten surfaces, not necessarily linear, i.e., which surfaces fulfill Jacobi condition:

$$H_u K_v - H_v K_u = 0. \tag{2.11}$$

The Jacobi condition yields an equation involving a 8th order polynomial in v with the coefficient of v^8 given by:

$$4a(u)^8 a'(u) b'(u)^8 \left(a'(u)b''(u) - a''(u)b'(u)\right). \tag{2.12}$$

Subsequently, as $a'(u), b'(u) \neq 0$, we must have $a'(u)b''(u) - a''(u)b'(u) = 0$, and thus $b'(u) = \rho_0 a'(u)$, $\forall u$, $\rho_0 \neq 0 \in \mathbb{R}$. Hence,

$$b(u) = \rho_0 a(u) + \rho_1, \quad \rho_1 \in \mathbb{R}. \tag{2.13}$$

Replacing now (2.13) in the previous expression (2.12), we get a 6th order polynomial in v and the term v^6 has the coefficient

$$\begin{aligned} 2\rho_0^6 a(u)^5 a'(u)^9 &\left[-4\rho_0\rho_1^2 a'(u)^3\right. \\ &\left.+(\rho_0 - 1)a(u)^2\left(-\rho_1^2 + 2\rho_0\rho_1 a(u) + (1 + 3\rho_0^2)a(u)^2\right)a'''(u)\right] \end{aligned}$$

and the constant term is:

$$\begin{aligned} -8\rho_0\rho_1^2\Big(\rho_0\rho_1 + (1+\rho_0^2)a(u)\Big)&\left[\rho_1^4 + 6\rho_0\rho_1^3 a(u) + 3(1+2\rho_0^2)\rho_1^2 a(u)^2\right. \\ &\left.+2\rho_0(2+\rho_0^2)\rho_1 a(u)^3 + (1+\rho_0^2)a(u)^4\right] a'(u)^{12}. \end{aligned} \tag{2.14}$$

Handling the 4th degree polynomial in $a(u)$ from the parenthesis $[\dots]$ in (2.14), the corresponding polynomial equation, i.e.,

$$\rho_1^4 + 6\rho_0\rho_1^3 a(u) + 3(1+2\rho_0^2)\rho_1^2 a(u)^2 + 2\rho_0(2+\rho_0^2)\rho_1 a(u)^3 + (1+\rho_0^2)a(u)^4 = 0,$$

yields $a(u)$ =constant, but we excluded this case. The same thing we get also from the parenthesis $(\dots)$ in (2.14), i.e., when

$$\rho_0\rho_1 + (1+\rho_0^2)a(u) = 0.$$

Hence, we must have $\rho_1 = 0$ and

- the coefficient of the 6th degree term becomes:

$$2\rho_0^6 a(u)^9 a'(u)^9 (\rho_0 - 1)(1 + 3\rho_0^2)a'''(u) = 0,$$

- the coefficient of the 4th degree term becomes:

$$2\rho_0^4 a(u)^9 a'(u)^9 (\rho_0 - 1)(\rho_0^2 + 1)(2 + 3\rho_0^2)a'''(u) = 0,$$

• the coefficient of the 2nd degree term becomes:

$$2\rho_0^2 a(u)^9 a'(u)^9(\rho_0 - 1)(\rho_0^2 + 1)^2 a'''(u) = 0.$$

Obviously, we can discuss two cases:

CASE 1 ($\rho_0 = 1$). Replacing $\rho_1 = 0$ and $\rho_0 = 1$ in (2.13) it follows that $a(u) = b(u)$. We get a flat surface parametrized by

$$X(u, v) = a(u)(v, \cos v, \sin v).$$

CASE 2 ($a'''(u) = 0$). It follows that $a(u) = a_0 + a_1 u + a_2 u^2$ and $b(u) = \rho_0 a(u)$. Then $K = 0$, hence we get again a flat surface.

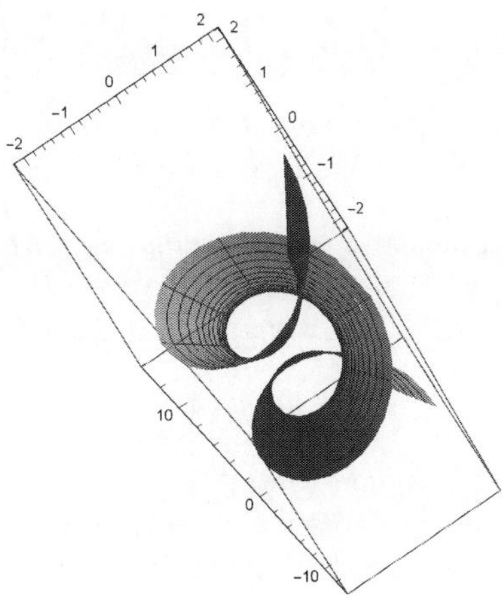

(A) Case 1. $a(u) = b(u) = u^3 + 1$

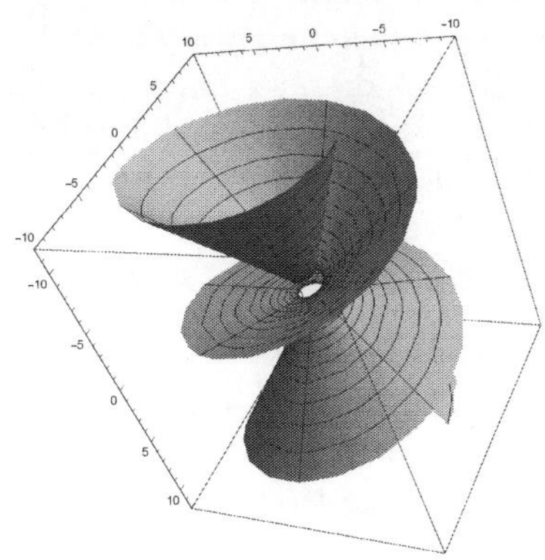

(B) Case 2. $a(u) = u^2 + 3u + 1$ and $\rho_0 = 0.2$

FIGURE 2. Weingarten surfaces parametrized by (2.6).

With this result we conclude the proof of the first theorem.
In the last subsection, we prove, in the same manner, the second theorem.

2.2. Proof of Theorem 1.2. The case when $c(u) = (u, y_0, z_0)$.

The surface is parametrized by

$$X(u,v) = (u,0,0) + (b(u)v,\ a(u)\cos v,\ a(u)\sin v)\,. \tag{2.15}$$

We point out two particular situations, which will be used in the sequel.

REMARK 2.3. If $a(u) = a_0 \in \mathbb{R} \setminus \{0\}$, then (2.15) becomes

$$X(u,v) = (u + b(u)v,\ a_0 \cos v,\ a_0 \sin v)\,,$$

which is a flat surface, hence a cylinder $y^2 + z^2 = a_0^2$.

REMARK 2.4. If $b(u) = b_0 \in \mathbb{R} \setminus \{0\}$, then (2.15) becomes

$$X(u,v) = (u + b_0 v,\ a(u)\cos v,\ a(u)\sin v)\,,$$

which is a helicoidal surface. As H and K depend only on u, the Jacobi condition (2.11) is satisfied, thus we have a Weingarten (not SW) surface.

We treat the same problems in the following using the same technique.

2.2.1. *Problem 1.* Constant mean curvature $H = H_0$ =constant. This time we obtain a 6th degree polynomial in v having the coefficient of the term v^6 given by

$$a(u)^4(1 - 4H_0 a(u)^2)b'(u)^6.$$

So, we should have $a(u)$ or $b(u)$ constants.

2.2.2. *Problem 2.* Constant Gaussian curvature $K = K_0$ =constant. This time we obtain a 4th degree polynomial in v having the coefficient of the term v^4 given by

$$-K_0 a(u)^4 b'(u).$$

So, we should have $b(u)$ constant (case excluded) or $K = 0$. If $K = 0$, from the coefficient of v^2 it follows that $a'(u)b''(u) = a''(u)b'(u)$ and thus $b(u) = \rho_0 a(u) + \rho_1$, $\rho_0 \neq 0$. From the coefficient of v we have $a'(u)b''(u) = 2a''(u)b'(u)$. Thus $a''(u)b'(u) = a'(u)b''(u) = 0$ and hence $a(u)$ and $b(u)$ are affine functions.

Thus, the surface is flat but not minimal if $a(u)$ (and consequently also $b(u)$) is not constant, but purely affine.

2.2.3. *Problem 3.* Special Weingarten surfaces: $\lambda H + \mu K = \zeta$

Analogous to the previous case in the above subsection, we obtain that there are no SW surfaces.

2.2.4. *Problem 4.* Let us study Weingarten surfaces, not necessarily linear, i.e., which surfaces fulfill the Jacobi condition (2.11).

We must have $a'(u)b''(u) - a''(u)b'(u) = 0$, and thus $b(u) = \rho_0 a(u) + \rho_1$, $\forall u$, $\rho_0, \rho_1 \in \mathbb{R}$, $\rho_0 \neq 0$. Furthermore, from the coefficient of v^5 we get that $a(u) = Au + B$, i.e., $a(u)$ is an affine function. Subsequently, from the coefficient of v^4: $-4A^{10}(Au + B)^4 \rho_0^5 \rho_1^2$ we get that $\rho_1 = 0$ since $A \neq 0$ and $a \neq$constant.

Thus, $a(u) = Au + B$ and $b(u) = \rho_0(Au + B)$ and the surface parametrized by

$$X(u,v) = (u + \rho_0(Au + B)v, (Au + B)\cos v, (Au + B)\sin v) \tag{2.16}$$

is flat.

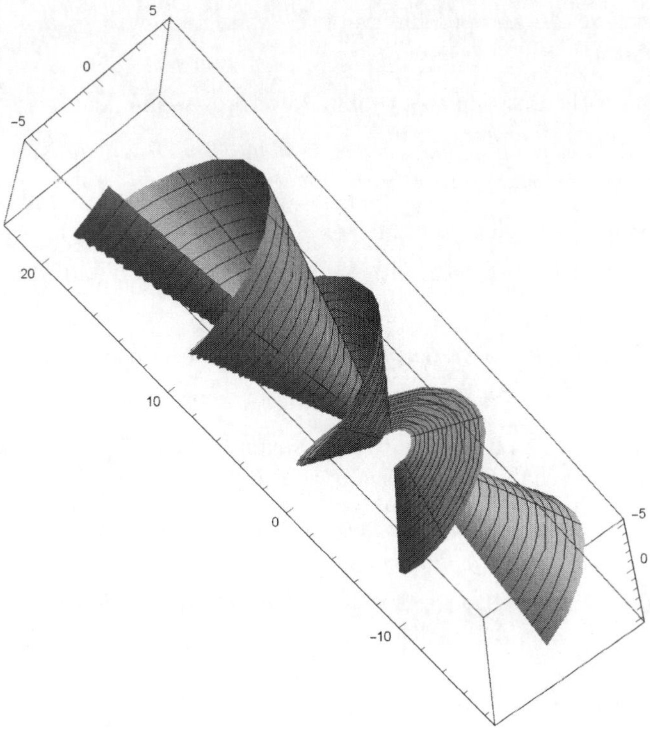

FIGURE 3. Weingarten surface for $A = 1$, $B = 3$ and $\rho_0 = 0.3$.

3. Appendix

This section is dedicated to the clarification of some technical details. Since the computations are quite long, they were assisted by Mathematica software. In order not to lose the train of ideas, we did not write here all the complete formulas.

The study of minimality and flatness reduces to the conditions $H_1 = 0$ in (2.4) and $K_1 = 0$ in (2.5), respectively. More precisely, these conditions can be written as:

$$H_1 = 0 \Leftrightarrow \sum_{j=0}^{3}[\mathbf{p}_j(u,v)\cos(jv) + \mathbf{q}_j(u,v)\sin(jv)] = 0, \quad \forall u, v,$$

and

$$K_1 = 0 \Leftrightarrow \sum_{j=0}^{4}[\mathbf{p}_j(u,v)\cos(jv) + \mathbf{q}_j(u,v)\sin(jv)] = 0, \quad \forall u, v,$$

where $\mathbf{p}_j(u,v)$ and $\mathbf{q}_j(u,v)$ are polynomials in v with coefficients depending on u.

In [**CC**] the authors solve a problem of linear independence for a set of cosine functions and sine functions. Related to it, we formulate and solve the following.

PROBLEM. Let $n > 0$ be a positive integer and $\alpha_1, \ldots, \alpha_n$ real numbers such that $|\alpha_i| \neq |\alpha_j|$ for all $i \neq j$. Let $p_j, j = 0, 1, \ldots, n$ and $q_j, j = 1, \ldots, n$ be polynomials of one real variable. If

$$p_0(x) + \sum_{j=1}^{n}(p_j(x)\cos(\alpha_j x) + q_j(x)\sin(\alpha_j x)) = 0, \tag{3.1}$$

for all $x \in I$, where I is an open interval in $\mathbb{R}$, then $p_0, p_1, \ldots, p_n$ and $q_1, \ldots, q_n$ are null polynomials.

Before giving the proof of this problem, we prove some lemmas.

LEMMA 3.1. *Let α be a nonnegative real number, P a polynomial of one real variable and $\lambda, \mu \in \mathbb{R}$ such that*

$$P(x) + \lambda \cos(\alpha x) + \mu \sin(\alpha x) = 0, \tag{3.2}$$

for all $x \in I$, where I is an open interval in $\mathbb{R}$. Then $\lambda = \mu = 0$ and P is the null polynomial.

PROOF. Taking successive derivatives (with respect to x in (3.2) one obtains

$$\alpha^k \big(\varepsilon_1 \lambda \cos(\alpha x) + \varepsilon_2 \mu \sin(\alpha x)\big) = 0, \quad \forall x \in I,$$

where $\varepsilon_1 = \pm 1$, $\varepsilon_2 = \pm 1$. It follows (use, for example, [**CC**]) $\lambda = \mu = 0$. Returning to (3.2) we conclude that P is identically zero. $\square$

LEMMA 3.2. *Let α be a nonnegative real number, P a polynomial of one real variable and $\lambda, \mu,\ \rho \in \mathbb{R}$ such that either*

$$P(x) + \lambda \cos(\alpha x) + (\mu x + \rho) \sin(\alpha x) = 0, \tag{3.3}$$

or

$$P(x) + (\mu x + \rho) \cos(\alpha x) + \lambda \sin(\alpha x) = 0, \tag{3.4}$$

for all $x \in I$, where I is an open interval in $\mathbb{R}$. Then $\lambda = \mu = \rho = 0$ and P is the null polynomial.

PROOF. We prove this Lemma only for the the case given in relation (3.3). Let

$$f(x) = P(x) + \lambda \cos(\alpha x) + (\mu x + \rho) \sin(\alpha x).$$

We have

$$\begin{aligned} f'(x) &= P'(x) + \alpha(\mu x + \rho) \cos(\alpha x) + (\mu - \alpha\lambda) \sin(\alpha x) \\ f''(x) &= P''(x) + \alpha(2\mu - \lambda) \cos(\alpha x) - \alpha^2 (\mu x + \rho) \sin(\alpha x). \end{aligned}$$

It follows $\alpha^2 f(x) + f''(x) = \big(\alpha^2 P(x) + P''(x)\big) + \alpha(\alpha\lambda + 2\mu - \lambda) \cos(\alpha x) = 0, \forall x \in I$.

Using Lemma 3.1 we deduce that $\alpha^2 P + P''$ is identically zero and $\alpha(\alpha\lambda + 2\mu - \lambda) = 0$. This implies that P is identically zero. Consequently, we have

$$\lambda \cos(\alpha x) + (\mu x + \rho) \sin(\alpha x) = 0,$$

$$\alpha(\mu x + \rho) \cos(\alpha x) + (\mu - \alpha\lambda) \sin(\alpha x) = 0, \quad \forall x \in I.$$

Combining the two equations we obtain

$$\big[\lambda(\alpha\lambda - \mu) + \alpha(\mu x + \rho)^2\big] \cos(\alpha x) = 0, \forall x \in I.$$

It follows $\mu = 0$ and $\lambda = \rho = 0$. $\square$

LEMMA 3.3. *Let α be a non-negative real number, P_0, P_1, Q_0 polynomials of one real variable such that*

$$P_0(x) + P_1(x) \cos(\alpha x) + Q_1(x) \sin(\alpha x) = 0, \tag{3.5}$$

for all $x \in I$, where I is an open interval in $\mathbb{R}$. Then P_0, P_1, Q_1 are the null polynomials.

PROOF. Let $f(x) = P_0(x) + P_1(x)\cos(\alpha x) + Q_1(x)\sin(\alpha x)$. Compute

$$\begin{aligned} f'(x) &= p_0'(x) + \big(P_1(x) + \alpha Q_1(x)\big)\cos(\alpha x) + \big(Q_1'(x) - \alpha P_1(x)\big)\sin(\alpha x) \\ f''(x) &= P_0''(x) + \big(P_1''(x) + 2\alpha Q_1'(x) - \alpha^2 P_1(x)\big)\cos(\alpha x) \\ &\quad + \big(Q_1''(x) - 2\alpha P_1'(x) - \alpha^2 Q_1(x)\big)\sin(\alpha x). \end{aligned}$$

We write

$$\begin{aligned} \alpha^2 f(x) + f''(x) &= (\alpha^2 P_0(x) + P_0''(x) + \big(P_1''(x)\big) + 2\alpha Q_1'(x)\big)\cos(\alpha x) \\ &\quad + \big(Q_1''(x) - 2\alpha P_1'(x)\big)\sin(\alpha x) = 0, \quad \forall x \in I. \end{aligned} \tag{3.6}$$

The relation (3.6) has the form

$$\tilde{P}_0(x) + \tilde{P}_1(x)\cos(\alpha x) + \tilde{Q}_1(x)\sin(\alpha x) = 0, \forall x \in I, \tag{3.7}$$

where

$$\begin{cases} \tilde{P}_0(x) = \alpha^2 P_0(x) + P_0''(x), \\ \tilde{P}_1(x) = P_1''(x)) + 2\alpha Q_1'(x), \\ \tilde{Q}_1(x) = Q_1''(x) - 2\alpha P_1'(x). \end{cases} \tag{3.8}$$

Suppose that P_1 and Q_1 are not simultaneously constant polynomials. We distinguish three situations:

(i) $\deg P_1 = \deg Q_1 = d \geq 1$;
(ii) $\deg P_1 = d \geq 0$, $\deg Q_1 = d + k$, $k \geq 1$;
(iii) $\deg Q_1 = d \geq 0$, $\deg P_1 = d + k$, $k \geq 1$.

In case (i) we have $\deg \tilde{P}_1 = \deg \tilde{Q}_1 = d - 1$. We continue to write equations in the form $\alpha^2 f(x) + f''(x) = 0, \forall x \in I$. At each step, the degree of the two polynomials decreases by 1 and therefore, after a finite number of steps, it becomes zero. We are now in the hypothesis of Lemma 3.1. The relations between two successive steps have always form of (3.8). Let $\tilde{P}_0(x) = 0$, $\tilde{P}_1(x) = 0$ and $\tilde{Q}_1(x) = 0$, for all $x \in I$. Taking into account that we are working with polynomials, it follows that P_0, P_1' and Q_1' are identically zero. This is false, so the assumption is false. Thus P_1 and Q_1 are both constant.

In cases (ii) and (iii) we have
$\deg \tilde{P}_1 = d + k - 1$, $\deg \tilde{Q}_1 = d + k - 2$ or
$\deg \tilde{P}_1 = d + k - 2$, $\deg \tilde{Q}_1 = d + k - 1$.

The difference between $\deg \tilde{P}_1$ and $\deg \tilde{Q}_1$ is 1. As before, we continue to write equations in the form $\alpha^2 f(x) + f''(x) = 0, \forall x \in I$. At each step, $|\deg \tilde{P}_1 - \deg \tilde{Q}_1| = 1$ and $\max(\deg \tilde{P}_1, \deg \tilde{Q}_1)$ decreses by 1. hence, after a finite number of steps, either $\tilde{P}_1$ is constant and $\tilde{Q}_1$ has degree 1, or $\tilde{Q}_1$ is constant and $\tilde{P}_1$ gas degree 1. This contradicts Lemma 3.2. The assumption is again false, so the polynomials P_1 and Q_1 are both constant. Using Lemma 3.1 we get the conclusion. □

LEMMA 3.4. *Let $n > 0$ be a positive integer and $\alpha_1, \ldots, \alpha_n$ real numbers such that $|\alpha_i| \neq |\alpha_j|$, $\forall i \neq j$. Let P be a polynomial of a real variable and $\lambda_j, \mu_j \in \mathbb{R}$, $j = 1, \ldots, n$ such that*

$$P(x) + \sum_{j=1}^{n} \big(\lambda_j \cos(\alpha_j x) + \mu_j \sin(\alpha_j x)\big) = 0, \tag{3.9}$$

for all $x \in I$, where I is an open interval in $\mathbb{R}$. Then $\lambda_j = \mu_j = 0$, $\forall j = 1, \dots, n$ and P is the null polynomial.

PROOF. After a finite number of derivations with respect to x we get

$$\sum_{j=1}^{n} \alpha_j^h \big(\varepsilon_{j1}\lambda_j \cos(\alpha_j x) + \varepsilon_{j2}\mu_j \sin(\alpha_j x)\big) = 0, \forall x \in I,$$

where $\varepsilon_{j1} = \pm 1$, $\varepsilon_{j2} = \pm 1$. Using for example [**CC**], the coefficients for $\cos(\alpha_j x)$ and $\sin(\alpha_j x)$ vanish. We conclude with $\lambda_j = \mu_j = 0$, $j = 1, \dots, n$. It then follows that P is identically zero. □

MAIN PROOF. We will proceed by induction on n.

For $n = 1$ the problem is equivalent to Lemma 3.3. Let us assume that the statement is true for "$n - 1$".

Let

$$f(x) = p_0(x) + \sum_{j=1}^{n} (p_j(x)\cos(\alpha_j x) + q_j(x)\sin(\alpha_j x)) = 0, \forall x \in I. \tag{3.10}$$

We compute

$$\begin{aligned} f'(x) &= p_0'(x) + \sum_{j=1}^{n} \big[\big(p_j'(x) + \alpha_j q_j(x)\big)\cos(\alpha_j x) \\ &\qquad + \big(q_j'(x) - \alpha_j p_j(x)\big)\sin(\alpha_j x)\big] = 0, \\ f''(x) &= p_0''(x) + \sum_{j=1}^{n} \big[\big(p_j''(x) + 2\alpha_j q_j'(x) - \alpha_j^2 p_j(x)\big)\cos(\alpha_j x) \\ &\qquad + \big(q_j''(x) - 2\alpha_j p_j'(x) - \alpha_j^2 q_j(x)\big)\sin(\alpha_j x)\big] = 0. \end{aligned} \tag{3.11}$$

With lost of the generality, assume that the greatest degree of polynomials p_j, q_j $(j = 1, \dots, n)$ is attained for p_n or q_n. Let it be m. Combine (3.10) and the first equation of (3.11) we obtain $\cos(\alpha_j x)$ and $\sin(\alpha_j x)$.

Let us make some notations:

$$\begin{array}{ll} A_n = p_n q_n' - p_n' q_n - \alpha_n(p_n^2 + q_n^2), & \\ A_j = p_j q_n' - p_j' q_n - \alpha_n p_j p_n - \alpha_j q_j q_n, & j = 1, \dots, n-1, \\ A_0 = p_0 q_n' - p_0' q_n - \alpha_n p_0 p_n, & \\ B_j = q_j q_n' - q_j' q_n - \alpha_n p_n q_j + \alpha_j p_j q_n, & j = 1, \dots, n-1, \\ T_n = p_n' q_n - p_n q_n' + \alpha_n(p_n^2 + q_n^2), & \\ S_j = p_j p_n' - p_j' p_n + \alpha_n p_j q_n - \alpha_j p_n q_j, & j = 1, \dots, n-1, \\ S_o = p_0 p_n' - p_0' p_n + \alpha_n p_0 q_n, & \\ T_j = p_n' q_j - p_n q_j' + \alpha_j p_j p_n + \alpha_n q_j q_n, & j = 1, \dots, n-1. \end{array}$$

We deduce

$$A_n(x)\cos(\alpha_n x) = -A_0(x) - \sum_{j=1}^{n-1} \big(A_j(x)\cos(\alpha_j x) + B_j(x)\sin(\alpha_j x)\big),$$

$$T_n(x)\sin(\alpha_n x) = -S_0(x) - \sum_{j=1}^{n-1} \big(S_j(x)\cos(\alpha_j x) + T_j(x)\sin(\alpha_j x)\big).$$

Combine with the second equation of (3.11) to obtain

(3.12)

$$A_nT_n\,[p_0''+\sum_{j=1}^{n-1}(p_j''+2\alpha_jq_j'-\alpha_j^2p_j)\cos(\alpha_jx)+\sum_{j=1}^{n-1}(q_j''-2\alpha_jp_j'-\alpha_j^2q_j)\sin(\alpha_jx)]$$
$$-T_nA_0(p_n''+2\alpha_nq_n'-\alpha_n^2p_n)-A_nS_0(q_n''-2\alpha_np_n'-\alpha_n^2q_n)$$
$$-T_n(p_n''+2\alpha_nq_n'-\alpha_n^2p_n)\sum_{j=1}^{n-1}\left(A_j\cos(\alpha_jx)+B_j\sin(\alpha_jx)\right)$$
$$-A_n(q_n''-2\alpha_np_n'-\alpha_n^2q_n)\sum_{j=1}^{n-1}\left(S_j\cos(\alpha_jx)+T_j\sin(\alpha_jx)\right)=0,\quad \forall x\in I.$$

Let N be the greatest degree of a polynomials among all polynomials p_j and q_j, $j=1,\dots,n-1$. Without danger of confusion, we denote by the same letter "j" the index for which this maximum is attained (one can have several indices, and in this case we fix one of them). Let us analyze the coefficients for $\cos(\alpha_jx)$ and $\sin(\alpha_jx)$, respectively:

$$A_nT_n(p_j''+2\alpha_jq_j'-\alpha_j^2p_j)-A_jT_n(p_n''+2\alpha_nq_n'-\alpha_n^2p_n)-A_nS_j(q_n''-2\alpha_np_n'-\alpha_n^2q_n),$$

$$A_nT_n(q_j''-2\alpha_jp_j'-\alpha_j^2q_j)-B_jT_n(p_n''+2\alpha_nq_n'-\alpha_n^2p_n)-A_nT_j(q_n''-2\alpha_np_n'-\alpha_n^2q_n).$$

We check the degrees of these polynomials, observing first that $\deg A_n=\deg T_n=2m$. We distinguish three situations (for the index j fixed):

(i) $\deg p_j=\deg q_j=N$
(ii) $\deg p_j<\deg q_j=N$
(iii) $\deg q_j<\deg p_j=N$.

In case (i) we have $\deg A_j=\deg B_j=\deg S_j=\deg T_j=m+N$. The degree of the coefficients of $\cos(\alpha_jx)$, respectively $\sin(\alpha_jx)$ are equal to $4m+N$ and they are, respectively, given by the polynomials

$$\alpha_n^2p_j(p_n^2+q_n^2)^2(\alpha_j^2-\alpha_n^2)\quad\text{and}\quad\alpha_n^2q_j(p_n^2+q_n^2)^2(\alpha_j^2-\alpha_n^2).$$

In case (ii) we have $\deg A_j=\deg B_j=\deg S_j=\deg T_j=m+N$. The coefficient of $\cos(\alpha_jx)$ vanishes, and the degree of the coefficient of $\sin(\alpha_jx)$ is equal to $4m+N$; it is given by the polynomial $\alpha_n^2q_j(p_n^2+q_n^2)^2(\alpha_j^2-\alpha_n^2)$.

Case (iii) is analogue to case (ii).

Relation (3.12) represents the statement corresponding to the induction hypothesis. Thus, all the polynomials, coefficients for $\cos(\alpha_jx)$ and $\sin(\alpha_jx)$ must vanish. Yet, according to the previous computations, taking into account that $|\alpha_j|\neq|\alpha_n|$ and α_j are nonzero real numbers, it follows that $m=0$ and $N=0$, that is all polynomials p_j and q_j, $j=1,\dots,n$ are, in fact, constants. According to Lemma 3.4, all the polynomials must vanish identically. □

References

[BK] C. Baikoussis and T. Koufogiorgos, *Helicoidal surfaces with prescribed mean or Gaussian curvature*, J. Geom. **63** (1998), no. 1-2, 25–29, DOI 10.1007/BF01221235. MR1651561

[CC] O. Christensen and K. L. Christensen, *Linear independence and series expansions in function spaces*, Amer. Math. Monthly **113** (2006), no. 7, 611–627, DOI 10.2307/27642004. MR2252932

[Dar] G. Darboux, *Leçons sur la théorie générale des surfaces. III, IV* (French), Les Grands Classiques Gauthier-Villars. [Gauthier-Villars Great Classics], Éditions Jacques Gabay, Sceaux, 1993. MR1365962

[Del] G. Delaunay, *Sur la surface de revolution dont la courbure moyenne est constante*, J. Math. Pures Appl. Series 1,**6**(1841), 309-320.

[DD] M. P. do Carmo and M. Dajczer, *Helicoidal surfaces with constant mean curvature*, Tohoku Math. J. (2) **34** (1982), no. 3, 425–435, DOI 10.2748/tmj/1178229204. MR676120

[HR] L. Hitt and I. M. Roussos, *Computer graphics of helicoidal surfaces with constant mean curvature*, An. Acad. Bras. Ci., **63** (1991), 211-228.

[Lop] R. López, *Special Weingarten surfaces foliated by circles*, Monatsh. Math. **154** (2008), no. 4, 289–302, DOI 10.1007/s00605-008-0557-x. MR2425758

[R] I. M. Roussos, *Surfaces in E^3 invariant under a one parameter group of isometries of E^3*, An. Acad. Brasil. Ciênc. **72** (2000), no. 2, 125–159, DOI 10.1590/S0001-37652000000200003. MR1763990

Department of Mathematics and Informatics, Gheorghe Asachi Technical University of Iasi, Bd. Carol I, n. 11A, 700506, Iasi, Romania

Email address: ana.irina.nistor@gmail.com

Contemporary Mathematics
Volume **821**, 2025
https://doi.org/10.1090/conm/821/16443

Characterizations of rotational biconservative hypersurfaces

Álvaro Pámpano

ABSTRACT. We extend a result of Montaldo, Oniciuc, and Pámpano (J. Math. Anal. Appl. **518** (2023), no. 1, Paper No. 126697) to show that the profile curves of non-CMC rotational (of all types) biconservative hypersurfaces of space forms $N^n(\rho)$ are p-elastic curves, for a suitable rational number $p \in [1/4, 1)$ which depends only on the dimension n of the space form $N^n(\rho)$.

In addition, we exploit this characterization to relate non-CMC closed rotational biconservative hypersurfaces of $N^n(\rho)$ with generalized Willmore surfaces in the 3-dimensional round sphere $\mathbb{S}^3(\rho/4)$. We also describe a relation between rotational minimal hypersurfaces of $N^n(\rho)$ and non-CMC rotational biconservative hypersurfaces of $N^{3n-4}(\rho)$.

1. Introduction

In 1964, Eells and Sampson [**7**] suggested to study biharmonic maps between Riemannian manifolds as an extension of harmonic maps. Biharmonic maps are the critical points of the bienergy functional, whose first variation formula and associated Euler–Lagrange equation were first computed by Jiang [**14**]. In the particular case that we are dealing with isometric immersions, this Euler–Lagrange equation can be decomposed into its tangential and normal components [**2, 6, 18, 25**].

For arbitrary variational problems, Hilbert [**13**] described the associated stress-energy tensor as a symmetric 2-covariant tensor given by the property that it is conservative at the critical points. It turns out that the tangential component of the Euler–Lagrange equation associated to the bienergy can be interpreted as the condition that the stress-energy tensor for the bienergy is conservative. From the computations of [**15**] and [**19**], it follows that for an hypersurface M^{n-1} of a space form $N^n(\rho)$ this condition is

$$2S_\eta\,(\mathrm{grad}H) + (n-1)H\mathrm{grad}H = 0,$$

where η is a unit normal vector field, S_η is the shape operator, and H is the mean curvature function. A hypersurface satisfying this equation is called a biconservative hypersurface. Since they were first introduced in [**3**] their study has been very intense. A couple of recent surveys [**5, 8**] collect the main results regarding biconservative hypersurfaces (and submanifolds in general) as well as a very detailed

---

2020 *Mathematics Subject Classification.* Primary 53A04, 53A05; Secondary 53A10.

Key words and phrases. Biconservative hypersurfaces, generalized Willmore surfaces, minimal hypersurfaces, p-elastic curves.

list of references. We refer the reader to these works and their references for more details.

Closely related to the topic of the present paper, we mention here that biconservative surfaces of 3-dimensional space forms $N^3(\rho)$ were locally classified in [**3**, **9**, **12**] and globally in [**23**]. More precisely, it was shown that non-CMC biconservative surfaces must be rotational and a relation between their mean and Gaussian curvatures was stated (a similar result holds for hypersurfaces in $N^5(\rho)$ under the assumption of constant scalar curvature; see [**10**]). Using these results, in [**22**], a variational characterization of the profile curves of non-CMC biconservative surfaces was obtained. Indeed, these profile curves are p-elastic curves for $p = 1/4$. Later on, in [**20**], this variational characterization was extended for the profile curves of non-CMC biconservative rotational hypersurfaces of spherical type. In this case, the profile curves are p-elastic curves for $p = (n-2)/(n+1) \in [1/4, 1)$ where $n \geq 3$ is the dimension of the ambient space form $N^n(\rho)$. In Section 3, we extend this characterization further by showing that the same result holds for non-CMC biconservative rotational hypersurfaces of the three possible types of rotations: spherical, parabolic, and hyperbolic [**4**].

The p-elastic curves occupy a central position in the study of variational problems for curves. They arise as the critical points of the functionals

$$\mathbf{\Theta}_p(\gamma) := \int_\gamma \kappa^p \, ds,$$

where κ si the curvature of the curve γ and s is its arc length parameter. In a letter of 1738 from D. Bernoulli to L. Euler, the functionals $\mathbf{\Theta}_p$ were introduced and the investigation of their extrema suggested. Throughout their long and storied history they have been used as effective scientific models and have found applications, not only in Mathematics, but also in many different fields of the natural sciences, ranging from biology and physics to architecture and art. The introduction of [**11**] has several of these applications for special values of $p \in \mathbb{R}$ described and referred.

Here, we highlight the construction of Hopf tori in $\mathbb{S}^3(\rho/4)$ based on 2-elastic curves in $\mathbb{S}^2(\rho)$ employed by Pinkall [**29**] to obtain the first examples of non-conformally minimal Willmore surfaces. His result relies heavily on the Symmetric Criticality Principle of Palais [**26**] and on the existence of closed 2-elastic curves [**16**]. By changing the base curve, the construction of Hopf tori given by Pinkall can be used to obtain examples of more general functionals whose energy density depends on the mean curvature (see, for instance, [**27**]). In particular, if the base curve is a closed p-elastic curve, the corresponding Hopf torus is a p-Willmore surface. Several illustrations for different values of $p \in (0,1)$ can be found in the appendix of [**11**].

In Section 4 we consider this construction of Hopf tori based on closed p-elastic curves for $p = (n-2)/(n+1)$ to describe a relation between non-CMC closed rotational biconservative hypersurfaces of $N^n(\rho)$ and p-Willmore surfaces. Since non-CMC closed rotational biconservative hypersurfaces lie necessarily on $\mathbb{S}^n(\rho)$ [**20**], it follows that the profile curves are closed p-elastic curves for $p = (n-2)/(n+1)$ in $\mathbb{S}^2(\rho)$ and, hence, the corresponding p-Willmore Hopf tori are in $\mathbb{S}^3(\rho/4)$. We illustrate this relation in Figure 2.

Finally, in Section 5 we exploit again the variational characterization of profile curves of non-CMC rotational biconservative hypersurfaces to show that there is a one-to-one correspondence between nonisoparametric rotational minimal hypersurfaces of $N^n(\rho)$ and non-CMC rotational biconservative hypersurfaces of $N^{3n-4}(\rho)$.

2. Rotational biconservative hypersurfaces of space forms

In this section we introduce the basic information about rotational biconservative hypersurfaces of Riemannian space forms $N^n(\rho)$. For more details about biconservative hypersurfaces we refer the reader to [**20**] and references therein, while for the theory of rotational hypersurfaces in space forms we suggest the seminal work [**4**].

2.1. Space forms. Let $N^n(\rho)$ be a n-dimensional Riemannian space form of constant sectional curvature $\rho \in \mathbb{R}$. If $\rho = 0$, $N^n(\rho) = \mathbb{R}^n$ denotes the Euclidean space of dimension n; when $\rho > 0$, $N^n(\rho) = \mathbb{S}^n(\rho)$ is the round sphere of dimension n; and, for negative values $\rho < 0$, $N^n(\rho) = \mathbb{H}^n(\rho)$ represents the n-dimensional hyperbolic space.

Consider the Euclidean semi-space $\mathbb{R}^{n+1}_\nu$ of dimension $n+1$. That is, $\mathbb{R}^{n+1}$ endowed with the canonical metric of index $\nu = 0, 1$, namely,

$$g = \sum_{i=1}^{n} dx_i^2 + (-1)^\nu dx_{n+1}^2 ,$$

where $(x_1, \dots, x_{n+1})$ denotes the standard coordinates of $\mathbb{R}^{n+1}$. For simplicity, the metric g will be denoted by $\langle \cdot, \cdot \rangle$. The space forms $N^n(\rho)$ can be isometrically immersed in $\mathbb{R}^{n+1}_\nu$. The Euclidean space $\mathbb{R}^n$ is

$$\mathbb{R}^n = \{\mathbf{x} \in \mathbb{R}^{n+1}_0 \,|\, x_{n+1} = 0\} ,$$

where $\mathbf{x} = (x_1, \dots, x_{n+1})$. The round sphere $\mathbb{S}^n(\rho)$ is the hyperquadric

$$\mathbb{S}^n(\rho) = \{\mathbf{x} \in \mathbb{R}^{n+1}_0 \,|\, \langle \mathbf{x}, \mathbf{x} \rangle = 1/\rho\} .$$

And, finally, the hyperbolic space $\mathbb{H}^n(\rho)$ is the hyperquadric

$$\mathbb{H}^n(\rho) = \{\mathbf{x} \in \mathbb{R}^{n+1}_1 \,|\, \langle \mathbf{x}, \mathbf{x} \rangle = 1/\rho\, ,\, x_{n+1} > 0\} .$$

2.2. Biconservative hypersurfaces. An oriented hypersurface M^{n-1} immersed in an n-dimensional Riemannian space form $N^n(\rho)$ is called biconservative if

$$2S_\eta\left(\mathrm{grad} H\right) + (n-1) H \mathrm{grad} H = 0, \tag{2.1}$$

where η is a unit normal vector field to M^{n-1}, S_η is the shape operator, and $H = \mathrm{trace}\, S_\eta/(n-1)$ is the mean curvature function. This equation is the tangential component of the Euler–Lagrange equation associated to the bienergy. The decomposition of the Euler-Lagrange equation of the bienergy with respect to its tangential and normal components was obtained with contributions of [**2,6,18,25**]. In addition, equation (2.1) was first introduced in [**3**] as the condition that the stress-energy tensor (as described by Hilbert; see [**13**]) associated to the bienergy is conservative.

When the mean curvature function H is constant (2.1) trivially holds and, hence, constant mean curvature (CMC) hypersurfaces are examples of biconservative hypersurfaces. Consequently, the main interest is to investigate non-CMC biconservative hypersurfaces. Developments on the theory of non-CMC biconservative hypersurfaces and recent results are beautifully collected in the following couple of surveys [**5,8**].

Let M^{n-1} be a hypersurface with principal curvatures κ_i, $i = 1, \ldots, n-1$, i.e., the eigenvalue functions of the shape operator S_η. The functions κ_i are continuous on M^{n-1} for all $i = 1, \ldots, n-1$. The set of points where the number of distinct principal curvatures is locally constant is a set M_A that is open and dense in M^{n-1}. On a nonempty connected component of M_A, which is open in M_A and so in M^{n-1}, the number of distinct principal curvatures is constant, and so, on that connected component, the principal curvatures κ_i are smooth and S_η is (smoothly) locally diagonalizable (see [**24, 30, 31**]). For a hypersurface M^{n-1} with nowhere zero grad$H$, the corresponding M_A has no connected component made up only by umbilical points. In this setting, it follows from (2.1) that a biconservative hypersurface M^{n-1} with nowhere zero grad$H$ has a principal curvature $\kappa_1 = -(n-1)H/2$ of multiplicity one on each connected component of M_A and, hence,

$$3\kappa_1 + \kappa_2 + \cdots + \kappa_{n-1} = 0\,, \tag{2.2}$$

holds. As a consequence, if a non-CMC biconservative hypersurface M^{n-1} of a space form $N^n(\rho)$ has exactly two distinct principal curvatures then they satisfy

$$\kappa_2 = \cdots = \kappa_{n-1}\,, \qquad \kappa_1 = \frac{2-n}{3}\kappa_2\,. \tag{2.3}$$

2.3. Rotational hypersurfaces. We next follow [**4**, Section 2] to provide a definition of rotational hypersurfaces of $N^n(\rho)$. Denote by $\mathbb{P}^k$ a k-dimensional subspace of $\mathbb{R}^{n+1}_\nu$ passing through the origin and by $\mathcal{O}(\mathbb{P}^2)$ the set of all orientation preserving isometries of $\mathbb{R}^{n+1}_\nu$ that leave $\mathbb{P}^2$ pointwise fixed. Choose subspaces $\mathbb{P}^2$ and $\mathbb{P}^3 \supset \mathbb{P}^2$ such that $\mathbb{P}^3 \cap N^n(\rho) \neq \emptyset$. Let γ be a curve immersed in $\mathbb{P}^3 \cap N^n(\rho) = N^2(\rho)$ that does not meet $\mathbb{P}^2$. The orbit of γ under the action of $\mathcal{O}(\mathbb{P}^2)$ is called a rotational hypersurface M^{n-1} of $N^n(\rho)$ generated by the profile curve γ.

In the case that $N^n(\rho) = \mathbb{H}^n(\rho)$ is the hyperbolic space, then $\nu = 1$ and we have three different cases depending on the causal character of the subspace $\mathbb{P}^2$. If the metric induced on $\mathbb{P}^2$ is Lorentzian, we have spherical rotational hypersurfaces; if it is Riemannian, the rotational hypersurfaces are of hyperbolic type; and, finally, if the restriction of the metric of $\mathbb{R}^{n+1}_1$ is degenerate, we obtain parabolic rotational hypersurfaces. To the contrary, in both the Euclidean space $\mathbb{R}^n$ and in the round sphere $\mathbb{S}^n(\rho)$ we only have spherical rotational hypersurfaces. In order to treat all the cases in a unified manner, we define the constant $\epsilon = -1, 0, 1$ so that $\epsilon = -1$ if the rotational hypersurface under consideration is of hyperbolic type, $\epsilon = 0$ if it is of parabolic type, and $\epsilon = 1$ if it is spherical.

A rotational hypersurface M^{n-1} of $N^n(\rho)$ has, at most, two distinct principal curvatures (see [**4**, Proposition 3.2]). These principal curvatures are given by

$$\kappa_1 = \mu = \frac{u_1'' + \rho u_1}{\sqrt{\epsilon - \rho u_1^2 - (u_1')^2}}\,, \tag{2.4}$$

and

$$\kappa_2 = \cdots = \kappa_{n-1} = \lambda = -\frac{\sqrt{\epsilon - \rho u_1^2 - (u_1')^2}}{u_1}\,, \tag{2.5}$$

respectively. Here, u_1 represents a suitable coordinate. More precisely, $u_1 = x_1$ if $\epsilon\rho < 0$ or $\rho = 0$, $u_1 = x_3$ if $\epsilon\rho > 0$, and $u_1 = x_1 - x_3$ when $\epsilon = 0$.

Conversely, an arbitrary hypersurface M^{n-1} with exactly two distinct principal curvatures one of which has multiplicity $n-2$ must be a rotational hypersurface [**4**, Theorem 4.2].

2.4. Rotational biconservative hypersurfaces. From the discussion above, we deduce [**20**, Proposition 2.2]. For later use, we state it here.

PROPOSITION 2.1 ([**20**]). *Let M^{n-1} be a hypersurface of a space form $N^n(\rho)$ with nowhere zero* grad$H$ *and with exactly two distinct principal curvatures everywhere. Then, M^{n-1} is biconservative if and only if it is rotational and the principal curvatures μ (of multiplicity one) and λ (of multiplicity $n-2$) satisfy*

$$3\mu + (n-2)\lambda = 0\,. \tag{2.6}$$

3. Profile curves as p-elastic curves

In this section we extend the result of [**20**, Theorem 3.3] that characterizes the profile curves of non-CMC (spherical) rotational biconservative hypersurfaces of space forms as p-elastic curves where $p \in [1/4, 1)$ is a rational number depending on the dimension n of the ambient space $N^n(\rho)$. This result is extended here for any possible type of rotation.

3.1. Curvature energy functionals. Let $\gamma : I \subseteq \mathbb{R} \longrightarrow N^2(\rho)$ be a smooth immersed curve parameterized by the arc length parameter $s \in I$. We denote by $T(s) := \gamma'(s)$ the unit tangent vector field along the curve γ, where $(\,)'$ denotes the derivative with respect to the arc length parameter s, and define the unit normal vector field $N(s)$ along γ to be the counter-clockwise rotation of $T(s)$ through an angle $\pi/2$ in the tangent bundle of $N^2(\rho)$. In this setting, the signed (geodesic) curvature $\kappa(s)$ of $\gamma(s)$ is defined by the Frenet-Serret equation

$$\nabla_T T(s) = \kappa(s) N(s)\,,$$

where ∇ denotes the Levi-Civita connection on $N^2(\rho)$. From the Fundamental Theorem of Curves, the curvature $\kappa(s)$ completely determines the curve $\gamma(s)$, up to isometries of $N^2(\rho)$.

Curves whose curvature $\kappa(s)$ never vanishes have a preferred orientation such that $\kappa > 0$ holds everywhere and are called convex curves. We will always assume that such curves are oriented in this preferred way.

Denote by $\mathcal{C}^\infty(I, N^2(\rho))$ the space of smooth immersed curves $\gamma : I \subseteq \mathbb{R} \longrightarrow N^2(\rho)$ parameterized by the arc length $s \in I$ and consider the curvature energy functional

$$\mathbf{\Theta}(\gamma) := \int_\gamma P(\kappa)\,, \tag{3.1}$$

acting on $\mathcal{C}^\infty(I, N^2(\rho))$. Throughout this paper we assume that $P(\kappa)$ is a smooth function defined on an adequate domain of the curvature κ of γ.

Using compactly supported smooth variations and standard arguments involving integration by parts, it can be shown that, regardless of the boundary conditions, critical points for $\mathbf{\Theta}$ must satisfy the Euler-Lagrange equation

$$\dot{P}_{ss} + \dot{P}\left(\kappa^2 + \rho\right) - \kappa P = 0\,, \tag{3.2}$$

where $\dot{P}$ is the derivative of P with respect to κ. Here, for convenience, we are denoting the derivatives with respect to the arc length parameter s with sub-indexes.

When $\dot{P}_s \neq 0$, the Euler-Lagrange equation (3.2) can be integrated once, obtaining the conservation law

$$\dot{P}_s^2 + \left(\kappa\dot{P} - P\right)^2 + \rho\dot{P}^2 = d\,, \tag{3.3}$$

for some constant of integration $d \in \mathbb{R}$. Whenever $\rho \geq 0$ holds, the constant of integration d must be positive. However, if $\rho < 0$, it may be positive, negative or zero.

Let $\gamma(s) \subset N^2(\rho) \subset \mathbb{R}^3_\nu$ be an arc length parameterized curve with nonconstant curvature $\kappa(s)$. It then follows that the function $\kappa(s)$ satisfies the conservation law (3.3) if and only if there exists a coordinate system (which we take to be the standard coordinates of $\mathbb{R}^3$) such that:

(1) Case $\rho = 0$ (i.e., $N^2(\rho) = \mathbb{R}^2$ is the Euclidean plane),

$$\gamma(s) = \frac{1}{\sqrt{d}}\left(\dot{P}, \int \left(P - \kappa\dot{P}\right) ds, 0\right). \tag{3.4}$$

(2) Case $\rho\, d > 0$ (i.e., either $N^2(\rho) = \mathbb{S}^2(\rho)$ is the round sphere, or $N^2(\rho) = \mathbb{H}^2(\rho)$ is the hyperbolic plane and the curve has time-like momentum, that is, $d < 0$ also holds),

$$\gamma(s) = \frac{1}{\sqrt{\rho\, d}}\left(\sqrt{d - \rho\,\dot{P}^2}\,\cos\psi(s), \sqrt{d - \rho\,\dot{P}^2}\,\sin\psi(s), \sqrt{|\rho|}\,\dot{P}\right). \tag{3.5}$$

(3) Case $\rho\, d < 0$ (i.e., $N^2(\rho) = \mathbb{H}^2(\rho)$ is the hyperbolic plane and the curve has space-like momentum, that is, $d > 0$ also holds),

$$\gamma(s) = \frac{1}{\sqrt{-\rho\, d}}\left(\sqrt{-\rho}\,\dot{P}, \sqrt{d - \rho\,\dot{P}^2}\,\sinh\psi(s), \sqrt{d - \rho\,\dot{P}^2}\,\cosh\psi(s)\right). \tag{3.6}$$

(4) Case $d = 0$ (i.e., $N^2(\rho) = \mathbb{H}^2(\rho = -1)$ is the hyperbolic plane and the curve has light-like momentum, that is, $d = 0$ holds),

$$\gamma(s) = \frac{1}{2\sqrt{2}\dot{P}}\left(2\left[2\widetilde{\psi}^2(s) - 1\right]\dot{P}^2 + 1, 4\sqrt{2}\,\widetilde{\psi}\,\dot{P}^2, 2\left[2\widetilde{\psi}^2(s) + 1\right]\dot{P}^2 + 1\right). \tag{3.7}$$

In the parameterizations given above the functions $\psi(s)$ and $\widetilde{\psi}(s)$ are defined by the condition that s is the arc length parameter of γ. Hence, they are given, respectively, by

$$\psi(s) := \sqrt{|\rho\, d|}\int \frac{\kappa\dot{P} - P}{d - \rho\,\dot{P}^2}\, ds\,, \qquad \widetilde{\psi}(s) := \frac{1}{2}\int \frac{\kappa\dot{P} - P}{\dot{P}^2}\, ds\,. \tag{3.8}$$

This characterization of curves whose curvature is a solution of (3.3) is classical in the theory and it follows from standard computations involving Killing vector fields along curves [**16**]. A detailed proof for the case $d > 0$ can be found in [**20**, Proposition 3.2], while the remaining cases can be shown adapting this result.

Suitably rotating these curves in $N^n(\rho)$ we obtain rotational hypersurfaces M^{n-1}. It turns out that the type of rotation is closely related to the sign of the constant of integration $d \in \mathbb{R}$. On one hand, in both $\mathbb{R}^2$ and $\mathbb{S}^2(\rho)$, $d > 0$ holds and the only possible type of rotation in both $\mathbb{R}^n$ and $\mathbb{S}^n(\rho)$ is spherical. On the other hand, in $\mathbb{H}^2(\rho)$ due to the parameterizations (3.5)–(3.7) it follows that the causal character of the subspace $\mathbb{P}^2$ pointwise fixed by the rotation is restricted. Indeed, we have that $\epsilon = \text{sign}(d)$, where we are understanding that $\text{sign}(0) = 0$. This relation in the 3-dimensional case was shown in [**1**, Proof of Proposition 5.3]. The higher dimensional case can be obtained adapting this result.

3.2. Characterization of profile curves. We are now in the right position to prove the main result of this section.

THEOREM 3.1. *Let M^{n-1} be a non-CMC rotational hypersurface in $N^n(\rho)$. Then, locally, M^{n-1} is biconservative if and only if its profile curve $\gamma(s) \subset N^2(\rho)$ satisfies the Euler-Lagrange equation associated to the curvature energy functional*

$$\boldsymbol{\Theta}_p(\gamma) := \int_\gamma \kappa^p\,, \tag{3.9}$$

where $p = (n-2)/(n+1) \in [1/4, 1)$. (Of course, it must be understood that the functional $\boldsymbol{\Theta}_p$ acts on the subspace of $\mathcal{C}^\infty(I, N^2(\rho))$ composed by convex curves.)

PROOF. Let M^{n-1} be a non-CMC biconservative rotational hypersurface. Since M^{n-1} has nonconstant mean curvature, then the curvature of its profile curve γ is also nonconstant. Hence, locally, by the Inverse Function Theorem we can assume that the arc length parameter of γ is a function of the curvature. Therefore, we can find suitable functions $\dot{P}(\kappa)$ and constants $d \neq 0$ such that

$$u_1(s) = x_i(s) = \frac{1}{\sqrt{|d|}}\,\dot{P}\left(\kappa(s)\right), \qquad i = 1, 3, \tag{3.10}$$

if $\epsilon \neq 0$, or

$$u_1(s) = x_1(s) - x_3(s) = -\sqrt{2}\,\dot{P}\left(\kappa(s)\right), \tag{3.11}$$

if $\epsilon = 0$. We recall here that the sign of ϵ determines the type of rotational hypersurface. This together with the arc length parameterization of the curve γ shows that, locally, γ is parameterized as in one of the cases (3.4)–(3.7) and, hence, its curvature $\kappa(s)$ satisfies (3.3) as well as the Euler–Lagrange equation (3.2).

It remains to prove that the integral $P(\kappa)$ of the function $\dot{P}(\kappa)$ defined in (3.10) and (3.11) is as in the statement, namely, $P(\kappa) = \kappa^p$ for $p = (n-2)/(n+1)$. For this objective, observe that since M^{n-1} is a non-CMC biconservative rotational hypersurface its two principal curvatures (cf. [**4**, Proposition 3.2]) satisfy the relation (2.6). Employing the definitions (2.4) and (2.5) of these principal curvatures together with (3.10) we obtain

$$\begin{aligned}
3\mu + (n-2)\lambda &= 3\frac{\dot{P}_{ss} + \rho\dot{P}}{\sqrt{\epsilon|d| - \rho\dot{P}^2 - \dot{P}_s^2}} - (n-2)\frac{\sqrt{\epsilon|d| - \rho\dot{P}^2 - \dot{P}_s^2}}{\dot{P}} \\
&= 3\kappa\frac{P - \kappa\dot{P}}{\sqrt{\epsilon|d| - d + (\kappa\dot{P} - P)^2}} - (n-2)\frac{\sqrt{\epsilon|d| - d + (\kappa\dot{P} - P)^2}}{\dot{P}} \\
&= 0\,,
\end{aligned}$$

where in the last line we have used the Euler–Lagrange equation (3.2) and the conservation law (3.3) to simplify the expression. Therefore, since $\epsilon = \operatorname{sign}(d)$, we have an ordinary differential equation in $P(\kappa)$, which can be explicitly solved obtaining

$$P(\kappa) = c\,\kappa^{(n-2)/(n+1)}\,,$$

for some constant of integration c. Since any multiple of an energy gives rise to the same Euler–Lagrange equation, the result follows immediately.

The case where $d=0$ follows in the same way using (3.11) in the relation (2.6). In this case $\epsilon=0$ holds.

Conversely, assume that the profile curve γ of a non-CMC rotational hypersurface M^{n-1} satisfies the Euler–Lagrange equation associated to $\mathbf{\Theta}_p$ given in (3.9). Since the curvature κ of γ cannot be constant, it solves the conservation law (3.3) for $P(\kappa)=\kappa^p$ and, hence, it is parameterized as in one of the cases (3.4)–(3.7). From the definition of the principal curvatures (2.4) and (2.5) a similar computation as above shows that the relation (2.6) holds and, hence, from Proposition 2.1 the hypersurface M^{n-1} is biconservative. □

The statement of Theorem 3.1 coincides with that of [**20**, Theorem 3.3]. However, in [**20**] the authors are only considering spherical rotational hypersurfaces, that is, the case $d>0$. Here, we are extending the result by allowing all the possible types of rotational hypersurfaces. In other words, we also consider the cases $d=0$ and $d<0$.

Regardless of the boundary conditions, we call p-elastic curves to those curves whose curvature satisfies the Euler-Lagrange equation associated to $\mathbf{\Theta}_p$ (cf. (3.2) for $P(\kappa)=\kappa^p$). These curves in $\mathbb{R}^2$ have been geometrically described in [**17**]. Their shape imitates that of a catenary and, hence, they were called catenary-type (the "proper" catenary arises, precisely, when $n=5$). None of them have periodic curvature and so they are not closed. To the contrary, in [**20**], it was shown that all p-elastic curves for $p=(n-2)/(n+1)$ in $\mathbb{S}^2(\rho)$ have periodic curvature and, additionally, countably many of them are closed. To be more precise, for every pair of relatively prime natural numbers (l,r) satisfying $r<2l<\sqrt{2}\,r$, there exists a closed p-elastic curve for $p=(n-2)/(n+1)$ in $\mathbb{S}^2(\rho)$. For convenience, we refer to them as p-elastic curves of type (l,r). We point out here that this assertion holds for every possible dimension n of the ambient space $\mathbb{S}^n(\rho)$ where the rotational hypersurface lies. Finally, it was also shown in [**20**], that there are no closed p-elastic curves for $p=(n-2)/(n+1)$ in $\mathbb{H}^2(\rho)$.

For visualization purposes we will just illustrate in Figure 1 three closed (rotational) biconservative surfaces in $\mathbb{S}^3(\rho)$, after stereographic projection. Observe that biconservative surfaces in 3-dimensional space forms are automatically rotational [**3**].

FIGURE 1. The stereographic projection of the three "simplest" closed (rotational) biconservative surfaces in $\mathbb{S}^3(1)$ together with their corresponding profile curves, which are $1/4$-elastic curves of type $(2,3)$, $(3,5)$, and $(4,7)$, respectively.

4. Relation to generalized Willmore surfaces

In this section we will show a relation between non-CMC closed rotational biconservative hypersurfaces in space forms $\mathbb{S}^n(\rho)$ and p-Willmore surfaces in $\mathbb{S}^3(\rho/4)$. We begin recalling the variational problem associated to p-Willmore surfaces.

4.1. Generalized Willmore surfaces. Let Σ be an oriented, connected and closed (i.e., compact without boundary) surface and denote by $\mathcal{C}^\infty(\Sigma, \mathbb{S}^3(\rho/4))$ the space of weakly convex smooth immersions $X : \Sigma \longrightarrow \mathbb{S}^3(\rho/4)$, that is the space of smooth immersions with nonnegative Gaussian curvature K and strictly positive mean curvature H. The p-Willmore energy is the functional

$$\mathcal{W}_p(X) := \int_\Sigma H^p \, dA \,, \tag{4.1}$$

acting on the space $\mathcal{C}^\infty(\Sigma, \mathbb{S}^3(\rho/4))$. Here, $p \in \mathbb{R}$ is an arbitrary real number and dA denotes the area element.

A critical point of the functional $\mathcal{W}_p$ is called a p-Willmore surface. From standard arguments, critical points of $\mathcal{W}_p$ must satisfy the Euler–Lagrange equation

$$p\Delta\left(H^{p-1}\right) + 4(p-1)H^{p+1} + p\left(\rho - 2K\right)H^{p-1} = 0\,, \tag{4.2}$$

where Δ is the Laplacian of the surface.

To prove the main result of this section, we will next introduce the construction of Hopf tori given by Pinkall [**29**]. We identify $\mathbb{R}^4$ with $\mathbb{C}^2$ and consider the Riemannian metric

$$g_{\mathbb{C}^2}\left((z_1, z_2), (\omega_1, \omega_2)\right) = \Re\left(z_1\overline{\omega}_1 + z_2\overline{\omega}_2\right),$$

where z_i and ω_i, $i = 1, 2$, are complex numbers. In $\left(\mathbb{C}^2, g_{\mathbb{C}^2}\right)$, the round sphere $\mathbb{S}^3(\rho/4)$ is the hyperquadric $g_{\mathbb{C}^2}((z_1, z_2), (\omega_1, \omega_2)) = 4/\rho$. Define the map $\widetilde{\pi} : \mathbb{C}^2 \longrightarrow \mathbb{C}^2$ by

$$\widetilde{\pi}(z, \omega) = \frac{\sqrt{\rho}}{4}\left(|z|^2 - |\omega|^2, 2\overline{z}\omega\right).$$

Then, the restriction of $\widetilde{\pi}$ to $\mathbb{S}^3(\rho/4)$ gives the standard Hopf mapping $\widetilde{\pi} : \mathbb{S}^3(\rho/4) \to \mathbb{S}^2(\rho)$.

Let γ be a curve in $\mathbb{S}^2(\rho)$, the complete lift $X_\gamma := \widetilde{\pi}^{-1}(\gamma)$ is a flat surface in $\mathbb{S}^3(\rho/4)$ whose mean curvature H is given in terms of the curvature κ of γ by

$$H = \frac{1}{2}\left(\kappa \circ \widetilde{\pi}\right).$$

We call the surface X_γ the Hopf cylinder based on γ. Since the fibers of X_γ are diffeomorphic to $\mathbb{S}^1$, it follows that X_γ is closed if and only if the curve γ is closed. In this case, we refer to X_γ as the Hopf torus based on γ.

4.2. Relation. We now state and prove the relation between non-CMC closed rotational biconservative hypersurfaces and p-Willmore Hopf tori.

THEOREM 4.1. *There exists a one-to-one correspondence between non-CMC closed rotational biconservative hypersurfaces of space forms $N^n(\rho)$ and non-CMC p-Willmore Hopf tori in $\mathbb{S}^3(\rho/4)$ for $p = (n-2)/(n+1)$.*

PROOF. Let M^{n-1} be a non-CMC closed rotational biconservative hypersurface of a space form $N^n(\rho)$. From [**20**, Proposition 5.1], then $N^n(\rho) = \mathbb{S}^n(\rho)$ is the n-dimensional round sphere. In addition, it follows from Theorem 4.1 (see also

FIGURE 2. Left: The stereographic projection of a non-CMC (rotational) biconservative surface M^2 in $\mathbb{S}^3(1)$ together with its profile curve γ. Center: The 1/4-elastic curve γ of type (3, 5) in $\mathbb{S}^2(1)$. Right: The stereographic projection of a non-CMC 1/4-Willmore Hopf torus X_γ in $\mathbb{S}^3(1/4)$ together with two horizontal lifts of the base curve γ.

[**20**, Theorem 3.3]) that its profile curve γ is a p-elastic curve with nonconstant curvature in $\mathbb{S}^2(\rho)$ for $p = (n-2)/(n+1)$. Moreover, γ must be closed.

We next construct the Hopf torus X_γ based on γ. Since γ is a closed p-elastic curve we deduce from the Symmetric Criticality Principle of Palais [**26**] that X_γ is a p-Willmore surface in $\mathbb{S}^3(\rho/4)$. Details on how to apply the principle for the present case can be obtained particularizing [**27**, proof of Theorem 3.1].

The converse follows reversing the argument, since both Theorem 4.1 and [**27**, Theorem 3.1] are equivalences. □

In the particular case $n = 3$, Theorem 5.1 was proven in [**27**, Corollary 6.4]. In this case, the non-CMC closed (rotational) biconservative surfaces as well as the non-CMC 1/4-Willmore Hopf tori lie in round spheres of dimension three and, hence, they are the ideal candidates to be illustrated after suitable projections. In Figure 2 we show one of these examples.

5. Relation to rotational minimal hypersurfaces

In this section we will employ the variational characterization of profile curves of non-CMC rotational biconservative hypersurfaces given in Theorem 3.1 to describe a relation between these hypersurfaces and rotational minimal hypersurfaces.

5.1. Rotational minimal hypersurfaces. A hypersurface M^{n-1} of a space form $N^n(\rho)$ is said to be minimal if its mean curvature function H is identically zero. That is, if

$$H = \frac{1}{n-1}\operatorname{trace} S_\eta = \frac{1}{n-1}\left(\kappa_1 + \cdots + \kappa_{n-1}\right) = 0\,, \tag{5.1}$$

holds. If, in addition, the hypersurface M^{n-1} is rotational, then the principal curvatures κ_i, $i = 1, \ldots, n-1$, are given in (2.4) and (2.5), respectively, and equation (5.1) is equivalent to

$$\mu + (n-2)\lambda = 0\,. \tag{5.2}$$

5.2. Relation. The following theorem gives the relation between minimal and non-CMC biconservative rotational hypersurfaces of space forms of suitable dimensions. A more general version of this result in the Euclidean ambient space $\mathbb{R}^n$ was shown in [**21**, Proposition 4.2], employing a different approach.

THEOREM 5.1. *There exists a one-to-one correspondence between rotational nonisoparametric minimal hypersurfaces of $N^n(\rho)$ and non-CMC rotational biconservative hypersurfaces of $N^{3n-4}(\rho)$.*

PROOF. A rotational hypersurface M^{n-1} of $N^n(\rho)$ is minimal if and only if equation (5.2) is satisfied where μ and λ are the principal curvatures of M^{n-1}. In addition, since M^{n-1} is nonisoparametric both μ and λ are nonconstant. Therefore, a nonisoparametric rotational hypersurface M^{n-1} of $N^n(\rho)$ is minimal if and only if its profile curve is a p-elastic curve in $N^2(\rho)$ for $p = (n-2)/(n-1)$. The proof of this assertion can be obtained adapting the proof of Theorem 3.1, just changing the equation (2.6) by (5.2). A more general result (which extends this one) will be presented in the future work [**28**] and, hence, we will avoid repeating it here.

Let $m \geq 3$ be a natural number (to be determined) and construct the rotational hypersurface M^{m-1} in $N^m(\rho)$ with profile curve γ. Since γ is a p-elastic curve in $N^2(\rho)$, it follows from Theorem 3.1 that M^{m-1} is biconservative if the following relation between $p = (n-2)/(n-1)$ and m holds

$$p = \frac{m-2}{m+1}\,.$$

Solving for m, we obtain

$$m = \frac{2+p}{1-p} = 3n - 4\,.$$

This finishes the proof. □

To put this result into context we consider here the Euclidean space $\mathbb{R}^3$ of dimension 3. In this case, the nonisoparametric minimal rotational surface is the catenoid which is generated by rotating a catenary. The catenary was known to Blaschke to be the nontrivial solution to the variational problem associated to $\mathbf{\Theta}_{1/2}$. Hence, from Theorem 4.1, it follows that the rotational hypersurface generated from the catenary M^4 is a non-CMC biconservative hypersurface of $\mathbb{R}^5$ (cf. [**21**]).

As a direct consequence of Theorems 4.1 and 5.1, the following corollary follows.

COROLLARY 5.2. *There exists a one-to-one correspondence between closed nonisoparametric rotational minimal hypersurfaces of $N^n(\rho)$ and non-CMC p-Willmore Hopf tori in $\mathbb{S}^3(\rho/4)$ for $p = (n-2)/(n-1)$.*

An illustration of this corollary for the case $n = 3$ can be found in [**27**, Figure 1]. In that figure the stereographic projection of a closed minimal surface in $\mathbb{S}^3(\rho)$ is shown together with its profile curve, which is a $1/2$-elastic curve in $\mathbb{S}^2(\rho)$. Additionally, it is also shown the stereographic projection of the associated $1/2$-Willmore Hopf torus in $\mathbb{S}^3(\rho/4)$.

Acknowledgments

The author would like to thank the referees for their valuable comments, which have helped to improve the manuscript.

References

[1] J. Arroyo, O. J. Garay, and A. Pámpano, *Constant mean curvature invariant surfaces and extremals of curvature energies*, J. Math. Anal. Appl. **462** (2018), no. 2, 1644–1668, DOI 10.1016/j.jmaa.2018.02.058. MR3774309

[2] A. Balmuş, S. Montaldo, and C. Oniciuc, *Biharmonic PNMC submanifolds in spheres*, Ark. Mat. **51** (2013), no. 2, 197–221, DOI 10.1007/s11512-012-0169-5. MR3090194

[3] R. Caddeo, S. Montaldo, C. Oniciuc, and P. Piu, *Surfaces in three-dimensional space forms with divergence-free stress-bienergy tensor*, Ann. Mat. Pura Appl. (4) **193** (2014), no. 2, 529–550, DOI 10.1007/s10231-012-0289-3. MR3180932

[4] M. do Carmo and M. Dajczer, *Rotation hypersurfaces in spaces of constant curvature*, Trans. Amer. Math. Soc. **277** (1983), no. 2, 685–709, DOI 10.2307/1999231. MR694383

[5] B.-Y. Chen, *Recent developments in biconservative submanifolds*, In: Geometrical Spaces and Related Topics, 2024.

[6] B.-Y. Chen, *Total mean curvature and submanifolds of finite type*, Series in Pure Mathematics, vol. 1, World Scientific Publishing Co., Singapore, 1984, DOI 10.1142/0065. MR749575

[7] J. Eells Jr. and J. H. Sampson, *Harmonic mappings of Riemannian manifolds*, Amer. J. Math. **86** (1964), 109–160, DOI 10.2307/2373037. MR164306

[8] D. Fetcu and C. Oniciuc, *Biharmonic and biconservative hypersurfaces in space forms*, Differential geometry and global analysis—in honor of Tadashi Nagano, Contemp. Math., vol. 777, Amer. Math. Soc., [Providence], RI, [2022] ©2022, pp. 65–90, DOI 10.1090/conm/777/15628. MR4410183

[9] Y. Fu, *Explicit classification of biconservative surfaces in Lorentz 3-space forms*, Ann. Mat. Pura Appl. (4) **194** (2015), no. 3, 805–822, DOI 10.1007/s10231-014-0399-1. MR3345666

[10] Y. Fu, M.-C. Hong, D. Yang, and X. Zhan, *Biconservative Hypersurfaces with Constant Scalar Curvature in Space Forms*, Preprint, `arXiv:2110.03438 [math.DG]`, (2021).

[11] A. Gruber, Á. Pámpano, and M. Toda, *Instability of closed p-elastic curves in* $\mathbb{S}^2$, Anal. Appl. (Singap.) **21** (2023), no. 6, 1533–1559, DOI 10.1142/S0219530523500173. MR4663707

[12] Th. Hasanis and Th. Vlachos, *Hypersurfaces in* E^4 *with harmonic mean curvature vector field*, Math. Nachr. **172** (1995), 145–169, DOI 10.1002/mana.19951720112. MR1330627

[13] D. Hilbert, *Die Grundlagen der Physik* (German), Math. Ann. **92** (1924), no. 1-2, 1–32, DOI 10.1007/BF01448427. MR1512197

[14] G. Y. Jiang, *2-harmonic maps and their first and second variational formulas* (Chinese), Chinese Ann. Math. Ser. A **7** (1986), no. 4, 389–402. An English summary appears in Chinese Ann. Math. Ser. B **7** (1986), no. 4, 523. MR886529

[15] G. Y. Jiang, *The conservation law for 2-harmonic maps between Riemannian manifolds* (Chinese), Acta Math. Sinica **30** (1987), no. 2, 220–225. MR891928

[16] J. Langer and D. A. Singer, *The total squared curvature of closed curves*, J. Differential Geom. **20** (1984), no. 1, 1–22. MR772124

[17] R. López and Á. Pámpano, *Classification of rotational surfaces in Euclidean space satisfying a linear relation between their principal curvatures*, Math. Nachr. **293** (2020), no. 4, 735–753, DOI 10.1002/mana.201800235. MR4089078

[18] E. Loubeau and S. Montaldo, *Biminimal immersions*, Proc. Edinb. Math. Soc. (2) **51** (2008), no. 2, 421–437, DOI 10.1017/S0013091506000393. MR2465916

[19] E. Loubeau, S. Montaldo, and C. Oniciuc, *The stress-energy tensor for biharmonic maps*, Math. Z. **259** (2008), no. 3, 503–524, DOI 10.1007/s00209-007-0236-y. MR2395125

[20] S. Montaldo, C. Oniciuc, and A. Pámpano, *Closed biconservative hypersurfaces in spheres*, J. Math. Anal. Appl. **518** (2023), no. 1, Paper No. 126697, 16, DOI 10.1016/j.jmaa.2022.126697. MR4489868

[21] S. Montaldo, C. Oniciuc, and A. Ratto, *Proper biconservative immersions into the Euclidean space*, Ann. Mat. Pura Appl. (4) **195** (2016), no. 2, 403–422, DOI 10.1007/s10231-014-0469-4. MR3476680

[22] S. Montaldo and A. Pámpano, *On the existence of closed biconservative surfaces in space forms*, Comm. Anal. Geom. **31** (2023), no. 2, 291–319. MR4685022

[23] S. Nistor and C. Oniciuc, *On the uniqueness of complete biconservative surfaces in 3-dimensional space forms*, Ann. Sc. Norm. Super. Pisa Cl. Sci. (5) **23** (2022), no. 3, 1565–1587, DOI 10.2422/2036-2145.202010_037. MR4553530

[24] K. Nomizu, *Characteristic roots and vectors of a differentiable family of symmetric matrices*, Linear and Multilinear Algebra **1** (1973), no. 2, 159–162, DOI 10.1080/03081087308817014. MR335536

[25] Y.-L. Ou, *Biharmonic hypersurfaces in Riemannian manifolds*, Pacific J. Math. **248** (2010), no. 1, 217–232, DOI 10.2140/pjm.2010.248.217. MR2734173

[26] R. S. Palais, *The principle of symmetric criticality*, Comm. Math. Phys. **69** (1979), no. 1, 19–30. MR547524

[27] Á. Pámpano, *Critical tori for mean curvature energies in Killing submersions*, Nonlinear Anal. **200** (2020), 112092, 18, DOI 10.1016/j.na.2020.112092. MR4135581

[28] Á. Pámpano, M. Samarakkody, and H. Tran, *Characterization of rotational linear Weingarten hypersurfaces*, Preprint.

[29] U. Pinkall, *Hopf tori in* S^3, Invent. Math. **81** (1985), no. 2, 379–386, DOI 10.1007/BF01389060. MR799274

[30] P. J. Ryan, *Hypersurfaces with parallel Ricci tensor*, Osaka Math. J. **8** (1971), 251–259. MR296859

[31] P. J. Ryan, *Homogeneity and some curvature conditions for hypersurfaces*, Tohoku Math. J. (2) **21** (1969), 363–388, DOI 10.2748/tmj/1178242949. MR253243

Department of Mathematics and Statistics, Texas Tech University, Lubbock, Texas 79409

Email address: alvaro.pampano@ttu.edu

Contemporary Mathematics
Volume **821**, 2025
https://doi.org/10.1090/conm/821/16444

Einstein warped-product manifolds and the screened Poisson equation

Alexander Pigazzini, Luca Lussardi, Magdalena Toda, and Andrew DeBenedictis

Abstract. We study a particular type of Einstein warped-product manifold where the warping function must satisfy the homogeneous version of the screened Poisson equation. Under these assumptions, we show that the dimension of the manifold, the (constant negative) Ricci curvature and the screened parameter are related through a quadratic equation.

1. Introduction and preliminaries

1.1. Einstein warped product manifolds. Warped-product manifolds, first introduced by Bishop and O'Neill [**1**], have gained significant importance in the field of differential geometry and are widely researched in the domain of General Relativity, particularly in relation to generalized Friedmann–Robertson–Walker spacetimes. A wealth of properties for warped product manifolds and submanifolds have been presented in various studies [**2**], [**3**] and [**4**].

To construct a warped-product manifold, consider two semi-Riemannian manifolds (B, g_B) and (F, g_F), and let τ and σ be the projections of $B \times F$ onto B and F respectively. The warped-product $M = B \times_f F$ is the manifold $B \times F$ with the metric tensor $g = \tau^* g_B + f^2 \sigma^* g_F$, where * denotes the pullback and f is a positive smooth scalar function on B, known as the warping function.

In more explicit terms, if X is tangent to $B \times F$ at (a, b) (where a is a point on B and b is a point on F), then:

$$\langle X, X \rangle = \langle d\tau(X), d\tau(X) \rangle + f^2(a) \langle d\sigma(X), d\sigma(X) \rangle.$$

In this context, B is referred to as the *base-manifold* of $M = B \times_f F$ and F is the *fiber-manifold.* If $f = 1$, then $B \times_f F$ simplifies to a semi-Riemannian product manifold. The leaves $B \times b = \sigma^{-1}(b)$ and the fibers $a \times F = \tau^{-1}(a)$ are Riemannian submanifolds of M. Vectors tangent to leaves are termed horizontal and those tangent to fibers are termed vertical. The orthogonal projection of $T_{(a,b)}M$ onto its horizontal subspace $T_{(a,b)}(B \times b)$ is denoted by $\mathcal{H}$, and $\mathcal{V}$ denotes the projection onto the vertical subspace $T_{(a,b)}(a \times F)$ [**5**].

2020 *Mathematics Subject Classification.* Primary 53C25, 53C21.
Key words and phrases. Warped-product manifolds; constant negative curvature; screened Poisson equation; time-independent Klein-Gordon equation.

If M is an n-dimensional manifold with metric tensor g_M, the Einstein condition gives $Ric_M = \lambda g_M$ for some real constant λ, where Ric_M represents the Ricci tensor of g_M. An Einstein manifold with $\lambda = 0$ is considered Ricci-flat.

With this understanding, we find that a warped-product manifold $(M, g_M) = (B, g_B) \times_f (F, g_F)$ (where (B, g_B) is the base-manifold, (F, g_F) is the fiber-manifold), with $g_M = g_B + f^2 g_F$, is Einstein if and only if (refer to [**2**]):

$$Ric_M = \lambda g_M \Longleftrightarrow \begin{cases} Ric_B - \frac{m}{f} Hess(f) = \lambda g_B \\ Ric_F = \mu g_F \\ f\Delta f + (m-1)|\nabla f|^2 + \lambda f^2 = \mu, \end{cases} \tag{1.1}$$

where λ and μ are constants, m is the dimension of F, $Hess(f)$, Δf and ∇f are, respectively, the Hessian, the Laplacian (given by $tr(Hess(f))$), and the gradient of f with respect to g_B, with $f : B \to \mathbb{R}^+$ a smooth positive function. Regarding the constant μ, for the system to produce a manifold that obeys the Einstein condition $Ric_M = \lambda g_M$ the last expression in (1.1) must be constant. The constant μ is specifically associated with the Ricci curvature of the fiber manifold F. The third equation in the above system (which includes μ) ensures the consistency of the curvature conditions over the entire manifold M, relating the warping function f, its derivatives and the curvatures λ and μ.

Contracting the first equation of (1.1) we get:

$$R_B f - m\Delta_{g_B} f = nf\lambda, \tag{1.2}$$

where n and R_B are the dimension and the scalar curvature of B, respectively, then (1.1), contracting also the second equation, becomes:

$$R_M = \lambda(n+m) \Longleftrightarrow \begin{cases} R_B f - m\Delta_{g_B} f = nf\lambda \\ R_F = \mu m \\ f\Delta_{g_B} f + (m-1)|\nabla_{g_B} f|^2 + \lambda f^2 = \mu. \end{cases} \tag{1.3}$$

REMARK 1.1. It is a well-established fact that an Einstein warped-product manifold, equipped with a Riemannian metric and a Ricci-flat fiber-manifold (F), cannot admit positive Ricci curvature solutions (see [**6**]). It can only accommodate zero or negative solutions. This conclusion is derived from the Bonnet-Myers theorem. Essentially, this theorem states that if $Ric_M \geq c$ for some constant $c > 0$, then M is compact. However, if V is a vertical vector (that is, a tangent vector on the fiber-manifold), then the formula for the Ricci curvature of warped product manifolds (as outlined in [**7**]) yields:

$$Ric_M(V,V) = Ric_F(V,V) - |V|^2\left(\frac{\Delta f}{f} + (m-1)\frac{|\nabla f|^2}{f^2}\right).$$

If the fiber-manifold F is Ricci-flat, then it follows that at the point where the warping function f reaches its minimum on the base-manifold (which is guaranteed to exist due to compactness), the condition $Ric_M(V,V) \leq 0$ is satisfied.

In this paper we are only interested in cases $R_F = 0$ with $Ric_M < 0$.

1.2. Screened Poisson equation and Klein–Gordon equation. In the realm of mathematics, the screened Poisson equation is a specific type of partial differential equation, conventionally represented as $[\Delta - \lambda^2]\phi(r) = -\psi(r)$. Here, λ is a constant, ψ is a function of position (r) known as the "source function", and ϕ is the function to be determined. When λ is zero, this equation simplifies to the Poisson equation.

A class of solutions to the inhomogeneous screened Poisson equation for a general ψ can be found using the method of Green functions. In this context, the Green function G is defined by the equation $[\Delta - \lambda^2]G(r) = -\delta^3(r)$. This equation frequently appears in various fields of physics, including the Yukawa theory of mesons, where the Green function represents the Yukawa potential [**8**]. The screened Poisson equation also appears in plasma screening, for example in limits of the Debye–Hückel theory [**9**] or the Thomas-Fermi theory [**10**]. The equation also has applicability in granular fluid flow [**11**], [**12**] and also emerges in the Klein–Gordon equation, which, aside from its importance in physics, holds a significant role in the theory of integrable systems or geometric PDE. This is a rapidly growing subfield at the intersection of differential geometry, differential equations, and mathematical physics.

The Klein–Gordon equation can be transformed into the form of a Schrödinger equation (as two coupled differential equations, each of first order in time) [**13**]. However, the Klein–Gordon equation carries a distinct meaning. Typically, in a physics course setting, it is presented with separated time and space. For instance, with respect to natural units and the Minkowski metric (+, -, -, -) (up to a potential change in sign), and in the case of time-independence ($[\Delta - \frac{m^2c^2}{\hbar}]\phi(r) = 0$), the Klein–Gordon equation is equivalent to the homogeneous form ($\psi = 0$) of the screened Poisson equation: $[\Delta - \lambda^2]\phi(r) = 0$. Here, the constant $\lambda^2 = \frac{m^2c^2}{\hbar}$, where c and $\hbar$ represent the speed of light and Planck's constant, respectively.

2. Einstein warped-product manifolds and homogeneous form of the screened Poisson equation

THEOREM 2.1. *Let M be an Einstein warped-product manifold with constant negative Ricci curvature ($\lambda < 0$) and m-dimensional Ricci-flat fiber manifold (F), and let the base-manifold (B) be a 2-dimensional manifold with negative constant Gaussian curvature K. There exists a warping function f satisfying the homogeneous form of the Screened Poisson equation, i.e., $[\Delta_{g_B} - 1]f = 0$, if and only if $\dim(M) = 2 + m$ and the constant λ of curvature are related by the following quadratic equation:*

$$\lambda^2(2-m) + \lambda(1 + \frac{3m}{2} - \frac{m^2}{2}) + \frac{m^2}{2} + \frac{m}{2} = 0. \tag{2.1}$$

PROOF. Let us consider (1.3) for a 2-dimensional base-manifold, i.e., $n = 2$:

$$R_M = \lambda(2+m) \iff \begin{cases} R_B f - m\Delta_{g_B} f = 2f\lambda \\ R_F = 0 \\ f\Delta_{g_B} f + (m-1)|\nabla_{g_B} f|^2 + \lambda f^2 = 0. \end{cases} \tag{2.2}$$

Considering initially, for ease of calculation, that B is a surface with constant Gaussian curvature $K = -1$, it is simple to notice that a general approach to this problem can be traced back to the following system of equations:

$$\begin{cases} \Delta_{g_B} f = p(f)q(f) \\ |\nabla_{g_B} f|^2 = p(f)^2. \end{cases} \tag{2.3}$$

Consider two real-valued functions, $p > 0$ and q, defined over an interval $I \subset \mathbb{R}$. Our objective is to ascertain the existence of a solution, f, for the system $|\nabla f|^2 = p(f)^2, \Delta f = p(f)q(f)$ on a certain (nonempty) open subset of the Poincaré upper half-plane, which is a Riemannian surface exhibiting a Gauss curvature K that remains consistently equal to -1, commonly referred to as a pseudospherical surface. We assert that if and only if the functions p and q satisfy the following differential equation, then a solution f exists:

$$p(t)p''(t) - p'(t)^2 + 2q(t)p'(t) - p(t)q'(t) - q(t)^2 + 1 = 0. \tag{2.4}$$

Working with Cartan differential forms associated to a Cartan moving frame is the most natural setting here. Let's start by noting that if such f exists, then the metric g can be written in the form $g = {\omega_1}^2 + {\omega_2}^2$ where $\omega_1 = (\mathrm{d}f)/p(f)$ and $*\mathrm{d}f = p(f)\,\omega_2$. Here $*$ is the usual Hodge operator.

Since $\mathrm{d}(*\mathrm{d}f) = \Delta f\,\omega_1 \wedge \omega_2$, it follows that $\mathrm{d}(p(f)\,\omega_2) = p(f)q(f)\,\omega_1 \wedge \omega_2$. This relates to the equality $\mathrm{d}(p(f)) = p'(f)\,\mathrm{d}f = p(f)p'(f)\,\omega_1$, and thus we must obtain that $\mathrm{d}\omega_2 = \big(q(f) - p'(f)\big)\,\omega_1 \wedge \omega_2$. Further, since $\mathrm{d}\omega_1 = -\omega_{12} \wedge \omega_2$ and $\mathrm{d}\omega_2 = \omega_{12} \wedge \omega_1$, it follows that $\omega_{12} = \big(p'(f) - q(f)\big)\,\omega_2$. The expansion of the Gauss–Codazzi equation $\mathrm{d}\omega_{12} = K\,\omega_1 \wedge \omega_2 = -\omega_1 \wedge \omega_2$ allows us to obtain equation (2.4).

Conversely, assuming that p and q satisfy (2.4), we proceed to examine the following equations:

$$\omega_1 = \mathrm{d}f/p(f), \text{ and} \tag{2.5}$$

$$\mathrm{d}\omega_2 = \big(q(f) - p'(f)\big)\,\omega_1 \wedge \omega_2 = \big(q(f) - p'(f)\big)/p(f)\,\mathrm{d}f \wedge \omega_2. \tag{2.6}$$

Considering a positive function s (according to the linear Ordinary Differential Equation theory), defined on interval I (unique up to a constant multiple) we get:

$$s'(f) = s(f)\big(q(f) - p'(f)\big)/p(f). \tag{2.7}$$

The above equations (2.5) and (2.6) imply that $\mathrm{d}\big(\omega_2/s(f)\big) = 0$, therefore, if we assume the domain to be simply connected, for a certain function h we have $\omega_2 = s(f)\,\mathrm{d}h$.

Consequently, equations (2.4) and (2.6) indicate that the metric:

$$g = \left(\tfrac{\mathrm{d}f}{p(f)}\right)^2 + (s(f)\,\mathrm{d}h)^2$$

on $I \times \mathbb{R}$ (with coordinates f and h), maintains a constant Gaussian curvature equal to -1, thereby enabling an isometric immersion into a domain within the Poincaré upper half-plane.

Therefore we showed that, if satisfied, equation (2.4) determines the existence of a solution for f such that M is an Einstein warped-product manifold with negative constant Ricci curvature, m-dimensional Ricci-flat fiber-manifold F and the base-manifold (B) a 2-dimensional hyperbolic manifold (i.e., with Gaussian curvature $K = -1$).

Our aim is to show, in a more general case, (that is, considering B a 2-dimensional manifold with negative constant Gaussian curvature, not necessarily $K = -1$), that if the warping function f, defined on B, satisfies $\Delta_{g_B} f = f$, then between the dimension of M and the constant λ, there is a second order equation that relates them.

In case the given metric g_B has a constant Gaussian curvature $K < 0$, rescaling the metric $\bar{g}_B = (-K)\,g_B$, we obtain the corresponding Gauss curvature in the new metric, namely $\bar{K} = -1$. So, from (2.3), we will have:

$$\begin{cases} \Delta_{\bar{g}_B} f = (-K)^{-1} \Delta_{g_B} f \\ |\nabla_{\bar{g}_B} f|^2 = (-K)^{-1} |\nabla_{g_B} f|^2, \end{cases} \tag{2.8}$$

and from this, now it is simple to determine $\bar{p}$ and $\bar{q}$. Therefore, in the most general scenario, we use (2.4) with $(\bar{p}, \bar{q})$.

That said, considering (1.3) with the settings expressed in *Theorem* 2.1, we obtain:

$$\begin{cases} R_B = 2\lambda + m \\ R_F = 0 \\ |\nabla_{g_B} f|^2 = -\frac{f^2(\lambda+1)}{m-1}, \end{cases} \tag{2.9}$$

and the second equation of (2.3) is equal to:

$$|\nabla_{g_B} f|^2 = \frac{f^2(-\lambda - 1)}{m - 1} = p(f)^2, \tag{2.10}$$

then

$$\nabla_{g_B} f = \frac{f\sqrt{-\lambda - 1}}{\sqrt{m-1}} = p(f), \tag{2.11}$$

therefore from (2.11) we deduce that λ must be ≤ -1.

Now, considering (2.8) (which corresponds to (2.3) with rescaled metric), we obtain:

$$\begin{cases} \frac{f}{-K} = \bar{p}(f)\bar{q}(f) \Longleftrightarrow \bar{q}(f) = \frac{\sqrt{m-1}}{\sqrt{-\lambda-1}} \frac{1}{\sqrt{-K}} \\ \frac{f\sqrt{-\lambda-1}}{\sqrt{m-1}} \frac{1}{\sqrt{-K}} = \bar{p}(f). \end{cases} \tag{2.12}$$

If we consider (2.4):

$$\begin{cases} \bar{p}''(f) = 0 \\ -\bar{p}'(f)^2 = \frac{\lambda+1}{m-1}(-K)^{-1} \\ 2\bar{q}(f)\bar{p}'(f) = 2(-K)^{-1} \\ \bar{q}'(f) = 0 \\ -\bar{q}(f)^2 = \frac{1-m}{\lambda+1}\frac{1}{K}. \end{cases} \tag{2.13}$$

Since $K = \frac{R_B}{2}$, from the first equation of (2.9), we get $-K = -(\lambda + \frac{m}{2})$, hence (2.4) is equivalent to:

$$\frac{\lambda + 1}{m - 1} + 2 + \frac{m - 1}{\lambda + 1} - \left(\lambda + \frac{m}{2}\right) = 0, \tag{2.14}$$

or

$$\lambda^2(2 - m) + \lambda\left(1 + \frac{3m}{2} - \frac{m^2}{2}\right) + \frac{m^2}{2} + \frac{m}{2} = 0. \tag{2.15}$$

□

The second degree equation (2.15) relates the dimension m of the fiber-manifold (F) and the constant λ (where $\lambda \in \mathbb{R}^-$), in order to define an Einstein warped-product manifold (M) with constant negative Ricci curvature and a 2-dimensional base-manifold (B) with constant negative Gaussian curvature, on which the defined warping function (f), satisfies the homogeneous form of Screened Poisson equation.

REMARK 2.2. More generally we can consider $[\Delta_{g_B} - \beta]f = 0$, with constant $\beta \in \mathbb{R}^+$, so (2.15) will be:

$$\lambda^2(2-m) + \lambda\Big(\beta + \frac{3m\beta}{2} - \frac{m^2\beta}{2}\Big) + m^2\Big(1 - \frac{\beta^2}{2}\Big) + m\Big(\frac{5\beta^2}{2} - 2\Big) = 0. \tag{2.16}$$

We have obtained that for this special family of Einstein warped-product manifolds, the second order equation (2.16) establishes a relation, in general, between the dimension (i.e., $2+m$), the constant of curvature (λ) and the "screened" parameter (β). These manifolds exist if and only if the equation (2.16) admits at least one negative real value as a solution for the dimension and screened parameter chosen.

EXAMPLE 2.3. If we consider that B is, for example, the upper Poincaré half-plane, then the line element of M, ds^2, will be:

$$\begin{aligned} ds^2 &= g_{B_{ij}} dx^i dx^j + f^2 g_{F_{uv}} dx^u dx^v \\ &= 4\frac{dx_1^2 + dx_2^2}{(1 - x_1^2 - x_2^2)^2} + f^2 g_{F_{uv}} dx^u dx^v \,, \end{aligned} \tag{2.17}$$

where $u, v = 3, 4, 5, \ldots$, and $g_{F_{uv}}$ represents the components of the (semi)-Riemannian Ricci-flat metric on F. Then

$$|g_B| = \frac{16}{(1 - x_1^2 - x_2^2)^4} \quad \text{and} \quad \sqrt{|g_B|} = \frac{4}{(1 - x_1^2 - x_2^2)^2}.$$

From $[\Delta_{g_B} - \beta]f = 0 \longrightarrow \nabla_i\partial^i f - \beta f = 0$, where

$$\nabla_i\partial^i f = \frac{1}{\sqrt{|g_B|}}\partial_i(\sqrt{|g_B|} g_B^{ij}\partial_j f),$$

we get:

$$\nabla_i\partial^i f = \frac{(1 - x_1^2 - x_2^2)^2}{4}\partial_i\Big[\frac{4}{(1 - x_1^2 - x_2^2)^2} g^{ij}\partial_j f\Big].$$

Now, after some straight-forward computation, we obtain:

$$\Delta_{g_B} f = \frac{(1 - x_1^2 - x_2^2)^2}{4}\Big[\frac{\partial^2 f}{\partial x_1^2} + \frac{\partial^2 f}{\partial x_2^2}\Big].$$

In summary, if and only if equation (2.16) will be satisfied, will there exist a function f such that it satisfies the following:

$$\frac{(1 - x_1^2 - x_2^2)^2}{4}\Big[\frac{\partial^2 f}{\partial x_1^2} + \frac{\partial^2 f}{\partial x_2^2}\Big] - \beta f = 0. \tag{2.18}$$

3. Open questions and future directions

An interesting future direction would be to extend these results to other cases of Einstein-warped product manifolds, wherever there is a well-understood corresponding family of integrable systems (geometric PDEs).

Acknowledgments

The authors would like to thank the anonymous referee for suggestions which have improved the manuscript. We also would like to thank the journal editors.

References

[1] R. L. Bishop and B. O'Neill, *Manifolds of negative curvature*, Trans. Amer. Math. Soc. **145** (1969), 1–49, DOI 10.2307/1995057. MR251664

[2] B.-Y. Chen, *Differential geometry of warped product manifolds and submanifolds*, World Scientific Publishing Co. Pte. Ltd., Hackensack, NJ, 2017. With a foreword by Leopold Verstraelen, DOI 10.1142/10419. MR3699316

[3] S. Kim, *Warped products and Einstein metrics*, J. Phys. A **39** (2006), no. 20, L329–L333, DOI 10.1088/0305-4470/39/20/L06. MR2238104

[4] D.-S. Kim and Y. H. Kim, *Compact Einstein warped product spaces with nonpositive scalar curvature*, Proc. Amer. Math. Soc. **131** (2003), no. 8, 2573–2576, DOI 10.1090/S0002-9939-03-06878-3. MR1974657

[5] B. O'Neill, *Semi-Riemannian geometry*, Pure and Applied Mathematics, vol. 103, Academic Press, Inc. [Harcourt Brace Jovanovich, Publishers], New York, 1983. With applications to relativity. MR719023

[6] B. Leandro, M. Lemes de Sousa, and R. Pina, *On the structure of Einstein warped product semi-Riemannian manifolds*, J. Integrable Syst. **3** (2018), no. 1, xyy016, 11, DOI 10.1093/integr/xyy016. MR3858737

[7] A. L. Besse, *Einstein manifolds*, Ergebnisse der Mathematik und ihrer Grenzgebiete (3) [Results in Mathematics and Related Areas (3)], vol. 10, Springer-Verlag, Berlin, 1987, DOI 10.1007/978-3-540-74311-8. MR0867684

[8] R. Machleidt, *Phenomenology and Meson Theory of Nuclear Forces.* In: I. Tanihata, H. Toki, T. Kajino,(eds) *Handbook of Nuclear Physics.* Springer, Singapore, (2023).

[9] P. Debye and E. Hückel, *Zur Theorie der Elektrolyte. I. Gefrierpunktserniedrigung und verwandte Erscheinungen*, Physik. Zeit. 24, 185-206, (1923).

[10] N. W. Ashcroft and N. D. Mermin, *Solid State Physics*, Thomson Learning, Toronto, (1976).

[11] K. Kamrin and G. Koval, *Phys. Rev. Let.*, 108, 178301, (2012).

[12] R. Tanios, S. El Mohtar, O. Knio, and I. Lakkis, `arXiv:1911.10944v1` [math.NA], (2019).

[13] W. Greiner, *Relativistic quantum mechanics*, 3rd ed., Springer-Verlag, Berlin, 2000. Wave equations; Translated from the second German (1987) edition; With a foreword by D. A. Bromley. MR1772212

MATHEMATICAL AND PHYSICAL SCIENCE FOUNDATION, SIDEVEJ 5, 4200 SLAGELSE, DENMARK
Email address: `pigazzini@topositus.com`

DIPARTIMENTO DI SCIENZE MATEMATICHE "G.L. LAGRANGE", POLITECNICO DI TORINO, C.SO DUCA DEGLI ABRUZZI 24, I-10129 TORINO, ITALY
Email address: `luca.lussardi@polito.it`

DEPARTMENT OF MATHEMATICS AND STATISTICS, 1108 MEMORIAL CIRCLE, MS 1042, TEXAS TECH UNIVERSITY, LUBBOCK TEXAS 79409-1042
Email address: `magda.toda@ttu.edu`

THE PACIFIC INSTITUTE FOR THE MATHEMATICAL SCIENCES AND FACULTY OF SCIENCE, SIMON FRASER UNIVERSITY, BURNABY, BRITISH COLUMBIA, V5A 1S6, CANADA
Email address: `adebened@sfu.ca`

SELECTED PUBLISHED TITLES IN THIS SERIES

821 **Bogdan D. Suceavă, Adara M. Blaga, Marian Ioan Munteanu, Cezar Oniciuc, Shoo Seto, Gabriel-Eduard Vîlcu, and Lihan Wang, Editors,** Recent Advances in Differential Geometry and Related Areas, 2025

818 **Howard S. Cohl, Roberto S. Costas-Santos, and Robert S. Maier, Editors,** Classical Hypergeometric Functions and Generalizations, 2025

817 **Mikhail Khovanov, Joshua Sussan, and Anton Zeitlin, Editors,** From Representation Theory to Mathematical Physics and Back, 2025

816 **Luca F. Di Cerbo and Laurenţiu G. Maxim, Editors,** Geometry and Topology of Aspherical Manifolds, 2025

815 **Daniel Orr and Joshua Jeishing Wen, Editors,** Macdonald Theory and Beyond, 2025

814 **Susan Montgomery and Siu-Hung Ng, Editors,** Quantum Groups, Hopf Algebras, and Applications, 2025

813 **Thomas Creutzig, Julia Plavnik, and David Ridout, Editors,** Quantum Symmetries, 2025

812 **Ross Geoghegan, Craig R. Guilbault, and Kim Ruane, Editors,** Topology at Infinity of Discrete Groups, 2025

811 **Krishnendu Gongopadhyay and Shashikant A. Katre, Editors,** Geometry, Groups and Mathematical Philosophy, 2025

810 **Robert J. Berman and Yanir A. Rubinstein, Editors,** Convex and Complex: Perspectives on Positivity in Geometry, 2025

809 **Fabrice Baudoin and Luca Rizzi, Editors,** New Trends in Sub-Riemannian Geometry, 2025

808 **Loring W. Tu, Editor,** Group Actions and Equivariant Cohomology, 2024

807 **Galina Filipuk, Editor,** Recent Progress in Special Functions, 2024

806 **Maximilian Engel, Hildeberto Jardón-Kojakhmetov, and Cinzia Soresina, Editors,** Topics in Multiple Time Scale Dynamics, 2024

805 **Anthony Iarrobino, Pedro Macias Marques, Maria Evelina Rossi, and Jean Vallès, Editors,** Deformation of Artinian Algebras and Jordan Type, 2024

804 **Mahir Bilen Can and Jörg Feldvoss, Editors,** A Glimpse into Geometric Representation Theory, 2024

803 **Peter Gothen, Margarida Melo, and Montserrat Teixidor i Bigas, Editors,** Moduli Spaces and Vector Bundles—New Trends, 2024

802 **Ralph M. Kaufmann, Martin Markl, and Alexander A. Voronov, Editors,** Higher Structures in Topology, Geometry, and Physics, 2024

801 **K. A. Brown, T. J. Hodges, M. Vancliff, and J. J. Zhang, Editors,** Recent Advances in Noncommutative Algebra and Geometry, 2024

800 **Avinoam Mann, Louis H. Rowen, David J. Saltman, Aner Shalev, Lance W. Small, and Uzi Vishne, Editors,** Amitsur Centennial Symposium, 2024

799 **Alberto A. Condori, Elodie Pozzi, William T. Ross, and Alan A. Sola, Editors,** Recent Progress in Function Theory and Operator Theory, 2024

798 **Marat V. Markin, Igor V. Nikolaev, and Carsten Trunk, Editors,** Advances in Functional Analysis and Operator Theory, 2024

797 **Sangita Jha, Mrinal Kanti Roychowdhury, and Saurabh Verma, Editors,** Recent Developments in Fractal Geometry and Dynamical Systems, 2024

796 **John Cremona, John Jones, Jennifer Paulhus, Andrew V. Sutherland, and John Voight, Editors,** LuCaNT: LMFDB, Computation, and Number Theory, 2024

795 **Michael A. Jones, David McCune, and Jennifer M. Wilson, Editors,** Mathematical Analyses of Decisions, Voting and Games, 2024

For a complete list of titles in this series, visit the
AMS Bookstore at **www.ams.org/bookstore/conmseries/**.